# Instructor Resources and Adjunct Guide

for

## *The Heart of Mathematics: An invitation to effective thinking*

Second Edition

by

Edward B. Burger, Williams College
Michael Starbird, The University of Texas at Austin
Deborah Bergstrand, Swarthmore College

 **Key College Publishing**
Innovators in Higher Education

1150 65th Street • Emeryville • California 94608

Edward B. Burger
Department of Mathematics
Williams College
Williamstown, MA 01267

Michael Starbird
Department of Mathematics
The University of Texas at Austin
Austin, TX 78712

Deborah Bergstrand
Department of Mathematics
Swarthmore College
Swarthmore, PA 19081

Key College Publishing was founded in 1999 as a division of Key Curriculum Press® in cooperation with Springer-Verlag New York, Inc. We publish innovative texts and courseware for the undergraduate curriculum in mathematics and statistics as well as mathematics and statistics education. For more information, visit us at www.keycollege.com.

Key College Publishing
1150 65th Street
Emeryville, CA 94608
(510) 595-7000
info@keycollege.com
www.keycollege.com

Development Editor: Allyndreth Cassidy
Production Director: McKinley Williams
Production Project Manager: Beth Masse
Copyeditor: Tara Joffe
Project Management: Elm Street Publishing Services, Inc.
Compositor: Interactive Composition Corporation
Art and Design Coordinator: Kavitha Becker
Printer: ePAC
Cover Designer: Marilyn Perry
Cover Photo Credit: *background:* Albano Guatti/Corbis. *center inset:* NASA. *top inset:* Craig Tuttle/Corbis. *top right inset:* Ergin Cavusoglu/Veer. *bottom right inset:* Hurewitz Creative/Corbis. *bottom left inset:* Joseph Sohm; Visions of America/Corbis. *top left inset:* Richard Hamilton Smith/Corbis.

Editorial Director: Richard J. Bonacci
General Manager: Mike Simpson
Publisher: Steven Rasmussen

Printed in the United States of America
10 9 8 7 6 5 4 3 2    08 07 06

1-931914-53-2

# Contents

# Introduction and Global Suggestions

## Description of Contents of *Instructor Resources and Adjunct Guide*

1. Introduction containing general suggestions and remarks for instructors and a sample course syllabus.
2. Overviews of each chapter of the text.
3. Commentary and suggestions for each section including:

   - A statement of the section's themes

   - A section called *Mathematical Underpinnings* which offers additional background for teachers

   - A box called *Our Experience* that lists

     - time spent on the section

     - emphasized items

     - items treated lightly

     - dangers

     - favorite topics

     - remarks

     - sample homework assignments

   - A collection of potential class activities designed to keep students actively engaged in each topic.

   - Some sections include template pages from which overhead transparencies can be made to support the presentation of the material. The template pages are found at the end of the chapter. Color versions of the templates can be found on the Instructor CD-ROM.

   - Sample test or quiz questions. These questions are also reproduced electronically on the Test Bank CD-ROM.

- Sample lecture notes that are also produced electronically on the Instructor CD-ROM.

**4.** Solutions to all the Mindscapes in the text, except for those in the In Your Own Words sections.

*Goal of textbook.* Our goal is to inspire students to be actively engaged in mathematical thought. Throughout the course we want them to discover ideas on their own, grapple with challenging new concepts, and learn various techniques of thought through repeated exposure. Many students entering this course do not arrive with a deep understanding of college-level mathematics. So we have presented topics—many of which are rather advanced—in such a manner as to make them accessible, interesting, and enticing to students who do not have a firm knowledge of the vocabulary and symbolic representation of mathematics. Our goal is not to teach these students the vocabulary and symbols of math. Instead, the goal is to open students' minds to ideas and to help them learn innovative modes of thought to empower them to approach and conquer all sorts of issues within and outside mathematics.

*Too much for one semester.* This book contains many more topics than can be treated in one semester. Our desire is for each instructor to craft the appropriate course by selecting those topics which he or she believes will be the most interesting and at the appropriate level for his or her students. The omitted sections are excellent sources for independent student projects. Finding supplemental material for independent research projects can be a challenge; however, the student text includes extra material (with exercises) that students can read, work through on their own, and which could lead to either term papers, oral presentations, or posters for a class poster session.

*Poster session / research project.* We have had much success holding a poster session near the end of the semester. Students in groups of two or three are assigned the project of preparing a paper and a poster for the session. On the day of the poster session, each group displays their poster, fields questions, and examines the other groups' posters. Bring food! Students are also asked to submit, with their group research paper, individual personal statements. These statements are an invitation to the student to reflect on the process of learning a piece of mathematics on his or her own; to share the frustration and struggle and then (hopefully) the feeling of accomplishment. These individual personal statements do not have any mathematical content. Here are a few sample sentences from various student individual statements (from the University of Colorado at Boulder):

- "I now look at certain things as math where I once looked at them as simply objects."

- "I've learned that you don't need a math teacher to preach the importance of equations but you can just as well use your own head and not a dead man's equation."

- "I honestly did not know that I had it in me to solve these seemingly complex problems without the aid of a professor. I now know I can. I am proud of this newfound insight."

- "For me, the most beneficial things weren't what I answered, but instead what it caused me to think about and stir new thoughts in my head."

*One section ≠ one class session.* In creating your syllabus, we wish to point out an important fact: one section ≠ one class session. Some sections require more than one day, while a few require less than a full day. Many sections can be used at different levels to allow shorter or longer treatments. For each section in this manual, we have included our estimate of the number of class days required, but you may have a different experience. Other instructors have decided to block out two 50-minute classes per section.

*Teaching styles.* Many formats of teaching are viable with this book. Our presentation of the material has an implicit discovery learning aspect in that we present examples and scenarios that evoke ideas in students' minds before we present worked-out notions. One of the authors teaches the class by frequently giving a brief explanation of an idea followed by small group discussions. Another author uses a more interactive, lecture-oriented style. In either case, following the pattern of this book will encourage student involvement in constructing and refining ideas for themselves. In this *Instructor Resources and Adjunct Guide* we suggest numerous opportunities for small group or entire class involvement and participation. These various activities can easily be converted to a more traditional lecture format if desired.

*Class activities.* Each class activity suggested in this manual usually consists of a question for the class to discuss or experiment with either as a whole or in small groups. The students' ideas can then be harvested, follow-up questions can be asked, and then the ideas can be clarified and explained. This discovery process allows the students to develop some of the ideas for themselves and thus the material seems fresh and, in fact, their own.

*Getting students to respond.* One of the authors was teaching from this text and began the topic of the fourth dimension by asking, "What questions do you have about the fourth dimension?" Absolute silence followed. Then he asked them to work in pairs and write two questions which he would collect from each pair. The rich collection of responses he received follow below. This experience is a great example of how the technique of soliciting answers seemed to make a huge difference in the responsiveness of the class. We have included samples of questions and activities that we have used in class to help get students to think about and talk about the various topics. They are listed later in this *Instructor Resources and Adjunct Guide* section by section.

Questions generated in class about the fourth dimension:

1. What is a dimension?
2. How is the fourth dimension graphically represented (X, Y, Z, ?)?
3. What IS the fourth dimension?
4. Why is the fourth dimension important?
5. How does the fourth dimension relate to geometry?
6. How does the fourth dimension compare to the third dimension?
7. Who discovered/explained/popularized the fourth dimension?
8. Where, or in what context is the fourth dimension found in real life and applied?
9. How is the fourth dimension represented numerically?
10. Does the fourth dimension really exist, and why should we care?
11. How can something be four-dimensional?
12. Can you eat it?
13. What do shapes look like in the fourth dimension?
14. What is it used to solve or depict?
15. Is the fourth dimension time?

*Interactive Explorations CD-ROM.*   The Interactive Explorations CD-ROM packaged with each new copy of the student text contains close to 40 powerful Java-based activities. Encourage students to use the CD each time they encounter the purple and yellow CD icon in the text. We suggest displaying computer output in class if you have the necessary equipment.

*Students should read their textbook.*   We have put a great deal of effort into making the book actually readable by real students. We do not visualize this book being used merely to assign Mindscapes at the end of the sections and to ask the students to find and adapt model examples previously worked out in the text. We are aiming for student thought which includes reading and understanding the material before grappling with the Mindscapes. A basic homework assignment is for the students to read the text and discuss the ideas with their classmates, roommates, friends, and family. We have found, in our experience, that there are many sufficiently intriguing ideas that students actually discuss outside of class. These discussions can be lively and interesting and should be encouraged. Often in Mindscapes, we ask the students to explain a particular notion to a friend not in the class and report on the experience.

Each chapter opens with a short intriguing introduction through questions, ideas, examples, or scenarios. The sections vary as to their independence from previous sections and the dependencies are diagramed in this *Instructor Resources and Adjunct Guide.*

*Exercises, called Mindscapes.* The Mindscapes at the end of each section fall into five categories. The first group of Mindscapes (Developing Ideas) are straightforward and should be easily answered after students have read the section. The second group (Solidifying Ideas) are designed to get students to solidify their understanding of the basic ideas of the section. The third group of Mindscapes (Creating New Ideas) requires more synthesis of ideas. The fourth group (Further Challenges) are more conceptual and thought provoking. The student text contains hints and solutions to selected Mindscapes in the back of the book (those Mindscapes are marked by (H) or (S) following the title). Solutions to all the exercises appear at the end of this *Instructor Resources and Adjunct Guide.*

The last group of Mindscapes (In Your Own Words) is a collection of questions to have students think creatively and generally about the ideas of the section. These questions add a writing component to the course and encourage students to express mathematical ideas in ordinary language. Many students in the course may be liberal arts majors who are more comfortable with words than with equations or symbols. These last Mindscapes give students a chance to come to grips with the mathematical ideas in their own terms. They also provide students with an opportunity for writing across the curriculum. One of the questions asks students to relate the section's meta-mathematical lessons to real life situations beyond mathematics. Encouraging students to make such connections from time to time can be valuable to convince them that what they are doing in class may have meaning to their real lives.

*Journal.* Along the lines of having students express ideas in their own terms, you may elect to ask students to keep a journal in which they record their views of the major ideas, points of interest or intrigue, surprises, curiosities, or changes in their outlook on the world. Such a journal allows students to reflect on their intellectual journey through mathematics at a broad and personal level, thus helping them make these new ideas their own.

*Tests.* We have included sample test questions for each section. Some of the suggested ones come from the Mindscapes, and the Mindscapes are good sources for additional test questions. As a practical matter, it is difficult to grade many long answer questions if the class is large, so we have included short answer questions and longer answer questions among the sample test questions appearing in the *Instructor Resources and Adjunct Guide* in the section-by-section commentary. All test questions printed here also appear electronically on the Test Bank CD-ROM. Additionally, the CD contains actual tests solicited from users of the first edition of this text.

*Students being active.*   Our hope is that the reader of the student text will be an active participant. Thus, throughout the text, we ask the reader to stop and think before moving on. Any way you can encourage students to actually engage in the reading will increase the positive impact of the text. The *Welcome* contains a statement of how to use the book. Encourage your students to read and follow the advice given there. The chapter following *Welcome* is a fast-paced, intriguing overview of the entire book. We suggest that you ask your students to read these two parts at the very start of the course. In fact, some instructors have their students read this chapter after the first day and then have the students vote as to which topics they wish to cover in the course! We think this is a wonderful idea, but have not tried it yet ourselves.

*Resources for instructors.*   For each section, we've included mathematical underpinnings to enhance instructor understanding. This material is written in a more sophisticated, condensed style than the text, offering instructors some additional background. The text itself presents all topics clearly and in detail, so even unfamiliar material is accessible.

We also include sample lecture notes for each section. Instructors can use these notes in class or as a guide or inspiration in designing their own lectures. The sample lecture notes are generally designed for a 50-minute class, designated "one day." In a few cases, the sample lecture notes are marked for a half-day, one-and-a-half days, or two days.

The sample lecture notes are set up so instructors can remove the pages from this *Instructor Resources and Adjunct Guide* and bring the lecture notes to class. All sample lecture notes are also reproduced electronically on the Instructor CD-ROM so that they can easily be tailored to fit your course.

*Instructor CD-ROM.*   Many of the resources contained within these pages are also available electronically on the Instructor CD-ROM. In addition to offering the templates in color, the CD provides all of the sample lecture notes. You will also find presentations prepared with PowerPoint presentation software that follow the sample activities suggested in this guide. We hope you will find that the CD is a valuable extension of the print resources.

*Instructor Videos.*  Users of *The Heart of Mathematics* have found that attending lectures and workshops by Edward B. Burger and Michael Starbird help in preparing to teach a course with the text. Because not all users can meet the authors in person, we are bringing the authors to you in this set of videos on two CDs. The authors explain their course philosophy and present each section's material in short videos that we filmed during Michael Starbird's 2004 Chautauqua lecture and the two author's 2004 MAA MathFest Minicourse.

*Web site and WebCT.* Both instructors and students will find extensive resources on the Web site for *The Heart of Mathematics* (www.heartofmath .com). To gain full access to all parts of the site, students will need to register at www.keycollege.com/online and enter their unique access code printed on the card packaged with their new text. Instructors should contact their sales representative to receive the correct instructor validation code for *The Heart of Mathematics.*

In addition to resources that directly support the text material, instructors can create WebCT sites complete with a customized syllabus, quizzes for each chapter of the text, PowerPoint slide presentations, and much more.

*Resources abound.* To receive any of the electronic resources mentioned above (Instructor CD, Instructor Videos, or Test Bank), please contact your sales representative. To find your rep, click the rep finder button at www.keycollege.com.

*Course enjoyable to teach.* We have tried to make the text fun for the students, but we have also tried to make this course a pleasure for the instructor. We hope the instructional suggestions in this manual are useful and make it easier to teach this course. We have taught this course many times ourselves and realize the challenge of enticing nonquantitative students to the joy of mathematics. We have found that students respond well to the intellectually substantial topics in this book and that the presentation suggested here makes those ideas accessible to a large number of students. We hope you find pleasure in teaching this course just as the students find pleasure in learning the mathematics. Of course we would be delighted to hear your experiences and suggestions for the text and the course. Good luck and have fun.

*Edward B. Burger*
*Michael Starbird*
*Deborah Bergstrand*

# Sample Syllabus

Here is a sample syllabus which was used by Edward Burger teaching a course with *The Heart of Mathematics*.

## The Art of Mathematical Thinking:
## An Introduction to the Beauty and Power
## of Mathematical Ideas

**Instructor:** Professor E. Burger

**Text:** *The Heart of Mathematics: An invitation to effective thinking*, 2nd edition, by Edward B. Burger and Michael Starbird

**Course Description:** Here we will consider some of the greatest ideas of humankind—ideas comparable to the works of Shakespeare, Plato, and Michelangelo. The great ideas we will explore here are within the realm of mathematics. What is mathematics? Mathematics is an artistic endeavor which requires both imagination and creativity. In this course, we will experience what mathematics is all about by delving into some beautiful and intriguing issues. There are three basic goals for this course.

1. To attain a better understanding of some rich mathematical ideas.
2. To build sharper skills for analyzing life issues that transcend mathematics.
3. To develop a new perspective and outlook on the way you view the world.

We will cover roughly six different topics. Although you will be challenged, the overriding theme of the course is to gain an appreciation for mathematics and to discover the power of mathematical thinking in your everyday life. We will follow the text reasonably closely although we will not cover all the material in class.

The only prerequisites for this course are an open and curious mind and the willingness to put aside any preconceived prejudices or dislikes for mathematics. Very little mathematical background will be expected and hopefully this course should be (for the most part) "self contained."

**Examinations:** There will be three quizzes during the term.

**Homework:** Homework will be assigned regularly from the text, collected, and graded. *Clarity of exposition is important*, and one should strive for **well**

**written, polished solutions.** For the most part, collaboration on homework with other members of this class is allowed, although solutions must be individually written up and collaborators *should be acknowledged*. It will be made clear when collaboration is not permitted. There will also be several *short* writing assignments throughout the semester.

**Research Project/Poster Session:** The only way to really understand mathematics is to learn and discover it on one's own. Thus, students will select a mathematical topic outside of those covered in our class, read and teach themselves any necessary background to understand it, and then investigate the topic. Students are **strongly** encouraged to work together in groups of two or three on this project. By working together, the individuals can learn from each other and share the experience. Each group will write a final paper on their findings and present a poster display during a class poster session at the end of the semester. Also, each student will write a short individual statement regarding the experience. Various interim reports will be collected throughout the term. Students are invited and encouraged to discuss all phases of the project with me.

**Grading Policy:**

| | |
|---|---|
| Quiz 1 | 20% |
| Quiz 2 | 25 |
| Quiz 3 | 25 |
| Research Project | 20 |
| Class participation | ?? |
| Homework | 10 |
| TOTAL | 100 |

**Pacing of the course and homework assignments:** Each of the weekly homework assignments consists of reading from the text and doing about 10 Mindscapes. (Note to instructor: The grading policy was that each homework assignment was graded out of 10 points. They received 4 points if they attempted every question. Between 3 and 6 questions were graded and clarity and correctness both influenced the grading.)

The weekly assignments are listed here:

**Special assignment due second day**—Read Section 1.1 and start Stories 1, 3, 4, 5, 7, and 8 (You may read Section 1.2 (hints), but do not read Section 1.3 (solutions)).

**Week 1**—Finish Section 1.1 Stories 1, 3, 4, 5, 7, 8, and also do Stories 2 and 6. Read Section 1.4 and do Mindscapes 6, 7, 9, 14.

**Week 2**—Read Section 2.1 and do Mindscapes I.4; II.8, 15.

**Week 3**—Read Section 2.2 and do Mindscapes I.2; II.6, 7, 15; III.29, 30.

**Week 4**—In Section 2.2, do Mindscapes II.22, 24; IV.36, 37. Read Section 2.3, do Mindscapes I.2; II.7, 12, 14, 15, 24; III.32.

**Week 5**—In Section 2.3, do Mindscape III.35. Read Section 2.6 and do Mindscapes I.3; II.6, 10; III.30; IV.40. Read Section 2.7 and do Mindscapes I.2; II.7, 10, 20, 23, 25.

**Week 6**—Read Sections 3.1, 3.2, 3.3 and do Section 3.1 Mindscapes I.4; III.1; Section 3.2 Mindscapes I.3; II.14, 16; III.30, 32; IV.36; Section 3.3 Mindscapes I.4; II.11, 13, 14; III.19.

**Week 7**—In Section 3.3, do Mindscapes II.9; III.16, 17. Read Section 3.4 and do Mindscapes I.4; II.6, 13. Read Section 3.5 and do Mindscapes I.2; II.6, 9, 10; III.20.

**Week 8**—Read Section 4.1 and do Mindscapes I.2; II.8, 12, 15. Read Section 4.3 and do Mindscapes I.3; II.9, 12; III.16, 17, 20. Read Section 4.5 and do Mindscapes I.2; III.16, 17.

**Week 9**—Read Section 4.7 and do Mindscapes I.1; II.7, 12, 14; III.16, 18. Read Section 5.1 and do Mindscapes I.4; II.6, 9, 10, 11, 12.

**Week 10**—Read Section 5.2 and do Mindscapes I.3; II.6, 8, 9, 25; III.33. Read Section 5.3 and do Mindscapes I.2; II.7, 9, 13; III.26; IV.40.

**Week 11**—Read Sections 6.1 and 6.2 and do Section 6.2 Mindscapes I.2; II.8, 12, 13; III.27, 28. Read Section 6.3 and do Mindscapes I.3; II.6, 13, 14, 20, 21.

**Week 12**—In Section 6.3 do Mindscapes II.23, 24, 25; III.26, 27; IV.40. Read Sections 7.1 and 7.2 and do Section 7.2 Mindscapes I.2; II.7, 8, 9, 12.

**Week 13**—In Section 7.2 do Mindscapes II.18, 19, 20; III.28, 30; IV.40. Read Section 7.3 and do Mindscapes I.2; II.23; III.26, 30, 32.

**Week 14**—Read Section 7.5 and do Mindscapes I.2, 4; II.9; III.17. Read Section 8.1 and do Mindscapes I.2; II.8, 10, 16. Read Section 8.2 and do Mindscapes I.3; II.9, 11, 13.

# Welcome!
# Surfing the Book
# Chapter 1   Fun and Games:
# An Introduction to Rigorous Thought

**1.1**   **Silly Stories Each with a Moral** [Conundrums evoking thought]
**1.2**   **Nudges** [Leading questions and hints]
**1.3**   **The Punch Lines** [Solutions and further commentary]
**1.4**   **From Play to Power** [Discovering strategies of thought for life]

## I.  Overview

Welcome! is a short introduction that sets the tone for what is ahead. We encourage you to have your students read the Welcome! to help solidify goals and expectations. Surfing the Book poses intriguing questions designed to pique students' interest in every chapter of the book. Our hope is that students will read a question and flip to the relevant pages out of curiosity. Some instructors have actually had their students read Surfing the Book and then vote on which chapters they wish to cover in the course—an intriguing idea.

Chapter 1, "Fun and Games," consists of nine conundrums or stories. Two are general logic puzzles, and the remaining seven are designed to encourage students to discover or grapple with one of the basic ideas from the subsequent chapters.

| | |
|---|---|
| That's a Meanie Genie | Logic puzzle |
| Damsel in Distress | Geometry |
| The Fountain of Knowledge | Number theory |
| Dropping Trou | Topology |
| Dodge Ball | Infinity |
| A Tight Weave | Fractals |
| Let's Make a Deal | Probability |
| Rolling Around in Vegas | Voting theory |
| Dot of Fortune | Logic puzzle |

*Our Experience*

*Time spent:* The first two or three 50-minute sessions of the course (although other instructors have spent considerably more time with this chapter)

*Items omitted or treated lightly:* One of us discusses only some of the nine stories.

*Dangers:* Students sometimes have difficulty interpreting what they read. A Tight Weave is difficult to understand because it involves an infinite process, so you may want to give help. Dot of Fortune also generates some controversy.

*Personal favorite topic or interactions:* Having students get acquainted while working on some entertaining mathematical questions

*Remarks:* We ask two students to play Dodge Ball and to explain the strategy. We then mention that the strategy of this simple game is a central idea for studying infinity. This example helps students see that these lighthearted stories are actually related to serious ideas.

*Sample homework assignment:* Write solutions to four of the stories and four of the Mindscapes. One of us assigns all nine stories for homework. Although hints and problem-solving techniques for the Mindscapes appear at the end of Chapter 1, students should use them very sparingly. Once students arrive at their solutions, they may read the hints to see if there are other ways of resolving the issue.

### Sample Class Activities

*Getting acquainted.* The stories in Chapter 1 can evoke lively discussion. Have groups of students get acquainted through choosing a story to discuss and working together on mathematics. Break the class into groups of three or four to work out some of the easier Mindscapes. The one rule: *no one can write anything down!* In this way, each person in the group must talk, contribute, or ask questions. No one goes off to work by him- or herself. This rule also allows for lively discussions. After a few minutes, make students switch groups and switch Mindscapes. In the last 15 minutes of class, solicit answers from the different groups and discuss with the entire class.

*Relevance of stories to future work.* Ask students to guess which stories are pertinent to which chapters or ideas. This discussion allows you to discuss some of the relationships of the stories to future chapters and encourages students to look for connections between ideas. Chapter 1 is intended to provide a playful introduction to the course and to set the tone that even playful-sounding stories can reveal substantial and serious ideas.

### Sample Test Questions

(The Mindscapes are another good source of test questions. All test questions below are available electronically on the Test Bank CD-ROM.)

- A voter with a rowboat finds herself in an unusual situation. On one side of a river is Bob Dole, Bill Clinton, and a big bag of Burger King Whoppers. The voter must get Dole, Clinton, and the bag of burgers across the river to the other side. Her boat, however, is large enough for her and just one other item or person. If she leaves Dole alone with Clinton, then Dole (an old war veteran) will beat up Clinton. If she leaves Clinton alone with the bag of Whoppers, then Clinton (a big burger lover) will devour the entire contents of the bag. Of course, Dole, Clinton, and the bag are all incapable of rowing the boat across the river. Is it possible for this voter to get all her cargo across the river? If so, carefully explain her method; if not, carefully explain why not.

- You have two measuring cups: one holds exactly 5 ounces of water and the other holds exactly 3 ounces of water. There are no markings on the cups, and you are not able to mark the cups in any way. You are given a huge bucket of water. Is it possible to measure and place exactly 4 ounces of water into the big cup? If so, carefully explain your method; if not, carefully explain why not.

- Explain the winning strategy for the game Dodge Ball (as described in this chapter). Would the strategy work for a different-sized board? For an infinitely big board?

# Chapter 2   Number Contemplation

## I.   Chapter Overview

This chapter introduces students to some of the fundamental ideas of number theory and its applications. It explores and develops a sense of number—both integers and reals—as having distinctive characters and features. By the end of the chapter, students should have a sense of a hierarchy of number—natural numbers, integers, rationals, and reals. They should see the significance of intriguing patterns among numbers through their applications to error-checking codes and perhaps cryptography. They should understand the real numbers and how the rationals and irrationals sit on the real number line. We hope students will see numbers as a more varied and interesting body of ideas than they first expected. Beyond the mathematics, we hope students will begin to learn how to develop new ideas and concepts.

---

### Dependencies
Section 2.5 (RSA coding) requires both 2.3 (Prime numbers) and 2.4 (Modular arithmetic). Section 2.6 (Irrational numbers) requires the notion of prime factorization found in Section 2.3. A portion of Section 2.7 (Real number line) uses the idea of irrationals from Section 2.6. Sections 2.1 and 2.2 are independent.

---

### Sections to Assign as a Reading Assignment or Term Project
Section 2.5 (RSA coding) is this chapter's most difficult section, and we often do not cover it during class. Sometimes, in the interest of time, we omit Section 2.4 (Modular arithmetic) and assign it either as a reading assignment or as a term group project.

*Reminders*

1. Class activities are suggested for each section. These can be done with students working in groups of two or three and then gathering their responses or as one global class discussion.
2. Before introducing a new topic, engage students with enticing questions, which they quickly discuss with neighbors; solicit responses within a minute. Use the responses as the springboard for the introduction.
3. We have included more sample class activities than you will probably want to use. Choose only the ones that are right for you and your class.
4. Each section can be treated in more or less depth. It is frequently a good idea to omit difficult technicalities in order to treat the main ideas well and then move on.

## II.  Section-by-Section Instructional Suggestions

**Section 2.1  Counting** [Pigeonhole principle]. This section introduces the natural numbers. The power of quantification and estimation is illustrated, and the Pigeonhole principle is introduced. All numbers have distinctive characteristics, and the section ends with the whimsical theorem that every natural number is interesting. An important goal of this section is to encourage students to take a quantitative look at their world, particularly at parts of the world they do not customarily view in a quantitative way.

---

*General Themes*

Using estimation to move from qualitative to quantitative thinking and reasoning is a powerful skill.

---

*Mathematical Underpinnings.*  (*Caveat:* These mathematical overviews are usually for the instructor only. The symbols and detailed proofs are not suitable for class presentations because mathematical symbols and terminology often present a formidable barrier for students.)

**Pigeonhole principle:** *If we place n + 1 objects in n boxes, then there exists at least one box containing at least two objects.*

**Proof:** Suppose no box has more than one object. Because there are only $n$ boxes, this assumption implies there are at most $n$ objects. But we start with $n + 1$ boxes; thus, the assumption is false and the result holds.

This principle can also be proved using induction. As simple as the Pigeonhole principle is to state, prove, and understand, it can actually be used to prove fairly complex results. Here's an application of the Pigeonhole principle that is less than straightforward.

**Claim:** *In any list of 10 natural numbers, $a_1$, $a_2$, ... $a_{10}$, there exists a string of consecutive numbers whose sum is divisible by 10.*

**Proof:** Consider the 10 sums $a_1$, $a_1 + a_2$, $a_1 + a_2 + a_3$, ..., $a_1 + a_2 + \cdots + a_{10}$. If one of these sums is divisible by 10, then we are done. If not, then each has a nonzero remainder when divided by 10. Because there are only nine such remainders, by the Pigeonhole principle there must be two sums with the same remainder. Denote these sums $a_1 + a_2 + \cdots + a_i$ and $a_1 + a_2 + \cdots + a_j$, where $i < j$. Because these two sums have the same remainder when divided by 10, their difference will have remainder 0. But their difference can be computed as $a_j + a_{j-1} + \cdots + a_{i+1}$. Thus, we have a string of consecutive numbers with sum divisible by 10.

---

*Our Experience*

*Time spent:* One 50-minute session

*Emphasized items:* The hairy body example

*Items omitted or treated lightly:* One of us sometimes skips the Intrigue of Numbers Theorem. We usually do not discuss the Hardy and Ramanujan story in class.

*Dangers:* The whole class is lighthearted, and we do not utter the word "induction."

*Sample homework assignment:* Read the section and do Mindscapes I.4; II.8, 15; III.19.

---

### Sample Class Activities

*Estimation.* Ask students to estimate various quantities to show that they can and to develop various techniques. How many Ping-Pong balls could be put in the room? Let students come up with the technique of estimating how many balls could fit in a small volume and then multiplying. How many leaves are on a tree? How many blades of grass are on a football field? How much is the college budget? How many cars would be needed to make a line, bumper to bumper, from New York City to San Francisco? For each question, have students make an estimation. Discard the extremes, and take the average.

*Quantification.* Ask students to consider various everyday activities and state how they could view them quantitatively rather just qualitatively. Have them estimate some pertinent numbers. For example, given a student's college costs, what is the cost of cutting one lecture in one class?

*Johnny Carson.* Johnny Carson was the most watched person in human history. Estimate the total number of viewers who watched Carson over his 30-year reign on the *Tonight Show* (you should count the number of times any particular person watched him—that is, don't count a person once if they watched him 100 times). Sadly, you may first have to explain who Johnny Carson is.

*Homework.* Ask each student to bring one million of something (grains of sand, etc.) to class, and have them explain how they determined they had a million.

*Pigeonhole principle.* Why are there two trees on Earth with the exact same number of leaves? Why does every person have many temporal twins on Earth—that is, people who were born on the same day and will die on the same day?

*Homework.* Ask students to look for examples of the application of the Pigeonhole principle and of quantification in everyday life and to bring those examples to class. Use the more interesting examples to demonstrate that the quantitative view of life leads to new insights.

*All natural numbers are interesting.* Start with the number 1 and ask students why it is interesting. Proceed with 2, 3, . . . , and keep going as long as students have interesting comments to make about the number. This activity can actually be quite lively. Once you get to the biggest natural number about which the class has specific commentary, ask why the next integer is interesting, thereby helping students to understand the proof. This whimsical theorem can be used to briefly introduce the idea that in mathematics, statements are proved. Although it can be used to talk about induction, we do not recommend doing so. If you choose to mention induction, you could use the Towers of Hanoi puzzle as another example.

## Sample Test Questions

(The Mindscapes are another good source of test questions. All test questions below are available electronically on the Test Bank CD-ROM.)

• In a standard deck of 52 cards, what is the smallest number of cards you must draw to guarantee that you will have at least one pair? (Remember, there are 13 different cards in each suit.)

• In a standard deck of 52 cards, what is the smallest number of cards you must draw to guarantee that you will have 5 cards of one suit? (Remember, a standard deck of cards has four suits with 13 cards in each suit.)

• Eighty thousand people attended The University of Texas versus Texas A&M football game. The fans of both teams were so happy, they decided to organize a party each day for a year. They decided that on each day, anyone with their birthday on that day would return to the stadium at noon to celebrate. Why would at least one party have more than 200 people?

• Each box of animal crackers contains two servings; each serving consists of exactly 12 crackers. There are exactly 18 different shapes of crackers. Are there always two crackers of the same shape in each box? Explain why or why not.

## Sample Lecture Notes for Section 2.1, "Counting" (1 day)

Question of the Day (to be written on the board for students to think about and discuss before class begins): *How many Ping-Pong balls are needed to fill up the classroom?*

5 minutes—Begin discussion by asking for collections that are too large or too complicated to count. Mention the number of hairs on a person's body.

10 minutes—Ask students to estimate the number of Ping-Pong balls needed to fill the entire classroom. How could we estimate how many balls we would need? Have students discuss and contribute. Bring some Ping-Pong balls to class to help the estimation.

5 minutes—Pose the hairy body question: Are there two nonbald people on Earth who have the exact same number of hairs on their bodies? Can we answer this question for certain? Begin a discussion and collect ideas.

15 minutes—Move from the qualitative to the quantitative through the estimation analysis done on text pages 41–43. Have students work in teams of two to look at the edge of each other's scalps and count how many hairs they see in a 1/4-inch × 1/4-inch square of hair. Use students' numbers in the calculations. (If it seems there are students who might be uncomfortable with this exercise, modify it appropriately. We have used this exercise in the past, and students enjoyed it.)

10 minutes—Get students to use the body-hair analysis to formulate a clear statement of the Pigeonhole principle—namely, if we have a collection of objects that are to be placed in a collection of containers and if there are more objects than containers, then at least one container will contain at least two objects. Then illustrate with several examples: Suppose we have five closed tennis-ball cans of standard size (each can holds up to three balls). How can we show for certain—without looking in the cans—that two cans contain the same number of balls? Why are there two trees with the same number of leaves?

5 minutes—Close the class with the importance of looking at issues quantitatively, even when they involve quantities that appear too large to consider.

**Section 2.2 Numerical Patterns in Nature** [Fibonacci numbers]. This section begins with the discovery of a pattern in the numbers of various spiral counts in flowers, pinecones, and pineapples. Fibonacci numbers are found in nature and art, and students are encouraged to find them in the spiral counts of various plants. This section introduces the Golden Ratio, $\lim\limits_{n\to\infty}\dfrac{F_{n+1}}{F_n}$, and discusses the game of Fibonacci nim and a winning strategy. Finally, this section provides examples of patterns in abstract mathematics reflected in nature, which may be a new and powerful idea to some students and an eye-opening concept to others.

---

*General Themes*

Looking at simple things deeply, finding a pattern, and using the pattern to gain new insights provides great value.

---

*Mathematical Underpinnings.* (*Caveat:* These mathematical overviews are usually for the instructor only. The symbols and detailed proofs are not suitable for class presentations because mathematical symbols and terminology often present a formidable barrier for students. In particular, in our experience, proofs by induction should be avoided in class.)

The Fibonacci sequence is defined by $F_1 = 1$, $F_2 = 1$, and for $n > 2$, $F_n = F_{n-1} + F_{n-2}$. The sequence, which begins $1, 1, 2, 3, 5, 8, 13, 21, \ldots$, appears in the spiral counts of various flowers and fruit—the clockwise and counterclockwise spiral counts are consecutive Fibonacci numbers. The quotient of two adjacent Fibonacci numbers can be expressed in a surprisingly attractive way as a continued fraction:

$$\frac{F_{n+1}}{F_n} = 1 + \cfrac{1}{1 + \cfrac{1}{1 + \cfrac{1}{1 + \cfrac{\ddots}{\phantom{1}} + 1}}},$$

where the number of plus signs equals $n - 1$.

11

The proof of this fact can be seen by induction on $n$: Note that if $n = 1$, then $F_2/F_1 = 1/1 = 1$, with no plus signs. Now assume that the statement holds for $n = k$, for some $k > 1$. That is,

$$\frac{F_{k+1}}{F_k} = 1 + \cfrac{1}{1 + \cfrac{1}{1 + \cfrac{1}{1 + \cdot\cdot\cdot\cdot + 1,}}}$$

where the number of plus signs equals $k - 1$.

Now we consider

$$\frac{F_{k+2}}{F_{k+1}} = \frac{F_{k+1} + F_k}{F_{k+1}} = 1 + \frac{F_k}{F_{k+1}} = 1 + \cfrac{1}{\cfrac{F_{k+1}}{F_k}},$$

which, given our inductive hypothesis, has the correct form and contains $k$ plus signs, thus completing the proof.

If we now take the limit as $k$ approaches infinity, we see it converges to

$$1 + \cfrac{1}{1 + \cfrac{1}{1 + \cfrac{1}{1 + \cdot\cdot\cdot\cdot,}}}$$

which is the Golden Ratio. On text page 55, we prove that this quantity equals $(1 + \sqrt{5})/2$.

Finally, we note the following:

**Theorem:** *Given any natural number N, the number can be expressed as the sum of nonadjacent Fibonacci numbers.*

**Proof:** We proceed by induction. If $N = 1$, then we can simply write it as the Fibonacci number 1. Assume the result holds for all numbers up through $k$. We now consider $k + 1$. If $k + 1$ is a Fibonacci number, then we simply write it as the degenerate sum consisting of just that number. If $k + 1$ is not a Fibonacci number, then we let $F_n$ denote the largest Fibonacci number less than $k + 1$. Because $k + 1$ is not a Fibonacci number, we know that $k + 1 - F_n$ is a natural number smaller than $k + 1$. By our inductive hypothesis, we know that $k + 1 - F_n$ can be written as a sum of nonadjacent Fibonacci numbers. Adding $F_n$ to that sum will produce a sum of Fibonacci numbers equal to $k + 1$. Note that $F_{n-1}$ cannot appear in the sum, because then $F_{n-1} + F_n$ would be a Fibonacci number less than or equal to $k + 1$, contrary to our choice of $F_n$. Thus, we have written $k + 1$ as the sum of nonadjacent Fibonacci numbers.

In fact, it can be shown that such a decomposition is unique—there is only one way (up to reordering) to write a number $N$ as a sum of nonadjacent Fibonacci numbers. Also, we have the following:

**Corollary:** *If $N$ is not a Fibonacci number and $F_k$ is the smallest Fibonacci number appearing in the above sum, then $N - F_k > 2F_k$.*

**Proof:** Because $F_k$ is the smallest term in the sum of nonadjacent Fibonacci numbers, then the next smallest term in that sum could be no smaller than $F_{k+2}$. But we can rewrite $F_{k+2}$ as $F_{k+2} = F_{k+1} + F_k = F_k + F_{k-1} + F_k = 2F_k + F_{k-1}$. Thus, we see that $N - F_k \geq F_{k+2} > 2F_k$, which completes the proof.

This theorem and its corollary allow us to prove that the strategy outlined for Fibonacci nim on text page 56 is, in fact, a winning strategy.

---

*Our Experience*

*Time spent:* One-and-a-half 50-minute sessions

*Items omitted or treated lightly:* We quickly illustrate that every natural number is the sum of nonadjacent Fibonacci numbers, and then we play and discuss Fibonacci nim.

*Dangers:* The standard notation of $F_n$ for the $n$th Fibonacci number is a difficult one for many students. That is, the notion of an index and the notation of a generic—the $n$th—Fibonacci number are both foreign notions. Stress how these issues are, for the most part, notational and different from the underlying ideas. Depending on your students' mathematics background and strength, provide them with numerous examples to shore up the notation, or simply avoid the notation altogether.

*Personal favorite topic or interactions:* One of us brings in pinecones or pineapples, has the students count spirals, and then gathers the data.

*Remarks:* We note that the Golden Ratio will be a quantity that will be revisited in Chapter 4, "Geometric Gems."

*Sample homework assignment:* Read the section and do Mindscapes I.2; II.6, 7, 17; III.28, 30; IV.37.

## Sample Class Activities

*Discovering the Fibonacci pattern through spiral counts.* Make overhead transparencies of the flowers on text pages 47–49 (Templates 2.2.1a–c). As a class, count the spirals on the flowers. Similarly, one of us has passed out one pinecone to each student and then had students individually count the spirals along the pinecone's outer surface. Have students guess the pattern and the next few numbers in the sequence.

*Dubious class activity—Fibonacci sequence in rabbit growth.* This application is found in the Mindscape II.6. Note that in this activity, each student represents two bunnies or rabbits. Ask one student to come to the front of the class to represent the first bunny pair. Mark down the time. Indicate that one month has passed, and the bunny pair is now a rabbit pair. At the two-month mark, the student representing the rabbit pair chooses a member of the class to come forward and represent the first bunny-pair offspring. Put the student representing the bunnies in a special place—the bunny hutch. After one month, the bunny student moves to the rabbit side and now as a rabbit pair chooses another pair of bunny offspring represented by one student. At each month, each person representing a rabbit pair chooses a pair of bunny offspring and the bunnies move to the rabbit side. Point out that at each generation, the total number at the next moment equals the total of the bunnies plus rabbits.

*Numbers as sums of Fibonaccis.* Take a number, say 86, and ask how students would go about writing it as the sum of Fibonacci numbers. How could they make the most progress in one step? Obviously, take out the biggest Fibonacci number. Ask students to explain why the remainder must be less than the next smaller Fibonacci number. Ask why the sum they get will not contain consecutive Fibonacci numbers but must express the original number as a sum of Fibonacci numbers. Ask students to explain why a number that is one less than a Fibonacci number (54 or 88, for example) will be the sum of every other Fibonacci number without skipping any.

*Fibonacci nim.* Play the game with a student, record the moves, and win. Then reenact the game; at each stage, compute the Fibonacci sums at each of your moves and demonstrate why, on the student's turn, he or she cannot win because double the number of sticks you took cannot be as large as the next number in the Fibonacci sum. Invite students to beat students not in the class.

*Base 2.* Write down the sequence 1, 2, 4, 8, 16, 32, 64. Now take a number, say 88, and try to express it as a sum of these numbers. Will the idea of taking out the biggest number work? Yes; however, numbers may be consecutive.

***Templates for Transparencies*** *(located at the end of the chapter):*

- Templates 2.2.1 **a.** Daisy **b.** Overlay of daisy with spirals marked **c.** Overlay of daisy with reverse spirals marked

- Template 2.2.2 Sunflower spirals

## *Sample Test Questions*

(The Mindscapes are another good source of test questions. All test questions below are available electronically on the Test Bank CD-ROM.)

- List the first 10 Fibonacci numbers.

- If Fibonacci nim is to be played starting with 20 sticks, how many sticks should be removed in the first move using the winning strategy?

- To what number does the ratio of consecutive Fibonacci numbers, $F_{n+1}/F_n$, converge? Name the number and express it numerically. Why does the ratio $F_{n+1}/F_n$ approach that number?

- Recall that $F_n$ represents the $n$th Fibonacci number.

    **a.** Write the first eight numbers in the sequence where the $n$th term in the sequence is given by $F_n + F_{n+2}$.

    **b.** By examining the sequence of numbers in part a, give a method for finding the next number in the sequence (other than the relation described in part a).

## Sample Lecture Notes for Section 2.2, "Numerical Patterns in Nature" (2 days)

### Day One

Question of the Day (to be written on the board for students to think about and discuss before class begins): *What's the next term in the sequence: pinecone, pineapple, daisy, _____?*

A week before class, inform students that they will need to bring a pineapple to class for this lecture.

2 minutes—Take a class photograph of students holding their pineapples—great photo op.

5 minutes—Ask students to look at their pineapples and note as many observations about their fruit as they can. Walk around the class, and share some of the spiral observations with the entire class.

5 minutes—Invite the class to move from the qualitative to the quantitative whenever possible. In this case, they found two sets of spirals on the pineapple. How many spirals are in each set? Firmly suggest that the number of spirals in one direction should equal the number of spirals in the other—after all, it's the same pineapple. Show the class how to carefully count the spirals, and have them write their answers without showing the numbers to anyone.

5 minutes—Pick a student at random and announce that their numbers are 8 and 13. Everyone will be amazed that (1) you guessed the student's counts correctly and (2) those values were the numbers the other students found as well. Write the numbers 8, 13 on the board, and show the overhead transparency of the daisy.

5 minutes—Count the spirals of the daisy (Template 2.2.1a) and write those numbers—21, 34—on the board next to the previous numbers.

5 minutes—Have students study these numbers, which are found in nature, and discover a pattern. Have the class guess the next two numbers and then figure out what numbers must have come before the 8. When they get to 1, 1, inform them they have discovered the so-called Fibonacci numbers. Have them guess how many spirals the huge sunflower has (answer: 55, 89). (See Template 2.2.2.) Stress that by looking at simple things deeply, we can see hidden patterns in our world around us that, when abstracted, allow us to have new insights about our world.

15 minutes—Armed with the theme of consecutive pairs of Fibonacci numbers, ask students to explore the relative sizes of two adjacent Fibonacci numbers—the bigger is not quite two times the smaller. What else do the students see? Make a chart of quotients of consecutive Fibonacci numbers, as on text page 52, and then help students discover—through cases as on page 53—that the quotient can be expressed solely in terms of 1's—another surprising discovery.

10 minutes—Ask what happens as the Fibonacci numbers used to compute the quotient of 1's get larger and larger. What number would we head toward (answer: an "infinite" fraction of 1's)? Use this observation to derive the Golden Ratio, as done on text pages 54 and 55. Foreshadow the Golden Rectangle.

## Day Two

Question of the Day (to be written on the board for students to think about and discuss before class begins): *Because not every counting number is a Fibonacci number, is there any way to express non–Fibonacci numbers in terms of Fibonacci numbers?*

5 minutes—Lament how sad it is that not every counting number is a Fibonacci number. Then write the first few Fibonacci numbers on the board and ask students if there is any way to express 15 in terms of Fibonacci numbers.

5 minutes—Have students consider various examples to help them discover that every counting number can be expressed as a sum of nonadjacent Fibonacci numbers.

20 minutes—Introduce Fibonacci nim, as described on text page 56, and play a round with a student so you win. Then explain (without formal proof) the winning strategy and play a few rounds with the class for students to gain practice at the game and at the decomposition process.

**Section 2.3 Prime Cuts of Numbers** [Prime numbers]. One fundamental principle of this section is the idea of seeking elementary building blocks as a technique for understanding. Building up larger numbers by multiplication evolves into a discussion of factorization. The Division Algorithm is presented as an example of clarifying a familiar idea, long division, into a precise statement. The section contains proofs of the Prime Factorization Theorem and the theorem that there are infinitely many primes. It concludes with a discussion of the Prime Number Theorem, which concerns the distribution of primes, and a discussion of some famous unsolved problems.

---

*General Themes*

Examining the building blocks of a complex structure answers old questions, invites new questions, and leads to greater understanding.

---

*Mathematical Underpinnings.* (*Caveat:* These mathematical overviews are usually for the instructor only. The symbols and detailed proofs are not suitable for class presentations because mathematical symbols and terminology often present a formidable barrier for students.)

Under the general topic of prime numbers, this section presents ideas ranging from elementary arithmetic techniques to unsolved problems in modern number theory. The Division Algorithm, stated in the text without proof, reminds students about their elementary school experiences with long division and reinforces the compact quality of mathematical notation. The Fundamental Theorem of Arithmetic is stated and proved: Every integer greater than 1 can be written as a product of primes. Proving that the factorization is unique up to the order of the factors is more subtle than, but not central to, this section's emphasis. We chose not to mention this aspect of factorization in the text because the concept of uniqueness is often difficult for students. You may explore in detail both the Fundamental Theorem and the Division Algorithm with students who need review of elementary arithmetic.

The section's core is a proof that there are infinitely many primes. The text presents a method showing that for each natural number $m$ there is a prime number greater than $m$. For each $m$, consider the number $N = (1 \times 2 \times 3 \times \cdots \times m) + 1$. The Division Algorithm shows that $N$ is not divisible by any primes less than or equal to $m$. So, either $N$ is itself prime or $N$ has prime factors that are greater than $m$.

Another classic proof of the infinity of primes uses proof by contradiction. Assume there are only a finite number of primes, say $n$ of them. Label them $p_1, p_2, \ldots, p_n$. Let $M$ equal the product of all these primes plus one, so $M = (p_1 p_2 \ldots p_n) + 1$. Because $M$ is an integer greater than 1, it can be factored into a product of primes. Once again, the Division Algorithm establishes that for each $i = 1, \ldots, n$, the remainder when $M$ is divided by $p_i$ is 1. So, $M$ is not divisible by any of the primes $p_1, p_2, \ldots, p_n$. Thus, there must be at least one more prime number, contradicting the assumption that there are only $n$ of them.

Please note that in either of these arguments, students often mistakenly presume that $N$ (or $M$) is itself prime. A few counterexamples will clarify this error, as will Mindscape II.7 (text page 77).

The section includes a number of results and conjectures that provide a rich context for discussing the nature of mathematical research. Fermat's Last Theorem is, of course, an engaging topic from inception to resolution. The discussion of the Prime Number Theorem—as $n$ increases without bound, the number of primes less than or equal to $n$ approaches $n/\mathrm{Ln}(n)$—provides an opportunity to mention logarithms, as well as the impact of computers on exploring mathematics. If students ask, there is no known formula for the exact number of primes less than a given $n$. Whether such a formula even exists is an open question. The text presents two other open questions: The Goldbach Question asks whether every positive even number can be written as the sum of two primes. (As of June 2004, this result is known to be true for all even numbers less than $6 \times 10^{16}$.) The Twin Prime Question asks if there are infinitely many pairs of primes that differ by two. It's fun for students to realize that even simply stated questions may not yet have been answered.

---

*Our Experience*

*Time spent:* One-and-a-half 50-minute sessions

*Emphasized items:* Infinitude of Primes Theorem and its proof

*Items omitted or treated lightly:* Prime factorization is often omitted. We often don't mention the Division Algorithm. We do not discuss the Prime Number Theorem or Fermat's Last Theorem, except in passing.

*Personal favorite topic or interactions:* We do discuss the two open questions, using them to show students that many questions are unanswered.

*Sample homework assignment:* Read the section and do Mindscapes I.2; II.7, 12, 14, 15, 24; III.32, 35; IV.36.

---

## Sample Class Activities

*Infinitely many primes question.* (See Template 2.3.1 (The first 100 primes)) After talking about building blocks and developing the idea of primes, ask students, Are there infinitely many primes? Why or why not? Display the transparency with the first 100 primes.

*Division Algorithm.* Ask students to do some simple long division and to express their answers with remainders. Then have them express the same long divisions with other "remainders." Help students see that in dividing $n$ into $m$, they can always choose a remainder from 0 to $(n - 1)$. The Division Algorithm expresses a fact about numbers that students basically know; thus, this is a good opportunity to briefly encourage them to formulate a precise statement to capture their own knowledge.

*Introduction to primes—Sieve of Eratosthenes.* In this section, although we do not present the idea of the Sieve of Eratosthenes, it is a good idea to introduce this topic in class. Give students the numbers from 1 to 100. Ask them to start at 2 and cross out the multiples of 2 other than 2; then proceed from 2 to the next number not crossed out, 3, and cross out all multiples of 3 other than itself; then proceed to the next number that is not crossed out, 5, and cross out all multiples of it other than itself. After crossing out the multiples of 7 and noting that 8, 9, and 10 are already crossed out, ask students why the uncrossed-out numbers on their list are all the primes from 1 through 100.

*Prime Factorization Theorem.* Ask students to factor some large numbers as much as possible and develop for themselves a factoring method. Gather ideas to find a technique that always factors numbers into primes and therefore proves the Prime Factorization Theorem. This is a good example for explaining to students that a proof is simply an explanation or technique that works in all the cases covered by the hypothesis. In this instance, doing a specific case and then doing the general case are very similar, so the Prime Factorization Theorem is a nonthreatening introduction to the meaning of a proof.

(*Note:* The text does not explicitly present induction; however, if you do so in your class, the Prime Factorization Theorem is a good example of the $n$ to $n + 1$ formulation of induction not being sufficient.)

*Infinitude of Primes proof.* After doing some examples and going over the Infinitude of Primes proof given in the section, ask students to discuss the variation of the proof in which $N$ is one more than the product of the first $m$ primes, rather than one more than the product of the first $m$ natural numbers. Mindscape III.35 involves this approach. Students may conjecture that $p_1 p_2 p_3 p_4 \ldots p_m + 1$ is always prime, which is a good example of a reasonable conjecture having good supporting evidence for quite some time and yet being untrue in general. Not until $(2 \times 3 \times 5 \times 7 \times 11 \times 13) + 1$ do you get a nonprime.

*Goldbach Question.* The end of the section mentions several questions, two that had remained unsolved for a long time before being settled (the Prime Number Theorem and Fermat's Last Theorem) and two that are still unsolved (the Twin Prime Question and the Goldbach Question). These examples illustrate that mathematics is not finished or complete and that long-standing mysteries are sometimes solved. Ask students to experiment with the Goldbach Question by writing random even numbers as the sum of two primes.

*Fermat numbers.* You may wish to examine the Fermat numbers: $2^{2^n} + 1$. Fermat conjectured that all such numbers are prime. The numbers are prime for $n = 1, 2, 3, 4$, but Euler showed that when $n = 5$, the Fermat number equals $4{,}294{,}967{,}297 = 641 \times 6{,}700{,}417$. There are no other known Fermat primes. We do not usually mention Fermat numbers.

*Factoring.* Ask students how many divisions they would have to perform to factor a large number. You might mention that factoring numbers effectively is a difficult question—for example, given an arbitrary number with 300 digits, no computer now can factor it definitely, even though multiplying such numbers is easy for a computer. This example ties into the question of public key coding.

**Template for Transparency** (*located at the end of the chapter*):

- Template 2.3.1 The first 100 primes

**Sample Test Questions**

(The Mindscapes are another good source of test questions. All test questions below are available electronically on the Test Bank CD-ROM.)

- Let $p$ stand for a prime that is larger than 3—that is, $p > 3$. Can $p + 1$ ever be prime? Why or why not?

- Are there infinitely many nonprimes? Explain.

- Let $M = (1 \times 2 \times 3 \times \cdots \times 1000) + 1$.

  **a.** Is $M$ necessarily prime?

  **b.** What can you say about the prime factors of $M$? Why?

  **c.** Let $S = (2 \times 3 \times 5 \times 7 \times 11 \times \cdots \times 991 \times 997) + 1$—that is, $S$ is one more than the product of all the primes less than 1000. Is $S$ necessarily prime?

  **d.** What can you say about the prime factors of $S$? Why?

- Prove that there are infinitely many primes.

## Sample Lecture Notes for Section 2.3, "Prime Cuts of Numbers" (1 day)

Question of the Day (to be written on the board for students to think about and discuss before class begins): *Can you write 71 as a product of two smaller numbers?*

10 minutes—Write some numbers on the board, and ask students to write the numbers as products of smaller numbers other than 1. Give an example, such as $36 = 2 \times 18 = 2 \times 2 \times 9 = 2 \times 2 \times 3 \times 3$. Define "prime numbers." State the Prime Factorization of Natural Numbers Theorem. Ask students why the theorem makes sense, thus leading into the Division Algorithm, the notion of "divide and conquer." (Treat the Division Algorithm informally or formally, as you wish.)

5 minutes—Ask whether the factorization theorem implies an infinity of primes, hence, no largest prime. If some students think yes, ask them to explain their thinking. Point out that even with only a finite number of primes, they can create infinitely many numbers simply by using higher and higher powers of those primes.

15 minutes—State and prove that there are infinitely many primes (text pages 70–73). Emphasize the beauty of the result and cleverness of the proof (text page 71), and point out the power and importance of the theorem. (For instance, having no largest prime number is important in the development of encryption algorithms used to protect electronic financial transactions. See Section 2.5.)

10 minutes—Ask: What proportion of numbers less than 100 are prime? What proportion less than 1000 are prime? Less than 1,000,000? Display the table on text page 72 and discuss the Prime Number Theorem. Treat the logarithm only in as much detail as you wish. Point out that nearly a century passed between the first conjecture of the Prime Number Theorem and a proof. Encourage students to read about Fermat's Last Theorem (text pages 73–75), for which a 350-year gap existed between conjecture and proof!

10 minutes—State the Twin Prime Question and the Goldbach Question. Ask students to work out a few examples for each question. Have the class discuss how they would try to prove these conjectures. Discuss any ideas students have before revealing that the questions are still unanswered. Finish by acknowledging how much we still don't know, even about such seemingly simple structures as the natural numbers.

**Section 2.4 Crazy Clocks and Checking Out Bars** [Modular arithmetic]. This section deals with modular arithmetic and its applications to check digits on universal product codes (UPCs) and bank checks. Starting with the ordinary experience of clocks, students can be drawn to develop the ideas of modular arithmetic. The section presents several opportunities for introducing general habits of thought. One opportunity is abstracting a rather concrete idea, like clock arithmetic on an ordinary clock, to a more general setting. Another is developing an idea by looking at examples and formulating statements about it. Check digits and bar codes are good everyday examples of mathematics in ordinary life.

---

*General Themes*

Generalizing a simple idea (telling time on a clock) can lead to important applications.

---

*Mathematical Underpinnings.* (*Caveat:* These mathematical overviews are usually for the instructor only. The symbols and detailed proofs are not suitable for class presentations because mathematical symbols and terminology often present a formidable barrier for students.)

This section includes several examples of using check digits to detect errors in numerical codes for products or accounts. For a 12-digit UPC code (the numbers found under the bar code), the standard method works as follows. If the UPC code has digits $d_1 d_2 d_3 d_4 \ldots d_{11} d_{12}$, they must satisfy the equation

$$3d_1 + d_2 + 3d_3 + d_4 + \cdots + 3d_{11} + d_{12} \equiv 0 \ (\text{mod } 10) \ (*).$$

The last digit, $d_{12}$, is usually called the check digit. Given the values of $d_1$ through $d_{11}$, the value of $d_{12}$ is exactly the digit needed to make the given sum equivalent to 0 (mod 10).

**Claim:** *Suppose $d_1 d_2 d_3 d_4 \ldots d_{11} d_{12}$ is a valid UPC code. If a single digit is changed, then $3d_1 + d_2 + 3d_3 + d_4 + \cdots + 3d_{11} + d_{12}$ will not be equivalent to 0 (mod 10).* (Thus we have a simple arithmetic way to check for single errors in a 12-digit UPC code.)

**Proof:** Suppose the first digit is given incorrectly as $d_1'$, but the remaining digits are reported correctly. We evaluate $d_1' d_2 d_3 d_4 \ldots d_{11} d_{12}$ in equation (*) to get $3d_1' + d_2 + 3d_3 + d_4 + \cdots + 3d_{11} + d_{12}$. If we let $e = d_1' - d_1$, the error in the reporting of $d_1$, then $d_1' = d_1 + e$. We have

$$3d_1' + d_2 + 3d_3 + d_4 + \cdots + 3d_{11} + d_{12}$$

$$= 3(d_1 + e) + d_2 + 3d_3 + d_4 + \cdots + 3d_{11} + d_{12}$$

$$= (3d_1 + d_2 + 3d_3 + d_4 + \cdots + 3d_{11} + d_{12}) + 3e \equiv 0 + 3e \equiv 3e \pmod{10}.$$

Because $e$ is a nonzero number between –9 and 9, the value $3e$ will never be equivalent to 0 (mod 10).

---

*Our Experience*

*Time spent:* One 50-minute session

*Items omitted or treated lightly:* Unless you intend to do public key cryptography, this section can be light and entertaining.

*Remarks:* We sometimes omit this section in class, because it makes for an outstanding group research project, with good poster session possibilities.

*Sample homework assignment:* Read the section and do Mindscapes I.4; II.6, 7, 9, 13; III.26, 32; IV.37.

---

### Sample Class Activities

*Mod clock arithmetic.* After giving some examples of mod clock arithmetic, ask students to devise a method for figuring out the day of the week for any day next year. Help them see how they must figure out 30 (mod 7) or 31 (mod 7) to go from month to month or 365 (mod 7) to go from year to year. Of course, you must deal with February and leap years. A great challenge is for students to compute how many years will pass before the days of the week are back on the same cycle.

*Mod clock arithmetic, more advanced.* If your class is strong enough for some added precision, ask students to formulate a numerical statement about when $x \equiv y \pmod{12}$. Lead students to discover that $x \equiv y \pmod{12}$ if and only if 12 divides evenly into $x - y$. Show them the utility of that formulation in several settings—for example, in seeing that if $a \equiv b \pmod{n}$ then (1) $a + c \equiv b + c \pmod{n}$, (2) $ac \equiv bc \pmod{n}$, and (3) $a^m \equiv b^m \pmod{n}$. This last one requires $a - b$ to be a factor of $a^m - b^m$.

26

*Check digits.* After reviewing the examples on text pages 86–87 and possibly some presented in the Mindscape section, ask students to devise a check digit scheme with two check digits, perhaps combining two of the schemes given in the text. Ask them how accurate such a system would be.

*Fermat's Little Theorem.* If you are going to do public key cryptography in detail (which we do not usually choose to do), you might want students to look at numerical examples of $a^{(p-1)}$ (mod $p$) and seek a pattern. For example, have different students in class simultaneously compute $2^6$, $3^6$, $4^6$, $5^6$, $6^6$ (mod 7), and then gather the answers. Experiment with other primes and then do an example mod 6 to show the failure of the patterns with nonprimes.

## Sample Test Questions

(The Mindscapes are another good source of test questions. All test questions below are available electronically on the Test Bank CD-ROM.)

- Define: $a \equiv b$ (mod $n$). Read "$a$ is equivalent to $b$ modulo $n$." Illustrate your definition with an example.

- Fill in the blanks in the following congruences with any correct answers other than the number itself:

  _____ $\equiv 8$ (mod 6); _____ $\equiv 132$ (mod 140); _____ $\equiv -3$ (mod 5).

- July 9, 1949, was a Saturday. What day of the week was July 9, 1950?

- You started a long mathematics exam at 2:00 p.m. You were told that you could work as long as you liked. You worked 487 hours straight. At what time of day did you finish?

## Sample Lecture Notes for Section 2.4, "Crazy Clocks and Checking Out Bars" (1 day)

Question of the Day (to be written on the board for students to think about and discuss before class begins): *Today is Monday, March 10. On what day of the week will the Fourth of July fall this year?* (Fall classes can use Halloween or New Year's Eve instead of the Fourth of July.)

5 minutes—Discuss the Question of the Day to introduce clock arithmetic. Ask if anyone is from or has visited a country that uses a 24-hour clock. Perform examples of addition in both formats.

10 minutes—Introduce the general idea of modular arithmetic and equivalence mod $n$. Use a specific value, say $n = 5$, for examples, then switch to a larger value, such as $n = 60$.

5 minutes—Do a few examples that require simplifying more complicated expressions mod $n$, including one that anticipates the UPC application. (For a great opportunity to review some of the laws of exponents, include examples with large exponents, such as reducing $2^{300}$ (mod 7).)

10 minutes—What good is modular arithmetic? Show students a sample bar code and UPC number. (If you asked them to bring in samples, you can use those, as long as they have a total of 12 digits.) Ask someone to read off all the UPC digits except the last one. Show the class how you can determine the last digit (following the technique on text pages 86–87). Now ask someone to read off the digits of another UPC, but this time with one wrong digit. Show the class how the same technique will show that the given UPC number has a mistake. Have students pair up and try this themselves.

10 minutes—Discuss the general idea of a check digit and why the technique for 12-digit UPC numbers will always detect a single error. Ask students what happens if two digits are interchanged. See if they can find an example when such a switch would not be detected by this method.

10 minutes—Discuss and do some examples of another type of error-detection system, such as bank identification numbers (text example on page 88) or post office money orders (Mindscape III.28). Have students pair up to check the method themselves.

**Section 2.5 Public Secret Codes and How to Become a Spy** [RSA public key cryptography]. This is probably the most notationally, algebraically, and computationally challenging section in the book. It is also one of the most modern and interesting applications of number theory. This dichotomy makes the topic a great challenge for both students *and* instructors. Because the authors' experience and approaches have been so varied, we will omit the traditional Our Experience box. We offer instead four possible approaches to this section.

1. Omit the topic completely, as one of us does. Or, suggest this topic as independent research for a stronger group of students.
2. Discuss the basic idea of public key codes and present an overview of the strategy, using the Carson City Kid to illustrate the theme.
3. Present the method as an algorithm and illustrate the procedure with some examples—that is, code a word and then decode it to see that you return to the original message. In this case, you outline how to use the method but give no explanation (such as no discussion of Fermat's Little Theorem).
4. If your class is strong or if you are really enthusiastic, you may wish to treat or outline the entire issue, including some justification of why the RSA Coding Scheme works. One of us actually took this route, including background and proof of Fermat's Little Theorem, but spent almost one month doing it.

As you can see, the authors have taken the extreme routes—perhaps one of the more intermediate stances would be best.

---

> *General Themes*
>
> Looking at things that seem abstract and devoid of application today may be central in our daily lives tomorrow.

---

*Mathematical Underpinnings.* (*Caveat:* These mathematical overviews are usually for the instructor only. The symbols and detailed proofs are not suitable for class presentations because mathematical symbols and terminology often present a formidable barrier for students.)

This section's topic is cryptography. The mathematics are quite complex, but the motivation is both fascinating and pragmatic. Because the material is so technical, we actually recommend that this section *not* be covered formally in class. A very bright, motivated student could study the technical aspects of this section as part of an independent project, but for the class as a whole, we encourage you to discuss only the philosophical underpinnings so students understand the general ideas and issues of public key cryptography.

Students will certainly appreciate the value of sending secret messages. It's easy to convince them that classical techniques, such as Caesar cyphers, have serious drawbacks: they are easy to break and to infiltrate. If someone steals the codebook, he or she can not only read private messages but also send fraudulant messages. More fundamentally, with such a code, anyone who can code a message can also decode any other message created using that code. This last point seems impossible to overcome. A coding process should be reversible.

Public key coding provides an answer to this conundrum: it is a coding process that cannot be reversed. Everyone knows the key, but no one except the owner can easily decode it. Public information; private decoding.

The text includes all the mathematics underlying basic public key cryptography. The principal result being applied is known as Fermat's Little Theorem:

*If p is a prime number and n is any integer that does not have p as a factor, then $n^{p-1}$ is equivalent to 1 (mod p).*

This theorem is an excellent example of a result in pure, abstract mathematics. It was proved in the seventeenth century with no motivation or application other than furthering the understanding of numbers and arithmetic. Now, more than 350 years later, this theorem is critical to all private electronic communication. Students who see abstract mathematics as having little or no practical value will appreciate learning that the three mathematicians who used Fermat's result to devise a successful method for public key cryptography sold their company for $400 million.

### Sample Class Activities

We suggest that you use some of the appropriate Mindscapes from the section as class activities, depending on which approach you decide to take. It's a beautiful topic, if you dare.

### Sample Test Questions

(The Mindscapes are another good source of test questions. All test questions below are available electronically on the Test Bank CD-ROM.)

- What is the remainder when $23^{2639}$ is divided by 13?

- State Fermat's Little Theorem. Use it to find a number between 0 and 30 equivalent to $6^{32}$ (mod 31). You must find this congruence in your head and state how you found it.

- This problem outlines an inductive proof of the following theorem, which is basically the same as Fermat's Little Theorem.

  **Fermat's Little Theorem:** *If p is a prime, then $n^p \equiv n \pmod p$.*

  **a.** State and prove the base case, $n = 1$.

  **b.** Show that if $k^p \equiv k \pmod p$, then $(k + 1)^p \equiv (k + 1) \pmod p$. Please explain your steps verbally (in sentences) so that we can understand your reasoning. You may want to use the following fact about the expansion of polynomials—namely, $(x + 1)^p = x^p + px \cdots + 1$. Here are some examples:

  $(x + 1)^2 = x^2 + 2x + 1$

  $(x + 1)^3 = x^3 + 3x^2 + 3x + 1 = x^3 + 3(x^2 + x) + 1$

  $(x + 1)^5 = x^5 + 5x^4 + 10x^3 + 10x^2 + 5x + 1 = x^5 + 5(x^4 + 2x^3 + 2x^2 + x) + 1.$

- Suppose you are the sender of a message, 13, that is encoded using the public information that the modulus $n = 35$ and the other number is 5.

  **a.** What number do you send?

  **b.** Now suppose you are the receiver. How do you decode the message?

  **c.** Why does your process of decoding the message work?

  **d.** Suppose you were able to factor 35 and you knew the public numbers 5 and 35. Could you crack the code? How would you do it? Explain.

  (*Instructors:* Note that $1 = 5 \times 29 - 6 \times 24$, if that is relevant.)

## Sample Lecture Notes for Section 2.5, "Public Secret Codes and How to Become a Spy" (1/2 day)

Question of the Day (to be written on the board for students to think about and discuss before class begins): *Which is easier, multiplying or factoring?*

5 minutes—Ask if any students have ever taken cash out of an automated teller machine (ATM) or used a credit card to buy something online. Ask who feels confident that their bank records are accurate and safe. Ask someone who says "yes" to briefly justify his or her confidence. This discussion leads into the need for ways to encode information.

5 minutes—On a transparency, display a simple message encrypted with a Caesar cypher. (Example: "ZKK RXRSDLR ZQD ETMBSHNMHMF MNQLZKKX, CZUD" encodes the message "ALL SYSTEMS ARE FUNCTIONING NORMALLY, DAVE," using a simple shift of the alphabet: A→Z, B→A, C→B, and so on.) Quickly reveal the coding mechanism, and acknowledge how simple it is to break such a code. Point out that anyone who knows how to encode a message with such a code will also know how to decode it. (This quality applies to all codes used prior to the late 1970s, even to highly sophisticated and historically significant codes, such as the Enigma code used by Nazi Germany during World War II.)

5 minutes—Now ask students if they think it's possible to create a code with which anyone can send encrypted messages to the owner, but no one other than the owner can decode (called a *public key coding* system). The answer is yes, and the method involves products of prime numbers. Explain that the details of such codes are rather technical, but the underlying mathematical principles arise very simply from a theorem proved 350 years ago by Fermat.

10 minutes—Give students a number, say 6, and tell them it's the product of two primes. Anyone who determines the two primes has broken the code! Do the same for 77. It's a product of two primes—can students find them? If so, they just broke the code. Now try 187 ($11 \times 17$), 851 ($23 \times 37$), and 19,549 ($113 \times 173$). (If some students have calculators, have them work in groups to try factoring 187 and 851.) Acknowledge that factoring gets harder as the primes get larger. Now give the number 802,027,811 and announce that it's the product of two primes ($80,021 \times 99,991$). Would students like to factor that number? Point out that even the fastest computers would take hundreds of years to factor a 100-digit number obtained as the product of two 50-digit primes.

35

5 minutes—Announce that students have now seen the basic idea underlying public key cryptography. Multiply two large primes together to get a very big number. Publicize your big number, but keep its two prime factors secret. Although others will be able to encode messages using the big number, following the technique explained in the text, decoding requires the two prime factors, which only you know. Because multiplying is easy but factoring is very hard, even for computers, your secrets are safe. If computers get better at factoring, just choose larger primes. (There are infinitely many!) Close with a salute to Fermat, whose seemingly abstract theorem of long ago has made our modern, ATM, Web-shopping world possible and to the mathematicians who first invented public key coding and ultimately sold their creation for $400 million.

**Section 2.6  The Irrational Side of Numbers** [Irrational numbers]. This section and Section 2.7, "Get Real," present an introduction to the real numbers. This section begins by explaining that between every two rational numbers there is another rational number. The principal content of the section is a proof that $\sqrt{2}$ is irrational. The same proof is shown to apply to other cases, such as $\sqrt{3}$ and $\sqrt{2} + \sqrt{3}$. These proofs provide excellent examples of the technique of following the consequences of an assumption, in this case that $\sqrt{2}$ is rational, and finding that the assumption leads to a self-contradictory conclusion.

---

*General Themes*

Following assumptions to their logical conclusions can yield powerful results.

---

*Mathematical Underpinnings.* (*Caveat:* These mathematical overviews are usually for the instructor only. The symbols and detailed proofs are not designed for class presentations because mathematical symbols and terminology often present a formidable barrier for students.)

The main focus of this section is to prove the existence of irrational numbers by proving the following theorem.

**Theorem:** *The square root of 2 is irrational.*

**Proof 1:** Suppose $\sqrt{2}$ is rational; for example, $\sqrt{2} = c/d$, where $c$ and $d$ are relatively prime natural numbers. If we square both sides, we see that $2 = c^2/d^2$, or

$$2d^2 = c^2 \text{ (*)}.$$

We see that 2 divides into $c^2$, and because 2 is a prime, we must have that 2 divides into $c$. That is, $c = 2n$. Thus, identity (*) becomes $2d^2 = 4n^2$, or simply $d^2 = 2n^2$. We see that $d^2$ is evenly divisible by 2, which again implies that $d$ must be evenly divisible by 2. Hence, 2 is a factor of both $c$ and $d$, which contradicts the assumption that $c$ and $d$ are relatively prime. Therefore, $\sqrt{2}$ is irrational.

**Proof 2:** Suppose $\sqrt{2}$ is rational; for example, $\sqrt{2} = c/d$, where $c$ and $d$ are natural numbers. If we square both sides, we see that $2 = c^2/d^2$, or

$$2d^2 = c^2 \text{ (*)}.$$

Thus, the number of prime factors (counting repeated factors) appearing on the right side of identity (*) is even, because every factor of the right side appears twice. On the other hand, the number of prime factors on the left side is odd, because the number of factors of $d^2$ is even and the extra factor of 2 makes the total number of factors on the left odd. However, by the Fundamental Theorem of Arithmetic, we know we have unique factorization into prime factors; in particular, any two factorizations of the same number will have the same number of prime factors. Thus, we have arrived at a contradiction; therefore, $\sqrt{2}$ must be irrational.

This proof can be generalized to prove the following theorem:

**Theorem:** *If n is a natural number that is not a perfect square, then $\sqrt{n}$ is irrational.*

The next theorem gives another way to generate irrational numbers.

**Theorem:** *Let r and s be two relatively prime natural numbers satisfying $s > r > 1$. Then $\ln_r s$ is irrational.*

**Proof:** Suppose that $\ln_r s$ is rational; for example, $\ln_r s = a/b$, where $a$ and $b$ are natural numbers. Then we have $r^a = s^b$. Thus, if $p$ is a prime factor of $r$, it must be a factor of $r^a$ and thus a factor of $s^b$ and hence a factor of $s$. But this is a contradiction because $r$ and $s$ are relatively prime. Hence, $\ln_r s$ is irrational.

---

*Our Experience*

*Time spent:* One to one-and-a-half 50-minute sessions

*Emphasized items:* Irrationality of $\sqrt{2}$

*Items omitted or treated lightly:* Irrationality of logs, $\sqrt{2} + \sqrt{3}$

*Dangers:* Students learn the proof of the irrationality of $\sqrt{2}$ but do not clearly see the role of evenness in the proof. Doing $\sqrt{3}$ is a good step toward explaining the essential elements of the proof.

*Remarks:*

1. Show why the proof fails for $\sqrt{4}$.
2. Do $\sqrt{12}$ to demonstrate the correct view of the role of prime factorization.

*Sample homework assignment:* Read the section and do Mindscapes I.3; II.6, 10, 15; III.30; IV.40.

---

## Sample Class Activities

*Rationals between rationals.* The text postpones until Section 2.7 the general issue of the real numbers and how the rational numbers and irrational numbers are distributed. However, this section explains that it is possible to approximate any quantity as closely as one likes with a rational number and that between any two rational numbers there are infinitely many other rational numbers. In this section, students are asked to find a rational between 1001/1003 and 1002/1003 and then to devise a general method for finding a rational between two rationals. Determine whether students can describe a method that is rather precise. You may suggest alternative methods to those that students are likely to devise. For example, point out that it is always possible to add $1/10^n$ for some $n$ to the smaller rational and keep it smaller than the second rational. In other words, there is some specific distance between the rationals, and adding anything smaller than that distance will provide a number between the two numbers. This idea will be helpful for Section 2.7.

*Proof of irrationality of $\sqrt{2}$.* The central idea of this section is the proof that $\sqrt{2}$ is irrational. After discussing the proof in detail, there are several questions you can ask to deepen students' understanding of it. For example, ask them to do and discuss the variation of proving $\sqrt{3}$ is irrational, as is done in the section. Another variation is to ask students to determine what would happen if they forgot the reduction step. That is, suppose they assumed that $\sqrt{2} = a/b$, where $a$ and $b$ are natural numbers, but they omitted the step of reducing $a/b$ to lowest terms. Ask students to work through the rest of the proof to see what happens. Point out that they cannot conclude immediately that there is a contradiction but that they can continue the process of replacing a number they have shown to be even by $2n$, where $n$ is smaller. Explain that because they cannot continue to get smaller and smaller $n$'s forever, they have another proof of the irrationality of $\sqrt{2}$.

*Irrationality of $\sqrt{n}$.* Ask students to prove that $\sqrt{8}$, $\sqrt{27}$, or $\sqrt{12}$ is irrational. Once they see the significance of an odd exponent in the prime factorization, ask them to characterize the prime factorizations of $n$ for which $\sqrt{n}$ is irrational. Point out that they have just proved the irrationality of any square root of any natural number that is not a perfect square. This insight shows that there are many irrationals, not just $\sqrt{2}$, and it provides a segue to Section 2.7.

## Sample Test Questions

(The Mindscapes are another good source of test questions. All test questions below are available electronically on the Test Bank CD-ROM.)

- If the number $M$ is an irrational number, then $1/M$ must be an irrational number as well. True or False?

- Prove that $\sqrt{5}$ is irrational.

- Prove that $\sqrt{15}$ is irrational.

- Prove that $\sqrt{12}$ is irrational.

- Prove that $\sqrt[3]{9}$ is irrational.

## Sample Lecture Notes for Section 2.6, "The Irrational Side of Numbers" (1 day)

Question of the Day (to be written on the board for students to think about and discuss before class begins): *How can we prove that all numbers are rational (in other words, all numbers are fractions)?*

5 minutes—Review the rational numbers and state, as a matter of fact, that all numbers are rational. You may want to attribute this assertion to the Greeks.

10 minutes—Explain that this assertion means that every number can be expressed as a fraction. Now construct $\sqrt{2}$, as on text page 113, and claim it is a rational number. See where this claim takes the class.

15 minutes—Slowly proceed through the proof, as given on text page 114. Reach a contradiction and explain how all the steps were correct, so therefore the original assumption must not be true. Then dramatically place a huge X through the original assertion.

5 minutes—Introduce the idea of irrational numbers, and state the theorem that was just proven; namely, that $\sqrt{2}$ is irrational. Explain how counterintuitive this idea was to the Greeks and is to us. Give some fun history, as described on text page 117.

10 minutes—Consider $\sqrt{3}$, and have students prove it is irrational. (*Note:* Because students will want to use the "even" idea used for $\sqrt{2}$, you should be ready to discuss the idea of 3 dividing into a number evenly rather than just being even.)

10 minutes—Say you're thinking of a number $B$ with the property that $3^B = 10$. Could $B$ be rational? Assume so and proceed with an argument similar to the one on text page 116. Conclude with the thought that if we make an assumption and we follow the logical consequences of that assumption and it leads us to something that is impossible, then that assumption must be false.

**Section 2.7  Get Real** [Real number line]. This section describes the real numbers and how rationals and irrationals are characterized by their decimal expansions. One challenging idea presented is that $0.9999\ldots = 1$. Many students have a difficult time understanding that they cannot talk about the real number "right next to" another real. Another idea challenging to students is that between every two reals there are irrational and rational numbers. The section ends with the provocative idea that a randomly chosen real is irrational, thus leading to the question of whether it makes sense to distinguish the infinitude of the rationals from that of the irrationals. This section shows students something familiar in a new, more refined light, and it might encourage them to reexamine other familiar ideas in other subjects.

---

*General Themes*

Sometimes things that seem commonplace and ordinary are actually exotic and unusual, and things thought to be rare and exceptional are, in fact, the norm.

---

*Mathematical Underpinnings.* (*Caveat:* These mathematical overviews are usually for the instructor only. The symbols and detailed proofs are not suitable for class presentations because mathematical symbols and terminology often present a formidable barrier for students.)

This section begins with that profoundly elegant model for the real numbers: the number line. The decimal expansion of a number $x$ is introduced as an "address" for the point on the number line corresponding to $x$. Given a decimal number, students can identify the corresponding point on the number line in a (perhaps infinite) sequence of steps.

The notion of a *continuum* is informally discussed in this section. Students can easily grasp that between any two distinct numbers there must be another number—simply take the average of the two. This is true, of course, even when looking only at rational numbers. More precisely, the rational numbers are *dense* in the set of real numbers because every nontrivial open interval on the real line contains rational numbers. This property of the rationals becomes more interesting when students realize that there are fewer rationals than irrationals. Chapter 3 discusses the comparative "sizes" of the set of rational numbers and the set of all real numbers; in the context of Chapter 2, however, it suffices to examine the following question: if you select a real number $x$ at random, what is the probability that $x$ is rational? The text (pages 129–130) answers this question informally for a number $x$ between 0 and 1 selected randomly: The probability that $x$ is rational is zero. Recognizing that we haven't formally defined what probability means in this context, here's a more rigorous proof of this claim.

**Theorem:** *If $x$ is a number between 0 and 1 chosen at random, then the probability that $x$ is rational is zero.*

**Proof:** As the text explains, we create a random number $x$ between 0 and 1 by generating digits $d_i$ to the right of the decimal point independently and randomly from among 0, 1, ... , 9. This produces a number $x = 0.d_1 d_2 \ldots$ . So, for each decimal place of $x$, each of the digits 0, 1, ... , 9 appears with probability 1/10. (The value of each digit could be determined by rolling a fair, 10-sided die, for instance.)

Now suppose $x$ is rational. Then we know that the decimal expansion must end in a repeating pattern, possibly a string of all zeroes. So, there is an integer $k$, $k \geq 1$, such that a specific pattern of $k$ digits repeats forever after some point in the list $0.d_1 d_2 \ldots$ . Thus, after the first occurrence of the pattern, every digit that follows is completely determined. Because the digits of $x$ are generated at random, each of these predetermined digits would appear with probability 1/10. Because the digits are produced independently, the probability of a particular sequence of digits is the product of the probabilities of each digit. So, the probability that all the remaining digits of $x$ really do match the pattern must be $\lim_{n \to \infty} \left( \frac{1}{10} \right)^n$, which equals 0. (We take a limit as $n$ goes to infinity because there are infinitely many digits remaining in the decimal representation of $x$.) Therefore, the probability that $x$ is rational is zero.

The informal upshot of this result is that almost all real numbers are irrational. Picking a point at random from the number line will certainly correspond to an irrational number. More formally, the set of rational numbers is a *set of measure zero* in the set of real numbers. The whole numbers and fractions we think of as so common are, in fact, rare birds in the aviary of real numbers.

---

*Our Experience*

*Time spent:* One to one-and-a-half 50-minute sessions

*Emphasized items:* Proof that 0.9999 . . . = 1

*Items omitted or treated lightly:* Betweenness results

*Dangers:* The ideas that real numbers have no immediate neighbors and that rationals and irrationals are intertwined are eye-opening propositions to some students.

*Personal favorite topic or interactions:* Representing repeating decimals as rationals

*Sample homework assignment:* Read the section and do Mindscapes I.2; II.7, 10, 20, 23, 25; IV.36.

## Sample Class Activities

*Rationals everywhere.* Ask students why every interval on the line contains infinitely many rational numbers.

*Irrationals everywhere.* Ask students why every interval on the line contains infinitely many irrational numbers.

*Decimals.* Ask students to draw a real line and to label the integers. Suppose a decimal number has been smudged, so all that is legible is the tenths digit, which is 3: XXX.3XXXXXX .... Ask students to shade in all the possible locations for that point. Suppose a decimal number has been smudged, so all that is legible is the hundredths digit, which is 7: XXX.X7XXXXX .... Ask students to shade in all the possible locations for that point.

*Decimal representation of rationals.* Ask students to neatly write out the long division 7 into 45, with at least 14 places after the decimal point. Ask them why it is quick to see what the decimal answer is forever. Ask them to explain why any rational number must have a repeating decimal representation.

*Shuffling rationals.* Suppose you have two rationals represented as decimals, such as $0.a_1a_2a_3a_4a_5 \ldots$ and $0.b_1b_2b_3b_4b_5 \ldots$, and you shuffle their digits to get $0.a_1b_1a_2b_2a_3b_3a_4b_4a_5b_5 \ldots$. Ask students whether the shuffled number must be rational.

*Unshuffling rationals.* Ask students to take a rational number in its decimal form. Why is the decimal number constructed simply by using the digits in the odd positions still rational? That is, given that $0.a_1a_2a_3a_4a_5 \ldots$ is rational, why is $0.a_1a_3a_5a_7a_9 \ldots$ rational?

*Bag of 0's and 1's.* Suppose each student has a bag of infinitely many 0's and 1's. Ask them how they could use the numbers to wreck the rationality of any decimal number. Specifically, how could they insert 0's and 1's, but never consecutively, into the decimal expansion of a number to make certain that the result is not rational. In other words, give them a decimal number 0.XXXXXXX ... and ask them to change it by adding 0's and 1's to create a number like 0.1X0X1X0X0X1X0X0X1X1X ... that they can be certain is not rational.

*0.9999 . . . = 1.* Students often resist the fact that 0.9999 . . . = 1. Hold a class debate, with some students taking the side that the two numbers are different and others taking the side that the numbers are the same. We hope the "same" siders will win.

*Mostly irrational.* First point out to students that the repeating decimal characterization of rationals means that random decimals are essentially certain to be irrational. Ask students to discuss the hard question of how to reconcile that observation with the facts that between every two rationals there is an irrational and that between every two irrationals there is a rational. Point out that the fact that irrationals are more common than rationals is an instance of our familiarity with rationals biasing our intuition about what is actually common and what is rare.

### Sample Test Questions

(The Mindscapes are another good source of test questions. All test questions below are available electronically on the Test Bank CD-ROM.)

• Which of the following statements are true?

   **a.** Between every two rational numbers there is a rational number.

   **b.** Between every two irrational numbers there is an irrational number.

   **c.** Between every two rational numbers there is an irrational number.

   **d.** Between every two irrational numbers there is a rational number.

• If a real number has an infinitely long decimal expansion, then it must be an irrational number. True or False?

• Is there a real number between 0.123333333 . . . and 0.124? If so, give an example of such a number; if not, just say "no."

• Express the decimal number 3.563737373737 . . . as $n/m$, where $n$ and $m$ are integers. Don't worry about reducing to lowest terms. Show your work.

• Prove that 0.7999999 . . . = 0.800000 . . . .

• What characterizes the decimal expansion of a rational number?

## Sample Lecture Notes for Section 2.7, "Get Real" (2 days)

### Day One

Question of the Day (to be written on the board for students to think about and discuss before class begins): *Is there a number between 0.999 . . . and 1?*

5 minutes—Ask students to define rational and irrational numbers. Give some examples, reminding them about $\sqrt{2}$, from Section 2.6.

5 minutes—Announce that, taken together, all these numbers are called the *real numbers*. Acknowledge that although it's clear what distinguishes these numbers, we also want to know what unites them. Write a short list of numbers on the board, such as 17, 0, –5/2, $\sqrt{2}$, and ask students to arrange the numbers in increasing order from left to right. Point out that this process would work, in theory, for any list of numbers because the real numbers are *ordered*. Thus, it seems natural to use a line to represent the real numbers—the number line!

10 minutes—Introduce decimal expansion as a method by which points on the number line correspond to numbers themselves. Display a portion of the number line, as suggested below, and ask students to locate some points (such as numbers), such as 1.7, –0.4, 2.1:

Point out that the decimal expansion of the number is the "address" for the corresponding point on the number line. Have students locate some more points—0.71, 0.76, and so on—to illustrate how additional decimal places refine the position of the point. Display a "magnified" portion of the line if desired.

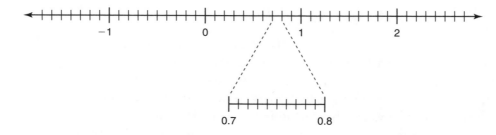

Conclude by noting that because addresses can be continually refined, there will always be a new number between any two distinct numbers on the number line. In particular, between any two distinct rational numbers, $a$ and $b$, is another rational number, $(a + b)/2$.

5 minutes—Ask students how they might locate an irrational number on the number line. Show them how to find $\sqrt{2}$, as in the text (pages 123–124).

15 minutes—Now ask how decimal expansions of rational numbers differ from those of irrational numbers. Find the decimal expansions of some rational numbers, some terminating and some repeating, such as 11/4, 1/3, 22/7. Suggest the result that rational numbers are exactly those with terminating or repeating decimal expansions but acknowledge the need to know that the process can always be reversed. Do some examples, such as transforming 7.636363 . . . and 12.34567567567 . . . into rational numbers, as on page 128.

10 minutes—Ask students if they think this decimal expansion is unique for each rational number. Have them discuss the Question of the Day among themselves, then take a vote. If someone thinks there is a number between 0.999 . . . and 1, ask them to describe it. If no one thinks such a number exists, ask someone to explain his or her thinking. Then give the algebraic proof that 0.999 . . . = 1 (text page 129). (*Note:* If you let them, your students may argue about this result for the entire class. This can be a very exciting conversation, or you may prefer to move on.)

**Day Two**

Question of the Day (to be written on the board for students to think about and discuss before class begins): *Look at the number 1.234567891011121314 15 . . . . How are the digits of this number created? Is it rational or irrational?*

10 minutes—Write the following numbers on the board and ask students which are rational and why: 1.25, 0.333 . . . , 17.3965, 4.121212 . . . . Remind students of the characterization of rational numbers as those with terminating or repeating decimal expansions. Pose the question: if a number is irrational, what must its decimal expansion look like? Discuss the Question of the Day. Ask students to create other examples of irrational numbers in decimal form.

15 minutes—Ask students to create a random decimal number. Have them call out digits at random for you to write on the board. Now ask them to imagine doing this forever. Will the resulting number be rational or irrational? Encourage students to discuss their thoughts briefly with each other, then conclude with the following claim: the probability that a randomly generated decimal number is rational is zero. (See the text discussion on pages 129–130.) This is a wonderful place for interesting discussion. If you like, have students use a die to generate a sequence of digits for a randomly generated decimal number. Note that it really doesn't matter that you'll only get digits from 1 to 6—the chances of creating a repeating decimal are still zero.

# Template 2.2.1a   Daisy

# Template 2.2.1b   Overlay of daisy with spirals marked

21 spirals

# Template 2.2.1c  Overlay of daisy with reverse spirals marked

34 spirals

# Template 2.2.2   Sunflower spirals

## Template 2.3.1    The first 100 primes

| | | | | | | | | | |
|---|---|---|---|---|---|---|---|---|---|
| 2 | 3 | 5 | 7 | 11 | 13 | 17 | 19 | 23 | 29 |
| 31 | 37 | 41 | 43 | 47 | 53 | 59 | 61 | 67 | 71 |
| 73 | 79 | 83 | 89 | 97 | 101 | 103 | 107 | 109 | 113 |
| 127 | 131 | 137 | 139 | 149 | 151 | 157 | 163 | 167 | 173 |
| 179 | 181 | 191 | 193 | 197 | 199 | 211 | 223 | 227 | 229 |
| 233 | 239 | 241 | 251 | 257 | 263 | 269 | 271 | 277 | 281 |
| 283 | 293 | 307 | 311 | 313 | 317 | 331 | 337 | 347 | 349 |
| 353 | 359 | 367 | 373 | 379 | 383 | 389 | 397 | 401 | 409 |
| 419 | 421 | 431 | 433 | 439 | 443 | 449 | 457 | 461 | 463 |
| 467 | 479 | 487 | 491 | 499 | 503 | 509 | 521 | 523 | 541 |

# Chapter 3  Infinity

## I.  Chapter Overview

Infinity is a wonderful topic for various reasons. It captures the imagination; many of the results go against what one would initially guess; and it beautifully illustrates the power of careful and logical analysis. Infinity allows for lively discussions and even passionate debates. Perhaps the central overall lesson for this chapter is the technique of taking a vague notion and making it precise. Students often have a mysterious, grand notion of infinity, but, as was the case for all humans for most of history, they do not have a means to think about infinity except in mystical terms. That a notion so grand can be tamed and developed into a logically meaningful system is startling. The technique of culling out a central definition, rather than relying on a vague feeling, is a powerful technique that students can use in many settings beyond mathematics.

Traditionally, topics of cardinality, and the arguments involved, have been viewed as too difficult for nonmathematics students; however, our experience with teaching these topics has been extremely positive. By building a strong sense of the idea of one-to-one correspondence, by avoiding as many formal definitions as possible, and by introducing some of the ideas through such means as the Dodge Ball game from Chapter 1, "Fun and Games," you can help students discover seminal ideas behind several of the core examples and proofs in this subject. In addition, infinity has an intrinsic fascination that helps keep interest high.

---

### Dependencies

Sections 3.1 and 3.2 are about one-to-one correspondence and are necessary for all subsequent sections of this chapter. Section 3.3 contains Cantor's proof that the reals have a larger cardinality than the natural numbers. Section 3.4, which includes Cantor's proof that the power set has larger cardinality than the set, requires Section 3.3. Section 3.5 is dependent only on Sections 3.1 and 3.2.

---

*Reminders*

1. Class activities are suggested for each section. These can be done with students working in groups of two or three and then gathering their responses or as one global class discussion.
2. Before introducing a new topic, engage students with enticing questions, which they quickly discuss with neighbors; solicit responses within a minute. Use the responses as the springboard for the introduction.
3. We have included more sample class activities than you will probably want to use. Choose only the ones that are right for you and your class.
4. Each section can be treated in more or less depth. It is frequently a good idea to omit difficult technicalities in order to treat the main ideas well and then move on.

## II. Section-by-Section Instructional Suggestions

**Section 3.1 Beyond Numbers** [Introduction to one-to-one correspondence]. This section introduces the central idea to the mathematical analysis of infinity. The lesson beyond mathematics for this section, and indeed for the chapter as a whole, is that the key to understanding the complex unknown is a deep understanding of the simple and familiar.

*General Themes*

Looking at familiar ideas in a different way can lead to new and important insights.

*Mathematical Underpinnings.* (*Caveat:* These mathematical overviews are usually for the instructor only. The symbols and formal definitions are not suitable for class presentations because mathematical symbols and terminology often present a formidable barrier for students.)

This section introduces the idea of a one-to-one correspondence as the measure of when two sets have the same size. More formally, given two sets A and B, a function $f$ from A to B is defined to be a *one-to-one correspondence* if $f$ is both one-to-one and onto. (Though "one-to-one correspondence" is the mathematical term, it's easier to use "one-to-one pairing.") Recall that a function is *one-to-one* if distinct elements in the domain must have distinct images in the range. Thus, if $x$ and $y$ are in A (the domain of $f$) and $x \neq y$, then $f(x) \neq f(y)$. This condition is often stated in the form of the contrapositive: if $x$ and $y$ are in the domain of $f$ and $f(x) = f(y)$, then it must be the case that $x = y$. Schematically,

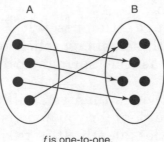

$f$ is one-to-one.

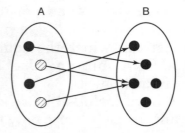

$f$ is not one-to-one.
(The striped elements of A have the same function value in B.)

The onto condition of the one-to-one pairing definition is defined as follows: A function $f$ from A to B is *onto* if every element in B is the image under $f$ of some element in A (the domain of $f$). Thus, for each $y$ in B, there is at least one $x$ in A such that $f(x) = y$. Schematically,

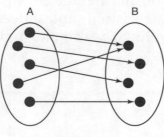

$f$ is onto.

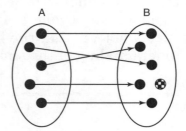

$f$ is not onto.
(The checkered element of B is not the image under $f$ of an element of A.)

Notice that a function that is both one-to-one and onto has an inverse function. That is, if $f$ is a function from A to B that is both one-to-one and onto, then there is a function $g$ from B to A such that for all $x$ in A, $g(f(x)) = x$, and for all $y$ in B, $f(g(y)) = y$. So, $g$ is the inverse of $f$, and vice versa. In this case, both $f$ and $g$ are one-to-one pairings.

### Sample Class Activities

*Provocative questions.* Ask students for their opinions about infinity. What is infinity? Students will have many vague notions. Lead them to the idea that one part of infinity refers to how many things you have when you have more than finitely many. In other words, part of the idea of infinity relates to an extension of counting and number.

*One-to-one correspondence.* Begin with a physical demonstration: bring in two jars filled with the same number of some objects, and ask how to determine if the two jars contain the same number of objects. The first answer you will probably get is to count the objects in each jar. Respond with the question, Suppose we couldn't count that high? Students will soon have the idea of pairing the objects; this idea naturally allows you to help students develop the idea of a one-to-one correspondence. Physically pair the objects and perform the correspondence (we suggest that the two sets have the same cardinality). Emphasize that without knowing how many things were in either jar, you have discovered that there are the same number of things in the two jars.

Chapter 3 uses the natural, and convincing, informal definition of a one-to-one correspondence instead of more formal definitions. The formal definitions of a map or function, domain, range, and onto can add a substantial layer of distance between students and the ideas in the cardinality proofs; however, you may prefer to include more formal definitions of these ideas.

*One-to-one correspondence with numbers.*   Illustrate the notion of a one-to-one correspondence with as many examples as you can. This variety will build a solid foundation and will help develop students' intuition before you move toward more foreign, abstract, and counterintuitive material. Examples such as the correspondence between {1, 2, 3, . . . , 9} and {2, 4, 6, . . . , 18} or between {1, 2, 3, 4, 5} and {2, 3, 4, 5, 6} are useful precursors to Section 3.2. Another useful example is having students discover a one-to-one correspondence between {–3, –2, –1, 0, 1, 2, 3} and {1, 2, 3, 4, 5, 6, 7}. After they comprehend one such correspondence, have them do another one with some of the numbers covered up. Don't reveal the hidden numbers until later. As an example, the sets could be {■, ■, . . . , ■, –2, –1, 0, 1, 2, ■, . . . , ■} and {1, 2, 3, ■, . . . , ■}. Ask students to find a pattern of one-to-one correspondence that could be maintained, no matter how many numbers are covered.

*Attribute vs. comparison.*   If your class is philosophically oriented, you may want to try this activity. Convince students that one-to-one correspondence is an excellent definition for two collections that have the same number of things. Stress the notion that one-to-one correspondence is a means of comparison. The philosophical distinction between "how many" versus "compare" can lead to interesting discussions. Ask the class for examples in regular life where we use comparisons more commonly than absolute statements. For example, when measuring the quality of tennis players, the method is to have one play the other rather than measuring various qualities of each, such as speed of serve or accuracy of return, and coming up with an absolute measure of quality.

### Sample Test Questions

(The Mindscapes are another good source of test questions. All test questions below are available electronically on the Test Bank CD-ROM.)

- Is there a one-to-one correspondence between the set {x, ?, &, m, @, A} and the set {1, 2, 3, 4, 5}? If there is, express it. If there is not, explain why not.

- Every seat in a theater has a row and seat number. Without counting the seats, how could you produce tickets so you know you will have the same number of tickets as seats?

## Sample Lecture Notes for Section 3.1, "Beyond Numbers" (1 day)

Question of the Day (to be written on the board for students to think about and discuss before class begins): *What comes to mind when you hear the word "infinity"? Write down your thoughts or questions.*

5 minutes—Call on several students or ask for volunteers to read their thoughts or questions about infinity.

5 minutes—Ask students what it means for a set to be finite. Promise a more formal discussion as part of Section 3.2 (see Lecture Notes for Section 3.2). For now, an intuitive sense of a finite set is sufficient—you can count the elements, there is some fixed number of elements in the set, and so on.

5 minutes—Ask how they compare the sizes of two sets. Display two boxes of donut holes and confess that you don't know how many are in each box because you nibbled at them on the way to class. Ask students how they would determine which box had more donut holes or whether both boxes contained the same number. Lead into the idea of one-to-one pairing. (Then pass around the donut holes!)

10 minutes—Provide many examples of one-to-one pairings, including small sets. Help students come to the realization that two sets have the same size precisely when their elements can be put into one-to-one pairing. Give a few examples of sets that are not the same size.

1 minute—Have students stand, raise their right hands, and solemnly swear that the only way two sets can be the same size is if their elements can be put into a one-to-one pairing.

**Section 3.2 Comparing the Infinite** [Examples of one-to-one correspondence]. This section introduces the basic definition of the subject of infinity—namely, that two sets have the same cardinality if they are in one-to-one correspondence. The notion of same cardinality is then explored through several examples. One example is a mental experiment involving Ping-Pong balls. Others involve one-to-one correspondences between the natural numbers and various sets, including {2, 3, 4, . . .}, the integers, and the rationals. These latter examples are presented as cases of seeking, in vain, sets with larger sizes of infinity. This section's key conceptual issue is for students to rely on the definition of same cardinality as one-to-one correspondence, rather than appealing to other intuitive notions. Getting a student to first think about the definition, rather than vague ideas, is not standard procedure outside of technical fields; therefore, developing the habit requires repetition and practice.

---

*General Themes*

Using a precise definition carefully can lead to counterintuitive discoveries that liberate our thinking and our view of the world.

---

*Mathematical Underpinnings.* (*Caveat:* These mathematical overviews are usually for the instructor only. The symbols and detailed proofs are not suitable for class presentations because mathematical symbols and terminology often present a formidable barrier for students.)

An infinite set is *countable,* or *countably infinite,* if it has the same cardinality as the natural numbers; otherwise, it is *uncountable.* Finite sets are also considered countable. One way to characterize countable sets is as follows: a set is countable if its elements can be listed. In the case of finite sets, this observation is obvious. But suppose $S$ is a countably infinite set, so we have a one-to-one correspondence with the natural numbers:

$$1 \quad 2 \quad 3 \quad 4 \quad 5 \quad \ldots$$
$$S_1 \quad S_2 \quad S_3 \quad S_4 \quad S_5 \quad \ldots$$

Notice that the one-to-one pairing establishes a listing of the elements of $S$: the $i$th element in the list is that element corresponding to $i$ in $\mathbb{N}$. So, for example, showing that the rationals have the same cardinality as $\mathbb{N}$ merely involves showing that it's possible to construct a list of the rationals. The real numbers are uncountable because Cantor's diagonalization argument shows that any claimed listing is missing at least one real number. (In fact, of course, such a claimed list would be missing an uncountable number of real numbers!)

61

One of the most interesting, but also most challenging, examples given in this section is the Ping-Pong Ball Conundrum (text pages 149–151). Some students find the parameters of the experiment *extremely* confusing. It's hard for some to grasp the idea that the experiment involves an infinite number of steps and yet is completed within a finite amount of time (1 minute). Although it's usually best to avoid introducing sums of geometric series or the idea of a limit, this topic does allow you to incorporate these topics. Because the first step in the experiment takes half a minute, and each successive step takes half as much time as its predecessor, the total amount of time for the experiment can be written as $1/2 + 1/4 + 1/8 + \cdots$, a geometric series that sums to 1 minute.

Another stumbling block for many students is the very reasonable observation that for each 10 balls going into the barrel, only 1 ball is being removed. Thus, it seems that there are 10 times as many balls going into the barrel as coming out. The hard part comes in recognizing that an infinite set that seems to be 10 times larger than another infinite set is actually the same size. However, it's easy to set up a one-to-one pairing between, say, one countably infinite set and another that appears to have 10 times as many elements. For instance, the sets A = {1, 2, 3, 4, . . .} and B = {10, 20, 30, . . .} can easily be shown to have the same cardinality. Thus, when the Ping-Pong ball experiment is actually completed, the number of balls removed from the barrel is exactly the same as the number that went in.

The Ping-Pong Ball Conundrum becomes much more transparent if you simply start with all the balls in the barrel. Removing them one by one, as the experiment dictates, clearly results in an empty barrel. However, we encourage you to offer this insight only after much class debate and discussion.

If you want to pose a more subtle question, consider what happens if the balls are not numbered. Even if you start with all the balls in the barrel, the outcome is not determined. When the experiment ends, there could be any number of balls in the barrel: zero, any finite number, or even an infinite number. To see how this could happen, suppose the balls are labeled as in the original conundrum but with invisible ink that can be seen only by the experimenter, who is using special glasses. As the experimenter, you could remove balls 1, 2, 3, . . . , leaving the barrel empty as in the original conundrum. Or you could remove balls 2, 3, 4, . . . , leaving only one ball in the barrel. Or you could remove balls 2, 4, 6, . . . , leaving an infinite number of balls in the barrel. All the while, to any observer, you are merely removing unlabeled balls from a barrel that started out with an infinite number.

Although the Ping-Pong Ball Conundrum can be very difficult for students, it can also be great fun for them to struggle with. If you're looking for a topic to stimulate lively class discussions, even arguments, this is it!

*Our Experience*

*Time spent:* One-and-a-half 50-minute sessions

*Emphasized items:*

1. Ping-Pong Ball Conundrum;
2. students writing one-to-one correspondences by making parallel lists—elements from one set beside or under elements from the other set

*Items omitted or treated lightly:* Countability of the rationals

*Dangers:* The idea of cutting time in half infinitely often is a source of student rebellion. Be prepared, but encourage this interest without being argumentative.

*Personal favorite topic or interactions:* If students think that any balls are left in the barrel at the end of the experiment, ask them to name which Ping-Pong ball(s) remains.

*Remarks:* Students are frequently apathetic about mathematics. If they can become interested enough to be argumentative, encourage that interest.

*Sample homework assignment:* Read the section and do Mindscapes I.3; II.14, 16; III.30, 32; IV.36.

### Sample Class Activities

*Ping-Pong balls.*   Describe the Ping-Pong ball activity. Remind students that the stopwatch will reach the end in 1 minute. Do the experiment. Have students vote for none, finitely many, or infinitely many balls left in the barrel. Ask doubters to name a ball left in the barrel so they can grudgingly convince themselves that the barrel is indeed empty.

*Bridge to infinity.*   Present the natural numbers as the first example of an infinite set. Ask students to compare the cardinality of the natural numbers with the natural numbers with 1 removed. Ask which set is larger. Allow the class to debate and discuss this question. Clarify the two opposing views. When and if the discussion moves off in some strange direction, bring it back on track by reminding students what it means for two sets to have the same cardinality—one-to-one correspondence. Stress that if a one-to-one correspondence can be constructed between two sets, then the two sets have the same cardinality. Make a connection to Section 3.1 by indicating how some of the one-to-one correspondences in that section did not actually depend on the sets being finite. Remind the class of the example from Section 3.1 in which they constructed a one-to-one correspondence between {1, 2, 3, 4, 5} and {2, 3, 4, 5, 6}. Ask students to produce a one-to-one correspondence between the set of natural numbers and the set of natural numbers greater than 1, and then have them explain the significance.

You may find that the class has many questions and arguments. A common response from students is that there is an extra number in the set {1, 2, 3, 4, . . .} not paired up with any natural number under the pairing $n \leftrightarrow n + 1$. Ask students to name such a number, and then show that it does not exist: there is no *last* number. These arguments and discussions make for an interesting class. Some students may be disturbed and even angry, but all such feelings are good indications that students are grappling with a challenging mathematical idea. Point out that these issues are counterintuitive and that mathematicians did not believe such theories at first. Celebrate the notion that it is hard to believe the truth—even when it is proven—when it goes against our preconceived notions and feelings. However, new ideas eventually become the basis for new intuition and belief.

*Some one-to-one correspondences work; some don't.*  Ask students to compare the cardinality of the even natural numbers with the cardinality of the natural numbers. Illustrate the notion that even though some pairings do not give a one-to-one correspondence, this does not imply that no one-to-one correspondence exists. In the example of the evens and the natural numbers presented in Section 3.1, ask students to find a one-to-one correspondence that pairs the evens with only some of the natural numbers. They will probably find the identity pairing in which the even number $2n$ is paired with itself, or the natural number $2n$. This pairing is not a one-to-one correspondence between the two sets; however, the fact that this particular pairing is not a one-to-one correspondence does not imply that the two sets do not have the same cardinality. The possibility for infinite sets coming up with a one-to-one pairing that does not work even when another pairing is a one-to-one correspondence is a difficult but important point. Ask students whether the same kind of phenomenon is possible with finite sets. This distinction between finite and infinite sets illustrates how our limited intuition and personal experience are not always reliable guides when dealing with infinite things.

*Seeking higher cardinalities, in vain.*  At this point, the goal is for students to seek a set that has a greater cardinality than the cardinality of the natural numbers. Make sure students understand what it means for one set to have greater cardinality than another: Given any correspondence (or pairing), it cannot be a one-to-one correspondence between the two sets. Consider the set of integers, which appears to have twice as many elements as the set of natural numbers. Ask students whether the integers and the natural numbers could have the same cardinalities. Then remind them of the pertinent exercise from Section 3.1 and see whether they can extend that example to construct the one-to-one correspondence. Next move to the set of rational numbers, and point out how between any two consecutive natural numbers there are infinitely many rational numbers. Thus, there appear to be infinitely many rational numbers for each natural number, which means there appear to be infinitely times as many rationals as natural numbers. Ask students whether they could construct a one-to-one correspondence between the rationals and the natural numbers. The text presents one example of a one-to-one correspondence between the natural numbers and the rationals (page 154). If you want to present another example, put the numerator and denominator in base 2 and then juxtapose the numerator and denominator into one number with a 3 inserted between the numerator and denominator. So the base 2 rational 110/111 would correspond to the base 10 natural number 1,103,111. This pairing convincingly shows that the rationals cannot be more numerous than the naturals. Of course, this pairing is not onto, so a further step needs to be done to show that any infinite subset of the naturals is in one-to-one correspondence with all the natural numbers.

Students have now seen that the apparently much more numerous rational numbers actually have the same cardinality as the natural numbers. This fact seems to support the (wrong) idea that all infinities have the same size. Students also have the idea that comparing the sizes of infinite sets involves constructing a one-to-one correspondence between the sets.

### Sample Test Questions

(The Mindscapes are another good source of test questions. All test questions below are available electronically on the Test Bank CD-ROM.)

- **a.** Define what is meant by "set A has the same cardinality as set B."

  **b.** Let $S$ be the set of square roots of all positive integers. Show that the set of natural numbers is a subset of $S$.

  **c.** Prove that $S$ has the same cardinality as $\mathbb{N}$, the set of natural numbers. (To receive full credit, you must include all details.)

- List the positive rational numbers in one list so that the pattern of the order is clear and so that all the positive rational numbers would eventually appear on the list.

- Let $\mathbb{R}$ be the set of real numbers and $\mathbb{R}^+$ be the set of nonnegative real numbers. Is the correspondence between these two sets $\mathbb{R} \to \mathbb{R}^+$ given by $n \to n^2$ a one-to-one correspondence? Explain your answer.

- **a.** Suppose you have two infinite sets $A$ and $B$, and you are told that there exists a pairing in which each element from $A$ is associated with exactly one element from $B$ and no two elements of $A$ are paired up with the same element of $B$. In this pairing, however, there are elements of $B$ that are *not* paired up with elements of $A$. Does this imply that the cardinality of $A$ and the cardinality of $B$ are not equal? Carefully explain why or why not.

- **b.** Suppose we replaced the word "infinite" in part a with "finite." Would that change your answer? Explain.

**Sample Lecture Notes for Section 3.2, "Comparing the Infinite"
(2 days)**

**Day One**

Question of the Day (to be written on the board for students to think about and discuss before class begins): *How much bigger is the set of rational numbers than the set of natural numbers? How many fractions are there between 0 and 1?*

10 minutes—To guide students' understanding of infinite sets, start with a careful look at finite sets. Ask students, What is the smallest number of elements a set can have? Someone should suggest zero, yielding the empty set, which is clearly a finite set. Now make the following definition very precisely: a nonempty set is *finite* if there is a natural number $n$ such that there is a one-to-one pairing between the set and the set 1, 2, . . . , $n$. Give some examples.

10 minutes—Define an *infinite set* to be a set that is not finite. Ask for examples: the natural numbers, the real numbers, the set of all possible poems, and so on. Have the class work together to find a rigorous proof that the set, $\mathbb{N}$, of all natural numbers is infinite. Try to let students discover for themselves the need for proof by contradiction: Suppose $\mathbb{N}$ is finite; then, there is some natural number $n$, and so on. Emphasize that the final conclusion ($\mathbb{N}$ is infinite) follows because they couldn't construct a particular kind of one-to-one pairing.

10 minutes—Define what it means for two sets to have the *same cardinality*. Point out that "cardinality" is really just a fancy word for size, but that the word "size" has too many other connotations to be used unambiguously with infinite sets. Be clear that the *cardinality* of a set is not the same as the *number of elements* in the set. Go through a variety of examples of infinite sets, letting the class discover one-to-one pairings as much as possible. (The text and Mindscapes provide many good examples.) Encourage students to try setting up their own one-to-one pairings, making corrections as they go.

5 minutes—Emphasize the importance of one-to-one pairing as the tool for comparing the "sizes" of infinite sets. Point out that many pairings for sets of integers and other numbers involve either a *shift,* as in comparing {1, 2, 3, . . .} with {2, 3, 4, . . .}, or a *shuffle,* as in comparing $\mathbb{N}$ to $\mathbb{Z}$ (text pages 151–152).

5 minutes—Suggest that the natural numbers form a sort of benchmark infinite set—if another infinite set has the same cardinality as $\mathbb{N}$, then the one-to-one pairing allows you to list the elements of the second set. Ask students to suggest sets that they think are "bigger" than $\mathbb{N}$, that is, have greater cardinality. Be sure that the set of rational numbers ends up as a candidate.

10 minutes—Show how to build a list of the rationals, thus establishing that **Q** and ℕ have the same cardinality. Here's one way that even a very naive first attempt can be modified to reach the desired goal:

First attempt:        1,   1/2,   1/3,   1/4,   1/5,   1/6, . . . .

Clearly, lots of rationals are missing, such as 2/3. Between each pair in the list above, insert all the missing fractions with numerator smaller than denominator (inserted numbers have boxes around them):

Insert more fractions: 1, 1/2, 1/3, 2/3 , 1/4, 3/4 , 1/5, 2/5, 3/5, 4/5 , 1/6 . . . .

But we're still missing all the fractions greater than 1. After each number in the list above, insert its reciprocal (again, newly inserted numbers are boxed):

Insert reciprocals: 1, 1/2, 2/1 , 1/3, 3/1 , 2/3, 3/2 , 1/4, 4/1 , 3/4, 4/3 , . . . .

What about all the negative rationals?

Insert negatives: 1, −1 , 1/2, −1/2 , 2, −2 , 1/3, −1/3 , 3, −3 , 2/3, −2/3 , . . . .

And, we can't forget 0!

Insert 0: 0 , 1, −1, 1/2, −1/2, 2, −2, 1/3, −1/3, 3, −3, 2/3, −2/3, . . . .

Conclude with a reminder that the list constructed above establishes a one-to-one pairing between **Q** and ℕ; thus, the answer to the Question of the Day is that there are exactly the same number of rational numbers as natural numbers. Explain how one-to-one pairings, as well as many other useful goals in life, are sometimes achieved only after numerous revisions.

## Day Two

Question of the Day (to be written on the board for students to think about and discuss before class begins): *If you have a barrel full of Ping-Pong balls and remove the balls one by one, how many balls are left when you're done?*

5 minutes—Begin by acknowledging that working with infinity and infinite sets can result in very counterintuitive scenarios, such as the result that there are as many rational numbers as natural numbers. Prepare students for a thought experiment, granting them imaginary powers to complete infinitely many tasks over the span of one minute.

10 minutes—Set up the Ping-Pong Ball Conundrum (text pages 149–151). Emphasize that the experiment has an infinite number of tasks, each taking half as long as the previous task. Be clear that the experiment lasts *exactly*

one minute; the infinite number of tasks will be completed in that time interval. It may help to create a table detailing the first few tasks of the experiment:

| Task Number | Time Left on the Clock at End of Task | Balls Added to the Barrel | Ball Removed from the Barrel |
|---|---|---|---|
| 1 | 1/2 min | 1–10 | 1 |
| 2 | 1/4 min | 11–20 | 2 |
| 3 | 1/8 min | 21–30 | 3 |
| 4 | 1/16 min | 31–40 | 4 |
| ⋮ | ⋮ | ⋮ | ⋮ |

10 minutes—Pose the big question: How many balls are left in the barrel at the end of one minute? Give students a few minutes to discuss the question among themselves. Take a vote on the suggested answers. If a student believes there are balls left in the barrel, ask him or her to name one. Ask other students to counter the claim.

10 minutes—Encourage lively discussion!

10 minutes—Acknowledge the profoundly counterintuitive quality of this example. Now ask students if their sense of the experiment changes if they start with all balls in the barrel. If you all feel ambitious, discuss the experiment as performed with unlabeled balls.

**Section 3.3   The Missing Member** [Cantor's diagonalization proof that $|\mathbb{N}| < |\mathbb{R}|$]. This section presents the proof that the cardinality of the real numbers is greater than the cardinality of the natural numbers. Cantor's diagonalization argument is a challenging concept that requires discussion and repetition; however, students have already been introduced to the idea in the Dodge Ball game from Chapter 1, "Fun and Games."

---

*General Themes*

Extending a new idea to its logical conclusion can lead to surprising and counterintuitive outcomes as well as a more accurate view of reality.

---

*Mathematical Underpinnings.* (*Caveat:* These mathematical overviews are usually for the instructor only. The symbols and detailed proofs are not suitable for class presentations because mathematical symbols and terminology often present a formidable barrier for students.)

This section is challenging. Many students will resist the idea that there are different sizes of infinity. Cantor's diagonalization argument requires focus, as does the more general idea of a proof by contradiction, which many students may have seen only in the proof that the square root of 2 is irrational (Section 2.6).

It helps to emphasize that if a set is the same "size" as $\mathbb{N}$, then its elements can be listed, as discussed in the Lecture Notes for Section 3.2. Thus, to show a set is larger than $\mathbb{N}$ boils down to showing that the elements of that set cannot be listed, that is, the set is not countable. Establishing such a negative property invariably requires a proof by contradiction: Assume the set is countable, examine an alleged list, and construct an element not on the list to yield a contradiction.

Cantor's diagonalization argument gives a technique for doing just this with the real numbers. Here's a compact version of the argument in the text (pages 164–167): Start by assuming the real numbers are countable—they have the same cardinality as $\mathbb{N}$. This assumption must yield a listing of the reals: $r_1, r_2, r_3, \ldots, r_i, \ldots$, which we may assume are represented in decimal expansion. Now construct a number $M$ that is not on the list. For simplicity, assume $0 < M < 1$. Now, choose for the $i$th decimal digit of $M$ a digit different from the $i$th decimal digit of the number $r_i$. Thus, $M$ will not equal $r_i$ for any $i = 1, 2, \ldots$, so $M$ cannot be on the list. Thus, any such list is incomplete, and the assumption that the reals are countable is false.

Some students may be tempted to fix the problem simply by adding $M$ to the list. There will likely be another student who points out that the argument could then be repeated to create another $M$ not on the list. Although this response is usually satisfying to students, it misses the point that the initial assumption guaranteed that the list was complete from the beginning, so the creation of the first $M$ is sufficient to derail the assumption.

Be alert to the possibility that students might, by accident, create an $M$ that ends with repeating 9's, such as $0.123999\ldots$. As discussed in Section 2.7 (text pages 128–129), such a number has a second decimal representation: $0.123999\ldots = 0.124$. In this case, the number $M$ might actually appear somewhere on the list in its alternate form. Because this scenario is potentially confusing, address it only if it comes up.

---

*Our Experience*

*Time spent:* One 50-minute session

*Emphasized items:* Students can readily learn the mechanical diagonalization construction. The challenge is getting them to understand what it implies. The idea that *if* a one-to-one correspondence between $\mathbb{N}$ and $\mathbb{R}$ existed, then you wouldn't be able to find a missing real is a difficult idea, requiring many repetitions and many chances for students to grapple with and express the idea individually.

*Dangers:* After finding a number not on the list, students often say, "Well, I'll put that number at the top of the list." Point out the flaw in this logic.

*Personal favorite topic or interactions:* Seeing students respond to the idea that there is more than one size of infinity.

*Sample homework assignment:* Read the section and do Mindscapes I.4; II.9, 11, 13, 14; III.16, 17, 19.

---

### Sample Class Activities

*Dodge Ball.* Remind students of the Dodge Ball game from Chapter 1, "Fun and Games." Play the game and ask students to generalize the game to more than six rows. After they have thoroughly grasped the strategy, ask them whether the game would work with an infinite number of rows.

*Real numbers.* Remind students that real numbers can be represented in their decimal expansion and that there are infinitely many digits in a decimal expansion of any real number. Show a decimal number in which all the digits are X's except for the digit in the sixth place, for example, 0.XXXXX7XXXXX . . . . Ask students to write a decimal number between 0 and 1 that differs from this number.

*Meaning of cardinality.* Ask students to write the natural numbers in a column on the left side of their paper, with a dot, dot, dot at the bottom to indicate that the list continues. Ask what kind of a list they would have to write on the right side of the page if the cardinalities of the natural numbers and real numbers were the same. Emphasize the idea that if every proposed list of reals in the right column failed to use up all the reals, then the cardinality of the natural numbers must not be the same as the cardinality of the reals.

*Cantor's diagonalization argument.* On a transparency, write a list of reals in the right column, but cover up all but the first line. Tell students that you will reveal the subsequent real numbers one at a time. It is their job to write down a decimal number as you go such that they are certain the number they are producing will not be anywhere on your list. Remind them of the Dodge Ball game and the 0.XXXXXX7XXXX . . . . question. Help them discover and then precisely formulate the criterion for what digit to put in each succeeding decimal place. Recapitulate how this diagonalization technique conclusively proves that the cardinality of the real numbers is not the same as the cardinality of the natural numbers and, therefore, that there is more than one size of infinity.

As you discuss Cantor's diagonalization procedure, several potential questions may arise. (1) Someone may say, "Well, that just shows that this particular correspondence doesn't work, so we have to try to find another." You might respond by saying how you could have done the argument using arbitrary decimal numbers, such as $a_{11}a_{12}a_{13}a_{14} \ldots a_{21}a_{22}a_{23}a_{24} \ldots$ . (2) Another comment you might hear is, "Well, I'll just pair up the number you created with some really big natural number." It is important to stress that your table only illustrates the procedure and that, in fact, the procedure is for *all* correspondences between the two sets, not just for the particular one you wrote on the transparency. It is also important to stress at the start that the whole list is fixed when you are working with it. If there were a one-to-one correspondence, there would be a fixed way of ordering the reals so that they all appeared. Therefore, the lists are etched in stone (we just don't know what to etch!).

It is worthwhile to move slowly but steadily and ask the class to help build the diagonalized number. At each stage, ask: No matter what digits we put on later, could this new number be the real number associated with the natural number 5? Solicit students' answers at each stage. This involvement will allow students to really grasp the issue.

*Understanding diagonalization.* Consider the set of all real numbers whose decimal expansion only has 1's and 7's. Ask students to produce a decimal number made of only 1's and 7's that is not on the list. Ask them why there are more decimal numbers made of 1's and 7's than there are natural numbers.

The diagonalization argument is the heart of the concept that infinities come in different sizes. The more firmly students grasp this idea, the more able they will be to press the ideas forward in Section 3.4. If you feel they have had enough, however, skip the next section with no loss of continuity. Simply mention that, in fact, there are infinitely many different sizes of infinity, and then invite them to read Section 3.4 on their own or as an extra credit or report assignment. However, if you have time and you have a strong class, Section 3.4 contains truly interesting ideas.

Students are often interested to hear about Cantor and his personal struggle in the mathematical community for understanding and acceptance, as well as his confinement in an asylum. The fact that his difficult notions were not even understood by professional mathematicians when the ideas were first presented allows for a certain sympathy and kinship between the students and the world of mathematics.

## Sample Test Questions

(The Mindscapes are another good source of test questions. All test questions below are available electronically on the Test Bank CD-ROM.)

- Following is a list of some decimal numbers corresponding to the first few natural numbers. Describe Cantor's diagonalization argument, and write down the first five digits of a decimal number that Cantor's argument produces.

$$1—0.436827384\ldots$$

$$2—0.728458988\ldots$$

$$3—0.433378349\ldots$$

$$4—0.444444444\ldots$$

$$5—0.222222222\ldots$$

- Suppose each natural number is associated with a decimal number between 0 and 1.

  **a.** Describe a decimal number between 0 and 1 that uses only the digits 3 and 5 and that is not on the list, that is, it is not associated with any one of the natural numbers. Illustrate your method by writing down five sample decimals associated with 1, 2, 3, 4, 5 and illustrating how you would start.

  **b.** In addition to the decimal number you described in part a, describe two other, different decimal numbers between 0 and 1 that use only the digits 3 and 5 and that are not on the list, that is, they are not associated with any one of the natural numbers.

- Prove that the cardinality of the set of real numbers is not the same as the that of the set of natural numbers.

- Let $S$ be the set of all real numbers between 0 and 1 with the property that their decimal expansions only have 0's and 7's. For example, the following numbers are elements of $S$:

$$0.7777007707070777707000\ldots$$

$$0.00000000700007777700000007\ldots.$$

  **a.** Show that there exists a rational number in $S$.

  **b.** Show that there exists an irrational number in $S$.

  **c.** Show that the cardinality of $S$ is not equal to the cardinality of the set of natural numbers.

  **d.** **Bonus:** Does the cardinality of $S$ equal the cardinality of the set of real numbers? Make a guess; no justification is required.

## Sample Lecture Notes for 3.3, "The Missing Member" (1 day)

Question of the Day (to be written on the board for students to think about and discuss before class begins): *Is there an infinite set "bigger" than* $\mathbb{N}$?

1 minute—Take a quick vote on the Question of the Day. Write the results on the board for later reference.

5 minutes—Recall that $\mathbb{N}$, $\mathbf{Z}$, and $\mathbf{Q}$ all have the same cardinality, so the elements of each set can be listed. To find a set of greater cardinality, be clear that you need to find a set whose elements cannot be listed.

5 minutes—Play the lottery. Describe a simple state lottery scenario, for example, picking four numbers from $1, 2, \ldots, 20$. Ask students what ticket-buying strategy would guarantee winning. Obviously, buying all possible tickets is a winning strategy. Point out that although buying one ticket for each possible choice of numbers (in this case, 4845 tickets) does guarantee you'll win, you might have to share the prize with other winners.

10 minutes—Now play Dodge Ball (text pages 8–9). Get to the point where everyone understands Player Two's winning strategy. Suggest revising the rules to allow Player One to list as many rows as she wishes as her first play of the game. Now ask what strategy will guarantee a win for Player One. Have students discuss this among themselves for a minute before confirming the "buy all the tickets" strategy: Player One will win if she simply writes down all possible rows of six 0's and 1's.

10 minutes—Extend Dodge Ball to an infinite grid, in other words, each row is an infinite sequence of 0's and 1's (acknowledge that such a game can be played only in the imagination). Ask whether Player One's winning strategy will work. Have students discuss in pairs whether Player One can simply "list" all possible infinite sequences of 0's and 1's. Ask how many think the answer is no. Ask for a volunteer to explain why, given any proposed list, it is always possible to construct a sequence that is not on the list.

5 minutes—Introduce Georg Cantor and his diagonalization argument. Clarify how the argument works for Infinite Dodge Ball, demonstrating that the set of infinite sequences of 0's and 1's cannot be listed. The class has found a set that has a larger cardinality than the natural numbers, thus resolving the Question of the Day.

10 minutes—Take a vote on whether the real numbers have a larger cardinality than $\mathbb{N}$. Claim that the class will now show that the real numbers do have a larger cardinality, that is, there is no way to construct a "list" that contains each real number. Start by assuming you do have such a list, as on text page 165. Now ask the class how to construct a number $M$ not on the list. Follow the Infinite Dodge Ball construction, except now you have a choice of nine other digits at each step to ensure that $M$ is different in the $i$th decimal place from the $i$th number on the alleged list.

5 minutes—Conclude with a brief discussion of Cantor's life and difficulties. Point out how deeply the entire mathematical community struggled with all these ideas and revelations about infinite sets. Thus, current students shouldn't feel bad as they struggle to get a handle on infinity.

**Section 3.4 Travels Toward the Stratosphere of Infinities** [Power Set Theorem]. This section contains the theorem and proof that there are infinitely many different sizes of infinity and that there is no largest infinite set. It uses the idea of the power set (the set of all subsets) of a set and then again uses Cantor's diagonalization argument to prove Cantor's theorem that the power set of a set always has a larger cardinality than the set. One theme of this section is to exploit ideas (in this case the diagonalization argument) that have been discovered and then use them to prove as much as possible. The section ends with the intriguing independence of the Continuum Hypothesis. The idea that there are statements that can be neither proved nor disproved is a disturbing thought to some students.

---

*General Themes*

Use a powerful idea or technique repeatedly to discover more fascinating results.

---

*Mathematical Underpinnings.* (*Caveat:* These mathematical overviews are usually for the instructor only. The symbols and detailed proofs are not suitable for class presentations because mathematical symbols and terminology often present a formidable barrier for students.)

This section presents a deep and fascinating theorem due to Georg Cantor: for every set $S$, the cardinality of $S$ is strictly less than the cardinality of the power set of $S$. If $S$ is a finite set, the result is easy to prove and understand. If $S$ is infinite, also called *transfinite,* then the proof is much more subtle. The outcome for transfinite sets also has enormous implications for our understanding of infinite sets.

The text presents a detailed proof, which, when done slowly and carefully, is accessible to students (pages 179–182). For fun, the text also gives a highly condensed, one-line version of the proof on page 182. Here's an expanded version of that proof, where $card(A)$ denotes the cardinality of the set $A$:

**Cantor's Power Set Theorem:** *If $S$ is a set, then $card(S) < card(\wp(S))$.*

**Proof:** We take it as obvious that $card(S) \le card(\wp(S))$. Let $f$ be an arbitrary function mapping the elements of $S$ to the elements of $\wp(S)$. By showing $f$ is not onto, we will establish that no one-to-one correspondence can exist between $card(S)$ and $card(\wp(S))$, so the inequality is strict.

To show that $f$ is not onto, we need only find one element of $\wp(S)$ that is not in the image of $f$, denoted $f(S)$. Consider the subset $M = \{x \in S \mid x \notin f(x)\}$. So, $M \in \wp(S)$. Now, suppose $M = f(y)$ for some $y \in S$. Is $y$ an element of $M$?

If $y \in M$, then $y \in f(y)$, because $M = f(y)$. But if $y \in M$, then $y \notin f(y)$, by the definition of $M$. Clearly this is a contradiction and $y$ cannot be in $M$. Thus, $y \notin M$, which means $y \notin f(y)$. But then $y$ satisfies the condition for membership in $M$, so $y \in M$, another contradiction.

Therefore, our assumption that $M = f(y)$ for some $y \in S$ must be false. So, $M \notin f(S)$, and $f$ cannot be onto. Because no one-to-one pairing exists between $S$ and $\wp(S)$, the result follows.

As impressive as Cantor's Power Set Theorem is, one of the most profound ideas in the section, if not the entire text, is the Continuum Hypothesis.

**Continuum Hypothesis:** *There is no set with cardinality between card($\mathbb{N}$) and card($\mathbb{R}$).*

Discussion of this topic is worth some serious class time. In 1940, Kurt Gödel proved that it is impossible to disprove the Continuum Hypothesis using standard mathematical axioms and proof techniques. In 1963, Paul Cohen proved that it is impossible to prove it. By now, students should be used to the idea that mathematics is full of conjectures, many of which have been or will someday be proven true or proven false. They may even believe that some conjectures are so hard to prove or disprove that we may never know if they're true or false. (Fermat's Last Theorem may have once been thought to be in this category.) But the idea of a conjecture that has been *proven* to be neither true nor false will likely boggle their minds. As the text states on page 184, "The Continuum Hypothesis is independent of our entire mathematical structure." This discussion provides a great opportunity to challenge students' images of mathematics as a complete, monolithic structure, where everything is so rigorously defined that all statements are either true or false, even if we don't always know which are which. The broader life lessons are valuable here as well. Even in the most highly structured system, there is room to ask questions that may not have answers.

There's one more great, confounding idea in this section: Russell's Paradox, which illustrates one of the limitations of set theory. Bertrand Russell considered the "set" $A$ of all sets that are not elements of themselves. Is $A$ an element of itself or not? If it is, then it violates the condition for membership in $A$, and so it cannot be an element of itself. If it is *not* an element of itself, then it satisfies its own membership condition, and so it *is* an element of itself. The only way to resolve this contradiction is to insist that

*A* cannot be a set in the first place. Thus, mathematicians have to place constraints on how sets are defined. The text asks students to grapple with this paradox in Mindscape III.19 in the form of Russell's barber's puzzle.

---

*Our Experience*

*Time spent:* One to one-and-a-half 50-minute sessions

*Emphasized items:* The idea of the power set is a first obstacle; however, students can learn the diagonalization procedure with finite examples. Generalizing to the infinite cases is harder. We sometimes only explain the statement of Cantor's Power Set Theorem and say that the Dodge Ball technique proves it, but we do not explain how in detail. Other times, we discuss the idea thoroughly. Either way, the ideas are interesting.

*Dangers:* This is an abstract and difficult topic. The notion of a set of subsets needs to be developed with many small examples.

*Personal favorite topic or interactions:* Having students vote on the truth of the Continuum Hypothesis and then telling them they are all wrong (or all right)

*Remark:* One of us has assigned this as outside reading with homework and test questions, without treating it in class. The other two think that's crazy.

*Sample homework assignment:* Read the section and do Mindscapes I.4; II.6, 13; III.16, 19.

---

### Sample Class Activities

*Posing questions.*   Now that students know that the cardinality of the reals is greater than the cardinality of the natural numbers, ask them to pose questions about infinity or infinities. Among their questions, encourage them to consider the four mentioned on text page 174:

1. Is there an infinity that is greater than the infinity of the set of natural numbers yet less than the infinity of the set of the reals? (Is the cardinality of the reals the "next" bigger infinity after the cardinality of the natural numbers?)
2. Is there an infinity greater than the infinity of the set of real numbers?
3. Are there infinitely many different sizes of infinity?
4. Is there a largest infinity—one that encompasses all others?

Ask students to speculate which of these questions is the most difficult or impossible to answer. Have students vote and tabulate their answers.

*Power set.*   Use several, very small examples to develop the notion of a power set. The notion of a set containing sets as elements is a new and sometimes difficult one for students. Using lots of examples and input from the class is helpful. One way to illustrate the idea is to ask three students to come forward and stand in a line. Then ask each possible subset to step forward. Developing a pattern for stepping forward will help them understand the formula for the number of subsets of a finite set. Record the eight subsets.

*Formula for number of subsets.*   Ask another student to come forward. Pose the question: Once we have written down all the subsets of three people, suppose a fourth person wanders up. How can we use our list of subsets of three to create a list of subsets of four? After resolving this issue, ask students to compute the cardinalities of the power sets of various finite sets and allow them to discover the general formula for the cardinality of the power set.

Point out that the power set of a finite set has many more elements than the set has.

*Note:* You must decide how much effort to spend on proving that the cardinality of the power set is always greater than the cardinality of the set itself, even for infinite sets, as was the case with finite sets. You can simply say that the proof is identical in nature to the Cantor diagonalization proof, and then assign the proof as a reading exercise. One of us does the proof in detail in class; however, this requires an entire class period to do it justice. It all depends on what is appropriate for your class, your schedule, and your personal preference. It is useful to state the theorem that the power set of any set has a larger cardinality than the set and to illustrate this theorem's powerful consequences. In particular, you can explain how this result settles questions 2, 3, and 4 from the list on the previous page.

*Power set of the natural numbers.*   Because the power set of the natural numbers is challengingly abstract, you may want to have students write examples of elements of the power set. Collect examples of subsets of ℕ, some that are finite and some that are infinite, and point out that each is a single element of the power set of ℕ.

*Power set of ℕ is bigger than ℕ.*   On a transparency, write the natural numbers in a column on the left, and ask students to give you subsets to write in a corresponding column on the right (making sure that some of the numbers chosen contain the corresponding number and some don't). Draw a large bracket sign containing underlines separated by commas: { __ , __ , __ , __ , . . .}. Under each underline write the natural numbers 1, 2, 3, . . . . Tell the class that their challenge is to construct a subset of $\mathbb{N}$ that is not on the list. They must tell you first whether 1 is or is not in this new set. Continue through five or six of the blanks, always filling in the number or writing "no" in the blank if that number is not in the set. Emphasize how those irrevocable decisions definitely tell them that the final set will not be the first set on the list, not the second, not the third, and so on. This idea requires students to articulate the idea to each other and themselves several times.

*Any power set is bigger.* Explain why ordering the elements of a set is not necessary. Suppose you have a set $S$ and a potential correspondence between the elements of $S$ and the elements of $\wp(S)$. The class goal is to construct a subset of $S$ that will not be on the list. Write down an arbitrary element $x$, and ask students how they would decide whether to put $x$ in the set they are constructing. Point out that their decisions are independent of the decisions made about other elements.

*Diagonalization practice.* Write several finite sets and correspondences between the elements of those sets and some elements of the power set. Ask students to construct a subset that they know will not be on the list.

*Infinitely many infinities.* Discuss the following: Suppose each student in the room took out a piece of paper and wrote an infinite set. How could they look at those papers and write down an infinite set that is larger than any one of the sets? Taking a union and then taking its power set is a likely answer.

*No biggest set.* Similarly, if someone proposes a biggest set, how could they come up with a still bigger set?

*Discussion.* A powerful way to end this section is to discuss, in broad, general terms, the Continuum Hypothesis and how this issue really pushes us to the very edge of mathematics as we know it. This independent result is always interesting for students. The idea that some issues can be shown to be neither true nor false is often eye-opening for them. Students learn that mathematics are not just black and white and begin to see the interesting and delicate hues. The idea that the axiomatic method has inherent limits is also an interesting notion for discussion. Such issues lead to a big finish for the abstract and technical concepts that students have been digesting.

## Sample Test Questions

(The Mindscapes are another good source of test questions. All test questions below are available electronically on the Test Bank CD-ROM.)

- Let $T = \{a, b, c, d, e\}$. How many elements does $\wp(\wp(\wp(\wp(T))))$ contain?

- Cantor proved that for any set $A$, the set of subsets of $A$ has a larger cardinality than $A$. Let $A = \{a, b, c, d, e\}$. Suppose you associate each element of $A$ with a subset of $A$ as follows:

$a$ is associated with $\{a, c\}$,

$b$ is associated with $\{a, c, d\}$,

$c$ is associated with $\{b\}$,

$d$ is associated with $\{a, b, c, d, e\}$,

$e$ is associated with $\{\ \}$.

There are many subsets of $A$ not associated with any element of $A$; however, what subset would you find using the construction that occurs in the proof of Cantor's Theorem?

- Let $S$ be an infinite set whose cardinality is larger than the cardinality of the real numbers. Can you construct a set whose cardinality is greater than the cardinality of $S$? If so, give a method of constructing one; if not, explain why not.

- In what theorem or theorems is Cantor's diagonalization argument used in the proof(s)? What do these theorems imply about infinity?

## Sample Lecture Notes for Section 3.4, "Travels Toward the Stratosphere" (2 days)

### Day One

Question of the Day (to be written on the board for students to think about and discuss before class begins): *Is there an infinity bigger than the reals?*

10 minutes—Start very slowly with some simple examples of "collections" and "subcollections." Avoid using the words *set* and *subset* until students have grasped the ideas, but use bracket notation. Have students create a list of all subcollections of {♠,♥,♦,♣}, as is done in the text. Do the same for other small, nonthreatening sets, such as {house, tree, dog}.

10 minutes—Have students discover that creating a subset is equivalent to doing the following: point to each element in the set and ask, "In it or not in it?" This yields the formula $2^n$ as the number of subsets of a set of size $n$. Confirm this formula with the previous examples, as well as with additional examples such as {Mac, PC} and {☺,★,↗}. During this discussion, make the transition from "subcollection" to "subset."

10 minutes—Define the power set. Emphasize that the elements of a power set are themselves sets, confirming with previous examples. Although this is a strange notion, it illustrates how broadly defined a set can be. (If you plan to discuss Russell's Paradox later, here's an opportunity to mysteriously acknowledge that even sets have limitations!) Now ask students to think about what the power set of the natural numbers is like. Write down a few elements, some finite and some infinite.

5 minutes—Observe that the collection of all subsets looks much larger than the original set. Claim that this is true for every set, even infinite sets, and state Cantor's Power Set Theorem.

10 minutes—Point out that the proof is simply a generalized version of Infinite Dodge Ball and the proof that the reals are "bigger" than the naturals. Follow the presentation in the text (pages 180–182).

5 minutes—Ask for a set with larger cardinality than the reals, $\mathbb{R}$. Students should respond with the power set of the reals, $\wp(\mathbb{R})$. Ask for a set with even larger cardinality, $\wp(\wp(\mathbb{R}))$, and continue until students catch on that there are an infinite number of different "sizes" of infinity and that there is no largest infinity.

**Day Two**

Question of the Day (to be written on the board for students to think about and discuss before class begins): *Is there an infinity between the cardinality of* $\mathbb{N}$ *and the cardinality of* $\mathbb{R}$?

5 minutes—Remind students that the cardinality of $\mathbb{R}$ is larger than the cardinality of $\mathbb{N}$. Ask them to guess at a set that has cardinality between these two. Have them vote on whether such a set exists.

10 minutes—State the Continuum Hypothesis as given on text page 184, and briefly discuss its status as "undecidable" within the standard axioms of set theory and logic. Point out that sometimes this hypothesis is included among the axioms of set theory. Be sure to mention the roles played by Kurt Gödel and Paul Cohen.

10 minutes—Discuss Russell's Paradox: consider the set $A$ of all sets that are *not* elements of themselves. (Be sure to recall the difference between elements and subsets.) Point out that $A$ is a strange set. The set of all integers is not itself an integer; the set of all students is not itself a student. But maybe the set of all wacky ideas is itself a wacky idea! Now ask students whether $A$ is an element of itself. Follow each case to its logical contradiction. Conclude by pointing out that mathematicians have in effect outlawed sets like $A$ precisely to avoid such a paradox.

**Section 3.5  Straightening Up the Circle** [Geometrical correspondences]. This section concludes the topic of infinity with some geometrical cardinality facts, such as a geometrical demonstration of one-to-one correspondences between a small line segment and a long line segment and between the open interval and the real line through the stereographic projection. These geometric correspondences can help solidify students' understanding of cardinality. The section also presents a one-to-one correspondence between the points of a filled-in square and the points of a line segment, thus showing that an increase in dimension does not necessarily mean an increase in cardinality.

---

*General Themes*

Don't be afraid to make mistakes as you try to apply a new idea or technique.

---

*Mathematical underpinnings.* (*Caveat:* These mathematical overviews are usually for the instructor only. The symbols and detailed proofs are not suitable for class presentation because mathematical symbols and terminology often present a formidable barrier for students.)

This section explores the cardinality of sets of points in familiar geometric objects, such as line segments and squares. Though still somewhat counterintuitive, as are so many infinite set cardinality comparisons, these examples are more accessible than the uncountable sets of Sections 3.3 and 3.4. Many of the one-to-one pairings are established geometrically and are therefore highly visual. Even spending just half a class on this material can greatly reinforce students' understanding of infinite cardinality. An added bonus is the text's emphasis on learning from failure.

Possibly the most fun example is looking at line segments of different lengths. Two such segments ($S$ for "small" and $B$ for "big") can be drawn as in Figure 1. Extending the lines determined by the endpoints of $S$ and $B$ creates a triangle with apex $A$ (Figure 2). Drawing a ray from $A$ pairs up each point on $S$ with a single point on $B$, because distinct, nonparallel lines intersect in exactly one point.

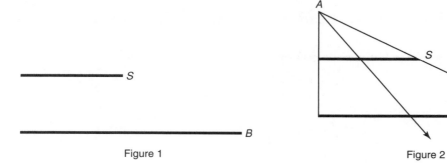

Figure 1                                        Figure 2

The most challenging and perhaps the most surprising example establishes that the cardinality of the set of points on a line segment is the same as the cardinality of the set of points inside a square. In this case, the argument is not geometric; instead, it cleverly exploits the decimal expansion of real numbers to create a one-to-one pairing between points in a square and points on a line segment. As described on text pages 196–199, each point $(x, y)$ in the square is paired with a point $z$ on the line segment by shuffling the decimal expansions of $x$ and $y$ into a single decimal expansion, yielding the number $z$.

The method seems simple and elegant, but it has a fatal flaw—the pairing as defined is not onto. The flaw is carefully explained in the text (pages 199–200). A more direct alternate approach uses the Schroeder-Bernstein Theorem: *Let A and B be sets. If there exists a one-to-one function from A into B and a one-to-one function from B into A, then there is a one-to-one correspondence between A and B.* If we let $A$ be our line segment and $B$ our square, then simply superimposing $A$ onto an edge of $B$ gives a one-to-one function from $A$ to $B$. Then, the coordinate shuffling operation gives a one-to-one function from $B$ into $A$.

---

### Our Experience

*Time spent:* One 50-minute session

*Emphasized items:* Cardinality of different-length line segments and stereographic projection

*Items omitted or treated lightly:* Cardinality comparisons of the solid square and the line segment

*Dangers:* Don't get bogged down in the technicalities of shuffling the digits in the square-to-line correspondence.

*Remarks:* We often assign this section to be read independently with homework, but not discussed in class.

*Sample homework assignment:* Read the section and do Mindscapes I.2; II.9, 10; III.17.

---

## Sample Class Activities

*Segments have the same cardinality.* After showing the correspondence in the section, ask students what would happen if they goofed and connected the wrong ends of the segments—that is, the lines cross between the intervals. Show how there is nevertheless a geometric one-to-one correspondence via all lines through the crossing point.

*Another method showing that a segment "equals" the real line.* Ask students if they can take an open interval and divide it into infinitely many subintervals. Lead them to produce increasingly shorter intervals going toward the ends. Now ask them to write the real line as infinitely many intervals. Again, lead them to say only the intervals between integers. Now ask them to show a one-to-one correspondence between the reals and the open interval by taking the increasingly shorter subintervals of the interval to the various intervals between integers on the real line.

*Projections.* Ask students to describe the points on the circle to which the integers correspond under the stereographic projection. Ask if they can show that an open semicircle is in one-to-one correspondence with the whole real line by a projection. Show them that projecting from the center of the circle would do the trick. Ask about other variations, such as how to project a **V**, with the top endpoints of the **V** removed onto the real line.

*Stereographic projection.* Ask students to describe a correspondence between a plane and a sphere with the north pole removed. Discuss how this projection relates to maps. Ask students what countries on a map are distorted under this projection. Draw an infinite straight line in the plane, and sketch the plane determined by that line and the north pole of the sphere. Note that the plane intersects the sphere in a circle. Ask students why a straight line in the plane always corresponds to a circle containing the north pole via the stereographic projection. Discussing the correspondences between polygons or countries in the plane and their corresponding positions on the sphere is useful for helping students figure out the projection. The stereographic projection is mentioned again in Chapter 5, where it is used in proving that there are only five regular solids using the Euler characteristic.

*Square equals segment.* Describe the shuffling correspondence between the points in the square and the points on the interval. The shuffling correspondence shows a surprising reality, namely, that the points in the square are not more numerous than the points in the segment. It is not a one-to-one correspondence, however, because there are points on the interval that are not hit. After showing the class this defect, describe the grouping and shuffling method that works. Now ask students some questions to help them understand the geometrical significance, or lack of significance, of these shuffling maps. Ask students to find places where nearby points in the square do not correspond to nearby points in the line segment. For example, $(0.499999\ldots, 0.4999999\ldots)$ goes to $0.449999\ldots$, whereas $(0.500000000001\ldots, 0.5000000000001\ldots)$ goes to $0.5500000000\ldots$. Thus, close points do not go to close points. In mathematical terms not used in the text, these questions are asking why the correspondence is not continuous. This activity will help students understand that the correspondences between the square and the interval differ from the previous correspondences involving segments—here we are doing something different from simply stretching.

### *Sample Test Questions*

(The Mindscapes are another good source of test questions. All test questions below are available electronically on the Test Bank CD-ROM.)

- Show geometrically that a circle of radius 2 has the same cardinality as a circle of radius 1.

- Let $C$ be the set of all circles in the $xy$-plane having their center at the origin. Does the cardinality of the set $C$ equal the cardinality of the integers $\mathbf{Z}$ or the cardinality of the real numbers $\mathbb{R}$? Prove your answer.

- Is there a one-to-one correspondence between the points on a 2-inch line segment and the points on a 5-inch line segment? If so, describe such a correspondence; if not, explain why not. Given your answer, what can you conclude about the cardinality of points on a 2-inch line segment compared with that of a 5-inch line segment?

- Show that the half-open interval $(0, 1)$ has the same cardinality as a circle.

- Show that the open interval $(-1, 1)$ has the same cardinality as the real line.

- Give a one-to-one correspondence between a sphere with the point at the north pole removed and the plane. (Recall that a sphere is the boundary (or surface) of a ball.)

- Prove that a filled-in square has the same cardinality as an interval.

90

## Sample Lecture Notes for Section 3.5, "Straightening Up the Circle" (1 day)

Question of the Day (to be written on the board for students to think about and discuss before class begins): *Are there more points in a square than on a line segment?*

8 minutes—Draw two line segments, $S$ and $B$, on the board. (Make the lines parallel, with the shorter one on top.) Ask students which line segment has more points. Regardless of answers, ask why the "straight projection" (see Figure 3) doesn't establish that the longer segment has more points. Finish with a bona fide one-to-one pairing, as in Figure 4.

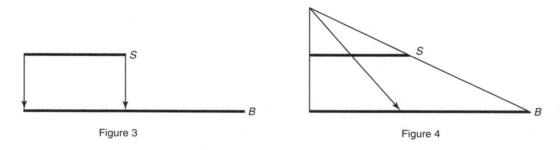

Figure 3                    Figure 4

7 minutes—Draw a circle with a point deleted and resting on the real line. Ask if the two objects have the same number of points. Have students create a one-to-one pairing (text pages 194–195).

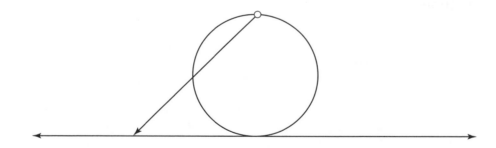

As an optional activity, have students work in pairs to construct a one-to-one pairing of the points on a sphere with the points on a plane. (This is Mindscape III.20.)

10 minutes—In preparation for discussing the Question of the Day, sketch a square on the plane and a line segment on the number line, as suggested in the following figures.

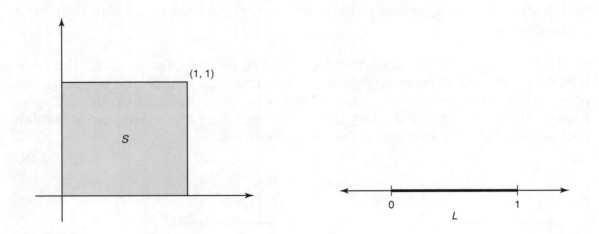

Ask students if there are more points in $S$ than there are on $L$. Describe the shuffling technique for pairing each point $(x, y)$ in $S$ with a point $z$ on $L$. For the shortest presentation, do only a few examples, and don't mention the tricky parts. To add some subtlety, point out the need to clarify how coordinates in $S$ will be written in decimal expansion: for any number with a decimal expansion ending in an infinite string of 0's, choose the alternate expansion ending in 9's. For example, use $0.714999\ldots$ instead of $0.75000\ldots$. Acknowledge that there is still a difficulty with the shuffle pairing, but refer interested students to the text.

# Chapter 4   Geometric Gems

## I.   Chapter Overview

This chapter presents a spectrum of geometrical insights from the ancient to the modern, from the plane to the fourth dimension. The chapter's overall goal is to help students comprehend the powerful consequences of visualization and geometric relationships. The sections vary in difficulty and in their proportion of mathematical challenge to pictorial interest. This chapter can be presented at a variety of mathematical levels to suit your needs.

---

### Dependencies
The sections in this chapter are largely independent, so any may be used or omitted. Section 4.5 on the regular solids is referred to in Section 5.3 (the Euler Characteristic), where it is proved that there are only five regular solids.

---

### Sections to Assign as a Reading Assignment or Term Project
Sections 4.2, 4.4, 4.5, and 4.6 are particularly well-suited for reading or term paper topics. Section 4.5 is also a good chapter to assign as a reading section to be covered on a test. Any omitted sections are good paper or extra credit topics.

---

### *Reminders*

1. Class activities are suggested for each section. These can be done with students working in groups of two or three and then gathering their responses or as one global class discussion.

2. Before introducing a new topic, engage students with enticing questions, which they quickly discuss with neighbors; solicit responses within a minute. Use the responses as the springboard for the introduction.

**3.** We have included more sample class activities than you will probably want to use. Choose only the ones that are right for you and your class.

**4.** Each section can be treated in more or less depth. It is frequently a good idea to omit difficult technicalities in order to treat the main ideas well and then move on.

## II.  Section-by-Section Instructional Suggestions

**Section 4.1  Pythagoras and His Hypotenuse** [Bhaskara's elegant proof]. This section states the Pythagorean Theorem and presents an elegant geometrical proof of it. The goal of this section is for students to attain a solid understanding of the Pythagorean Theorem and one of its proofs.

---

*General Themes*

Seemingly abstract ideas can be made tangible, and manipulating simple shapes can lead to profound results.

---

*Mathematical Underpinnings.*   (*Caveat:* These mathematical overviews are usually for the instructor only. The symbols and detailed proofs are not suitable for class presentations because mathematical symbols and terminology often present a formidable barrier for students.)

**Pythagorean Theorem:** *In a right triangle, the square of the length of the hypotenuse equals the sum of the squares of the lengths of the other two sides.*

This theorem has many proofs, one of which is in the text, another of which is outlined in the Sample Lecture Notes. The figure shows a way to combine the text proof with a little algebra for yet another proof:

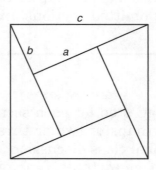

The figure is constructed using four identical right triangles, with side lengths as labeled. The area of the large square is clearly $c^2$. Alternatively, we can write this area as the sum of the areas of the four triangles plus the area of the small inner square. Thus,

$$c^2 = 4\left(\frac{1}{2}\right)ab + (a-b)^2$$

$$= 2ab + a^2 - 2ab + b^2$$

$$= a^2 + b^2.$$

Some students will discover this proof on their own.

---

*Our Experience*

*Time spent:* One 50-minute session

*Emphasized items:* Students physically doing the proof with the kit pieces

*Personal favorite topic or interactions:* Posing parts of the proof for students to work out at their seats using the kit pieces

*Remarks:*

1. Everyone knows the Pythagorean Theorem, but few know how to prove it. This proof is fun and brings the theorem to life.
2. An overhead projector is useful for demonstrating the proof with the pieces.

*Sample homework assignment:* Read the section and do Mindscapes I.2; II.6, 12, 15; III.18.

---

### Sample Class Activities

*Pythagorean Theorem.* Ask students to bring to class the right triangles and square from the kit. Have them assemble the pieces into a square, with each side being a hypotenuse. Then have them shift two triangles to make a figure consisting of two squares connected in the shape of a fattened **L**, demonstrating the theorem. Students should physically do this exercise to appreciate it.

*Alternative proof of the Pythagorean Theorem.* The Pythagorean Theorem has many interesting proofs. Leading the students to discover one of the alternative proofs is a good classroom exercise. It can show them that when thinking about why things are true, "the proof" should be changed to "a proof." For example, using the same four triangles from the previous activity, ask students to construct a square with outside dimensions $a + b$. Note that the inside space is a square of area $c^2$. Have them rearrange the triangles into two rectangles within the big square so that the remaining space is two squares—one $a^2$, one $b^2$. Examples of this proof and the area described in the first activity are provided on the Interactive Explorations CD-ROM packaged with each student text.

*Homework remark.* Bringing mathematics to others is a major step, and often a novel one, for students. Giving them incentives for truly explaining the Pythagorean Theorem and its proof to someone outside the class can be a formative experience for them.

### Sample Test Questions

(The Mindscapes are another good source of test questions. All test questions below are available electronically on the Test Bank CD-ROM.)

- State the Pythagorean Theorem completely, and indicate a proof of it by a picture and one or two sentences.

- What makes Bhaskara's proof of the Pythagorean Theorem so elegant?

- Take four copies of a right triangle. Arrange them so they outline a square with sides $a + b$, with a missing square of size $c^2$ in the middle. Rearrange the triangles into two rectangles within the outline of the same $a + b$ square, so that the remaining area is $a^2 + b^2$.

## Sample Lecture Notes for Section 4.1, "Pythagorean Theorem" (1 day)

Question of the Day (to be written on the board for students to think about and discuss before class begins): *A baseball diamond is really a square measuring 90 feet on a side. How far does a catcher have to throw the ball to get it from home plate to second base?*

5 minutes—Ask how many students remember the Pythagorean Theorem, and have someone state the theorem. Briefly discuss why the theorem might be an important result.

10 minutes—Ask for examples of whole-number side lengths of right triangles (many students will remember (3, 4, 5) or (5, 12, 13); another is (8, 15, 17)). Discuss how the Pythagorean Theorem works in reverse, so that students can determine whether a triangle with given side lengths is a right triangle.

5 minutes—Display the following figures on the overhead or on the board. Have students discuss in pairs how they might use the diagrams to prove the Pythagorean Theorem.

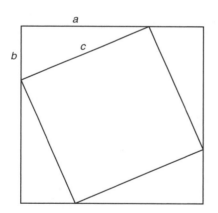 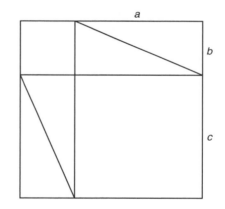

5 minutes—Complete the discussion of this first proof, emphasizing how an algebra formula can be interpreted as a claim about area. Emphasize that the particular right triangle used could be replaced with *any* right triangle. Suggest that it's time for students to discover for themselves another way of proving the theorem.

15 minutes—Have students use the four right triangles and the small square from their kit. (You may want to bring some extra sets to class!) If you are able to have a computer and projector in your classroom, consider bringing the Interactive Explorations CD-ROM to class as an alternative method of

demonstrating the proofs. Instruct them to create a single large square with side length equal to the hypotenuse. After establishing that the area of this square is $c^2$, ask students how they could use the five shapes to prove the theorem. Suggest the "fat L" shape, and work through the analysis to show that this shape comprises two squares with areas $a^2$ and $b^2$, respectively. (Throughout the construction process, encourage students who finish quickly to help others who get stuck.)

10 minutes—Reinforce the validity of a geometric proof for an algebraic formula. Discuss the generality of the arguments by asking students what the diagrams might look like using right triangles with different proportions. Point out that there are many other proofs of the Pythagorean Theorem (provide online references, if desired).

**Section 4.2  A View of an Art Gallery** [View-obstruction question from computational geometry]. In this section, the Art Gallery Theorem is stated and proved, and an unsolved problem in computational geometry in the plane is posed. The section shows plane geometry as a still living study that contains easily stated questions that remain unsolved.

```
General Themes

Playing with examples can lead to general results.
```

*Mathematical Underpinnings.* (*Caveat:* These mathematical overviews are usually for the instructor only. The symbols and detailed proofs are not suitable for class presentations because mathematical symbols and terminology often present a formidable barrier for students.)

**The Art Gallery Theorem:** *Suppose we have a polygonal closed curve in the plane with v vertices. Then there are v/3 vertices from which it is possible to view every point on the interior of the curve. If v/3 is not an integer, then the number of vertices we need is the biggest integer less than v/3.*

**Proof:** The theorem's proof has three major steps, detailed in the text (pages 224–226) and summarized here. Let $C$ be a polygonal closed curve with $v$ vertices.

1. Triangulate $C$ by adding "diagonal" edges from one vertex to another until the interior of $C$ is divided into triangles. (Be sure the new edges do not intersect inside $C$.)
2. Color each vertex of the resulting triangles with one of three colors so that each triangle has vertices of three different colors.
3. Notice that every point in each triangle is visible from each vertex of the triangle, and thus every point in the interior of $C$ is visible from the set of vertices of one color.

Because the color used least occurs at no more than $v/3$ vertices of $C$, placing "guards" at those vertices will suffice.

We can use mathematical induction to prove that $C$ can be triangulated. First, if $C$ has only three vertices, then $C$ is already triangulated. Now, assume that any polygonal closed curve on fewer than $n$ vertices can be triangulated, and consider $C$ on $n$ vertices. Drawing one diagonal edge in $C$ will divide $C$ into two polygonal closed curves with fewer than $n$ vertices. Because these two closed curves can both be triangulated by our induction assumption, we may conclude that $C$ can also be triangulated.

We can also prove, through a similar induction argument, that the vertices of the resulting triangles can be colored as required.

Once the vertices have been colored with red, yellow, and blue, for example, suppose there are $R$ red vertices, $Y$ yellow vertices, and $B$ blue vertices. Then $R + Y + B = v$. If all three values were greater than $v/3$, then their sum would be greater than $v$, which is a contradiction. Thus, some color occurs on at most $v/3$ vertices. Because every triangle has a vertex of each color, placing "guards" at those vertices will keep every interior point of the gallery under surveillance. (Note that this final argument is like a reverse Pigeonhole principle: If you put $v$ pigeons into three holes, then there must be a hole with at most $v/3$ pigeons.)

---

*Our Experience*

*Time spent:* One 50-minute session

*Items omitted or treated lightly:* We don't make a big deal of the fact that the interior of the polygon can be triangulated.

*Personal favorite topic or interactions:* Discussion of the open question to show that some simple questions are still unanswered

*Remark:* One of us omits this section in class, because it makes for an outstanding group research project with good poster session possibilities.

*Sample homework assignment:* Read the section and do Mindscapes I.3; II.9, 11; III.20.

---

### Sample Class Activities

*Triangulations.* Have students draw a polygonal closed curve and find a segment that spans across the interior and joins two vertices. They should continue to find such spanning segments as often as they can without crossing segments. Ask them what each piece is. They should observe that the interior of the original polygonal curve is divided into triangles with vertices at the original vertices of the polygonal curve.

*Art Gallery Theorem.* On their triangulated polygonal curve, ask students to label each vertex $R$ (for red), $Y$ (for yellow), or $B$ (for blue) so that each triangle has one vertex of each color. Ask them to find the color that appears least frequently. Point out that cameras placed at those vertices would be vertices of every triangle and therefore could see the whole interior of the curve.

*The mirrored gallery.* Ask students to draw a polygonal closed curve in the plane and then demonstrate how to find a place from which every point inside the curve would be visible if the walls were mirrored. Point out that no one knows whether it is always possible to find such a point in every polygonal closed curve. Invite them to submit a solution for extra credit.

### *Sample Test Questions*

(The Mindscapes are another good source of test questions. All test questions below are available electronically on the Test Bank CD-ROM.)

- **a.** State the Art Gallery Theorem.

   **b.** Demonstrate your understanding of the theorem's proof by using this figure. Simply draw and label things and show where the proof tells you to locate the cameras. Little to no explanation is necessary.

- In these figures, where would you place the guards so that the museum is protected with the fewest number of guards? In at least one example, two guards are enough. Determine which one(s).

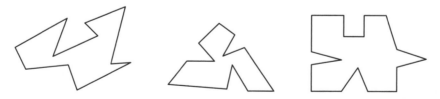

- Show how to divide the interiors of these polygons into triangles with vertices at vertices of the polygons.

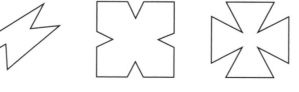

## Sample Lecture Notes for Section 4.2, "A View of an Art Gallery" (1 day)

Question of the Day (to be written on the board for students to think about and discuss before class begins): *Here's a floor plan for an art gallery. If you stand in the corner marked A, which walls or parts of walls can you see? What if you stood in the corner marked B? How many guards would you need, each standing at a corner, so that each wall could be seen by at least one of the guards? (The guards can turn their heads, but they cannot leave their corners.)*

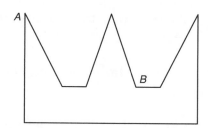

5 minutes—Ask if students have been to art museums. Acknowledge the need for security cameras or guards. Draw examples of different floor plans, such as the following, and introduce the idea of a polygonal closed curve.

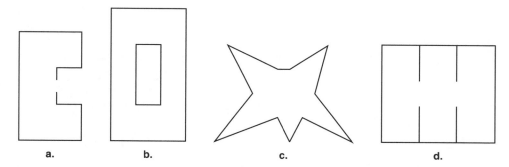

5 minutes—Have pairs of students draw a gallery and note the number of vertices. (Suggest they not make it too large or complicated!) Then ask them to choose to put guards at vertices so that every interior point is seen. Ask them to do so efficiently, using as few guards as possible and noting the number of guards required.

5 minutes—Now collect data from some or all students in a table on the board (add rows as necessary):

| Number of Vertices ($n$) | Smallest Number of Guards Needed | $n/3$ |
|---|---|---|
|  |  |  |

103

If you haven't already, answer the Question of the Day and include the data for the given gallery in the table.

5 minutes—Ask students what they notice about the middle and right columns of the table. Then state the Art Gallery Theorem. Illustrate the theorem with a comb gallery and perhaps one or two student-created galleries.

10 minutes—Introduce triangulation and do an example. Have students redraw their galleries, or draw a new one, and triangulate the galleries. They should then color the vertices of their gallery so that each triangle has three colors. Ask students to work in pairs to discuss how coloring the vertices relates to the Art Gallery Theorem.

10 minutes—Discuss more formally how coloring the vertices is used to prove the theorem (text pages 225–226).

10 minutes—Introduce related questions, such as mirrored galleries, galleries with "holes" or obstructions, galleries with only right-angle corners (Mindscape III.19), and so on. Ask for other ideas, acknowledging that many of these questions are still unanswered.

**Section 4.3 The Sexiest Rectangle** [The Golden Rectangle]. The Golden Rectangle is constructed with proportions equal to the Golden Mean and was introduced in Section 2.2 (Fibonacci Numbers). This section discusses the property that removal of the largest square leaves a rectangle of the same proportions, as well as the logarithmic spiral. Through pictures and photographs, students can be led to recognize a role that mathematics, and the Golden Rectangle in particular, have played in art, architecture, and nature.

---

*General Themes*

Examining the interplay between seemingly different disciplines and perspectives—mathematics and art, geometry and nature, the analytical and the aesthetic—can be very satisfying.

---

*Mathematical Underpinnings.* (*Caveat:* These mathematical overviews are usually for the instructor only. The symbols and detailed proofs are not suitable for class presentations because mathematical symbols and terminology often present a formidable barrier for students.)

This section defines and presents a construction for the Golden Rectangle. A rectangle with base $b$ and height $h$ is a Golden Rectangle if $b/h = \varphi$, which is the Golden Ratio. (We assume, without loss of generality, that $b \geq h$.) The section's main theorem proves the following: Cutting the largest possible square off the end of a Golden Rectangle leaves a smaller rectangle that is also a Golden Rectangle (text page 239). Here's a slightly different version of the proof:

**Claim:** *Given a Golden Rectangle with dimensions $b \times h$, removing a square of side length $h$ leaves a Golden Rectangle with dimensions $h \times (b - h)$.*

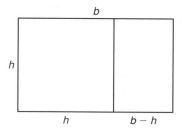

**Proof:** We need only show that $h/(b - h) = \varphi$. Notice $\dfrac{h}{b-h} = \dfrac{1}{\dfrac{b}{h}-1} = \dfrac{1}{\varphi-1},$

because the original rectangle is golden. Because $\dfrac{1}{\varphi-1} = \varphi$, however, we also have that the smaller rectangle is golden.

105

This result leads to a technique for constructing a Golden Rectangle. Starting with a square *abcd*, find the midpoint *m* of the side *ab*. Extend the side *ab* and drop the arc of a circle with center *m* and radius *mc* to find the point *e*:

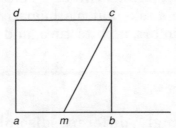 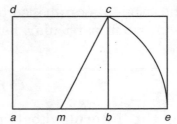

(See text pages 240–241.)

The section ends with a construction of the logarithmic spiral using nested Golden Rectangles. A number of claims are made, but the proofs lie beyond the scope of this course. In fact, the last result on tangent lines involves subtle arguments from calculus. This is a wonderful opportunity to mention calculus and its powerful applications.

---

### Our Experience

*Time spent:* One 50-minute session

*Emphasized items:* Regeneration feature or construction

*Dangers:* Students sometimes have difficulties with the proof that the construction of a Golden Rectangle works.

*Personal favorite topic or interactions:* Having students find the Golden Rectangle in pictures of art and architecture, such as in Mondrian's work or the Parthenon

*Remark:* Art, architecture, and snails make this section visually appealing. Students can debate whether the Golden Rectangle really is especially attractive.

*Sample homework assignment:* Read the section and do Mindscapes I.3; II.9, 12, 13; III.16, 17, 20.

---

### Sample Class Activities

*Golden Rectangles in art.* (Template 4.3.1) If possible, display some pictures of the Parthenon (see the template at the end of the chapter), Mondrian paintings, or other artworks. Ask students, working in groups, to find Golden Rectangles in each. For each picture, have a contest among groups to find as many Golden Rectangles as they can. Choose volunteers from each group to show where the Golden Rectangles are in each picture.

*Constructing the Golden Rectangle.* (Template 4.3.2) Ask students to construct the Golden Rectangle. Then have them verify, with a ruler and algebraically, that the remaining rectangle, after removal of the largest square, does indeed have the proportions of the original rectangle. Use the transparency template as necessary.

*Deducing the Golden Mean.* Tell students they have a rectangle with the property that upon removal of the largest square the remaining rectangle has the same proportions as the original rectangle. Ask them to deduce that the proportions of the original rectangle are indeed the Golden Mean and that therefore they must have started with a Golden Rectangle. Indeed, the Golden Rectangle is often defined as one with this self-replicating property.

*Logarithmic spiral.* (Template 4.3.3) Instruct students to draw or construct the logarithmic spiral. Ask them to locate the center point of the spiral by drawing enough of the spiral to locate the center. Then ask them to draw the diagonals of the biggest and second biggest rectangles to check that the intersection does indeed locate the center point. Discuss the reasons for this as a class.

Finally, ask each student to take a random point on the spiral and measure the angle between the line tangent to the spiral at that point and the line going to the center point of the spiral. Take a poll of the class to see if they all reached the same answer. Use the template as necessary.

### Templates for Transparencies (located at end of chapter):

- Template 4.3.1—Architecture with the Golden Rectangle

- Template 4.3.2—Golden Rectangle with a square inside

- Template 4.3.3—The logarithmic spiral in a Golden Rectangle

## Sample Test Questions

(The Mindscapes are another good source of test questions. All test questions below are available electronically on the Test Bank CD-ROM.)

- What is the area of a Golden Rectangle with height 2?

- Find three Golden Rectangles in this figure.

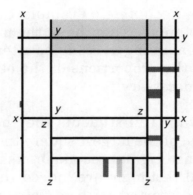

- Draw the logarithmic spiral in the following figure.

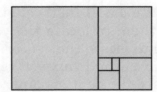

- Show that if you have a rectangle such that upon removal of the largest square you are left with a rectangle of the same proportions, then the original rectangle was a Golden Rectangle.

## Sample Lecture Notes for Section 4.3,
## "The Sexiest Rectangle" (1 day)

Question of the Day (to be written on the board for students to think about and discuss before class begins): *Imagine the ideal shape that comes to mind when you hear the word "rectangle." Draw the shape on the board and initial your rectangle.*

5 minutes—Ask several students to defend their choice of the ideal rectangle. (We hope that some look like a Golden Rectangle!) Choose a few representative samples, and have the class vote on its favorite.

5 minutes—Introduce the Golden Rectangle, emphasizing that the proportions, not the actual size or orientation, of the rectangle determine whether it's golden.

5 minutes—Discuss the Golden Rectangle in art and architecture. Display pictures of the Parthenon (Template 4.3.1) or artwork, such as the art in the text. Photos of buildings from your school's publicity materials may also provide good examples.

10 minutes—Draw a Golden Rectangle with $h = 1$ and a square marked off, and examine the base-to-height ratios of the larger and the smaller rectangles. Show that these two ratios are equal (text page 239).

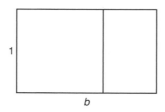

15 minutes—Present the construction of a Golden Rectangle and have each student create his or her own. You may bring in compasses and straightedges for students to use. Show that the construction does produce a Golden Rectangle as in the text (pages 240–241).

10 minutes—Show how to construct the logarithmic spiral. Alternatively, have students work in pairs on Mindscape II.13. (Encourage them to start with a very small square!)

**Section 4.4  Soothing Symmetry and Spinning Pinwheels** [Aperiodic tilings]. This section develops the idea of symmetry and presents an example and proof of an aperiodic tiling of the plane. Seeking patterns in the world is one of the themes that students can take from the course. In this section, students learn to see different types of organizational patterns, specifically, rigid symmetry and symmetry of scale.

---

*General Themes*

Making intuitive ideas more precise can lead to greater understanding and completely new creations.

---

*Mathematical Underpinnings.  (Caveat:* These mathematical overviews are usually for the instructor only. The symbols and detailed proofs are not suitable for class presentations because mathematical symbols and terminology often present a formidable barrier for students.)

To study symmetry, this section focuses on tiling the plane, a topic students may not have seen before. Some tiling patterns are simple and very old, but this section also includes the results of some fairly recent research, thus giving you an opportunity to show students that even ancient mathematical ideas still give rise to open questions and current research. The early material is very accessible and visual. Overhead transparencies are an ideal tool for examining rigid symmetries in various tilings of the plane or to construct a tiling using a particularly shaped tile.

The section begins with a general discussion of symmetry and then introduces the two types of symmetry found in plane tilings: rigid symmetry and symmetry of scale. The main question is, Must a pattern that has symmetry of scale also have rigid symmetry? The answer is no, as illustrated by a tiling called the Pinwheel Pattern discovered in 1994 by John Conway of Princeton University and Charles Radin of The University of Texas at Austin (text pages 253–254). The rest of the section describes the Pinwheel Pattern in detail, showing that it has symmetry of scale but no rigid symmetries. Constructing the Pinwheel Pattern and understanding its symmetry of scale is quite accessible. The proof that it has no rigid symmetries is trickier, but could easily be omitted.

In constructing the Pinwheel Pattern, you'll notice that the basic tile is a right triangle (called a Pinwheel Triangle) with one leg twice as long as the other. As the text notes (page 254), these proportions allow you to construct a large Pinwheel Triangle using a particular arrangement of five small Pinwheel Triangles. This arrangement has qualities that are essential to proving the aperiodicity of the Pinwheel Pattern. Here's a more generic way to use four congruent right triangles to create a larger, similar right triangle.

111

This construction works with any right triangle (see figure), but notice that the resulting tiling does have rigid symmetries.

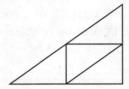

It's easy to find this pattern of four triangles within the Pinwheel Pattern, which is potentially confusing. You'll need to encourage students to focus on the five-triangle pattern, not the four-triangle pattern.

---

*Our Experience*

*Time spent:* One or two 50-minute sessions, depending on the approach

*Emphasized items:* The meaning of rigid symmetry and symmetry of scale

*Items omitted or treated lightly:* The proof of the uniqueness of scaling

*Dangers:* The proof of the uniqueness of super-tiles and the reason that this proof implies aperiodicity are difficult ideas.

*Personal favorite topic or interactions:* The pictures are intriguing.

*Remarks:*

1. One of us omits this section in class, because it is an outstanding group research project with good poster session possibilities.
2. This section can be treated in several different ways. Understanding the whole section in detail is challenging. Another approach is to take one 50-minute period to talk about symmetry, present the construction of the Pinwheel Pattern, and state that it is aperiodic, but omit the proof of its aperiodicity, that is, the proof and implications of the uniqueness of scaling. Sometimes we have done a thorough treatment, which might be done in approximately two 50-minute sessions.

*Sample homework assignment:* Read the section and do Mindscapes I.2; II.6, 8, 10, 12; III.16.

---

### Sample Class Activities

*Symmetry.* Ideally give each student a piece of paper with a square tiling and a transparency with the exact same tiling on it (see Template 4.4.1). Ask students to perform certain symmetries by sliding, rotating, and flipping the transparency. Have them mark a lattice point on the paper and transparency and describe and name all symmetries that leave that point fixed. Then ask them to perform some pairs of those symmetries in different orders. You may opt to help the class observe the idea of noncommutative properties of symmetries; however, the section deals neither with this issue nor with the question of the group structure of symmetries.

*Symmetry of scale—a different type of symmetry.* Ask students to group sets of four squares in their square tiling to create another square tiling with larger, four-unit super-tiles (see Template 4.4.2). Show how these super-tiles can be grouped into fours to create super-super-tiles, and so on. Discuss that there are different groupings of four square tiles to create the four-unit super-tiles. The possibility of different groupings is a central idea in proving the aperiodicity of the pinwheel tiling, later in the section.

*Generating tilings with symmetry of scale.* Ask students to take an equilateral triangle and surround it with three others to create a large equilateral triangle (see Template 4.4.3). Have them shade the original triangle in the middle. How can this idea be repeated to create a tiling for the whole plane? Ask students to describe the process.

*Super-tiles with 1, 2, $\sqrt{5}$ right triangles.* Ask students to take five identical 1, 2, $\sqrt{5}$ right triangles (see Template 4.4.4) and create a larger right triangle of the same proportions. They should then mark the central triangle. Ideally, using physical models, such as paper, can make this process concrete for students. Ask the class to use the process from the previous activity to build a tiling of the plane. Ask them to notice how this tiling, called the *Pinwheel Pattern,* automatically has a symmetry of scale because of its method of construction. Have them draw the first 25 tiles.

*Finding the super-tiles.* Give students a copy of part of the plane covered by the pinwheel tiling (Template 4.4.5). Pick a triangle at random, and ask students to draw the five-unit super-tile of which it is a part. Ask whether they had a choice about the group of five triangles that make up the five-unit super-tile.

The remaining activities pertain to the proof of the uniqueness of scaling and, therefore, the aperiodicity of the Pinwheel Pattern. Consequently, they may be skipped if you wish to treat this section more lightly. The proof of aperiodicity is challenging for students.

*Uniqueness of scaling.* Ask students to take a five-unit super-tile and a transparency copy of it (Template 4.4.6) and try to position the central triangle over any of the other four triangles. Notice that in every case, some edges cross; thus, alternative groupings are impossible. Contrast this situation with the case of equilateral triangle tiling or square tiling.

113

*No rigid symmetries, but large super-tiles can be matched.* Ask students to take a transparency copy of part of the Pinwheel Pattern (Template 4.4.5) and try to overlay it on a paper copy to produce a rigid symmetry. Ask them whether they can make five triangles match up. Then ask them to find a rigid symmetry that matches 25 triangles. In each case, point out how havoc arises outside the lined-up region. Then help them understand the proof that the uniqueness of scaling implies that there are no rigid symmetries.

***Templates for Transparencies*** *(located at end of chapter):*

- Template 4.4.1—Square tiling

- Template 4.4.2—Square tiling with grouping showing symmetry of scale

- Template 4.4.3—Equilateral triangle tiling

- Template 4.4.4—Five identical 1, 2, $\sqrt{5}$ right triangles, creating a larger one

- Template 4.4.5—The Pinwheel Pattern

- Template 4.4.6—Five-unit super-tiles, mismatching

***Sample Test Questions***

(The Mindscapes are another good source of test questions. All test questions below are available electronically on the Test Bank CD-ROM.)

- On the picture of a portion of the Pinwheel Pattern, draw to demonstrate that it has a symmetry of scale.

- Look at the picture of a portion of the Pinwheel Pattern. For each tile marked, draw the 5-unit super-tile and the 25-unit super-super-tile of which it is a part.

- Show how to surround an equilateral triangle with three identical equilateral triangles so that the whole configuration is a larger equilateral triangle.

- Show how to surround the Golden Triangle by four other identical Golden Triangles so that the whole configuration is a larger Golden Triangle.

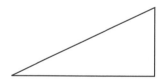

## Sample Lecture Notes for Section 4.4, "Soothing Symmetry and Spinning Pinwheels" (1 day)

Question of the Day (to be written on the board for students to think about and discuss before class begins): *What is the most symmetric shape you can think of?*

5 minutes—Acknowledge the affinity human beings have for symmetry. Take a sample of students' answers to the Question of the Day. (Presumably some will have answered "circle" or "sphere.") Connect the allure of the circle with early models of planetary motion, spinning tops, wheels, and so on. Refer to the description of the preference for a human face with features averaged by a computer (text page 250).

10 minutes—Ask students to describe common patterns in floor tilings (square, hexagonal, etc.). Which patterns have symmetry? What do we mean by "symmetry"? Hand out transparencies to introduce the idea of "a symmetry." Have students work in pairs to discover various *rigid symmetries*: shift, rotational, and flip (see Sample Class Activities on pages 112–114 of this *Instructor Resources and Adjunct Guide*). Point out that in this activity, students are taking a vague, intuitive idea and making it more precise.

10 minutes—Introduce *symmetry of scale*—grouping tiles to make super-tiles. Hand out the square tiling (Template 4.4.1), equilateral triangle tiling (Template 4.4.3), and hexagonal tiling transparencies. Have students work in pairs to find super-tiles, super-super-tiles, and so forth. Point out that there are different ways to define a super-tile for a given tile. For example, with square tiling, a super-tile might comprise four squares or maybe nine squares. Be sure someone notices that hexagonal tiling does *not* have symmetry of scale.

5 minutes—Make the following observation: If you can take a tile and surround it with copies of itself to create a super-tile of the same shape, then you have a method to tile the plane. When you use this method, you automatically get a symmetry of scale. Illustrate with the square tiling and equilateral triangle tiling.

5 minutes—Reinforce the two types of symmetry: rigid and scale. Point out that there are tilings (for example, the square tiling) that have both rigid symmetries and symmetry of scale. Point out that the hexagonal tiling has rigid symmetry but no symmetry of scale. Ask if there is a tiling that has symmetry of scale but no rigid symmetries. (Take a vote!)

10 minutes—Create the Pinwheel Pattern. The basic tile is a right triangle that has one leg twice as long as the other. Demonstrate that five such tiles can be arranged to form a super-tile of the same shape. (If you decide to allow more class time for this section, you could have students discover this construction themselves. See page 113 of this *Instructor Resources and Adjunct Guide*.) Display transparencies to show a super-super-tile and a detailed sample of a completed tiling of the plane (Template 4.4.5). Ask students to comment on the tiling (it's very twisted!).

5 minutes—Suggest that the Pinwheel Pattern has no rigid symmetries—that is, it is *aperiodic*. Remark that although this is difficult to prove formally, it can be illustrated informally. Use two transparencies made from Template 4.4.5 (you may want to enlarge this pattern). Highlight any particular super-tile in one transparency. Point out that any rigid symmetry would have to move this super-tile to another super-tile. Highlight a *different* super-tile in the second transparency and line up the two super-tiles as a rigid symmetry would have to do. Observe that the rest of the overlapped figures form a blur. Acknowledge that your example is not a proof.

*Note:* This material can be expanded to fill two class days. Have students spend more time discovering rigid symmetries and creating super-tiles. Much of the second day can be spent going over a proof that the Pinwheel Pattern is indeed aperiodic, as described in the text.

**Section 4.5  The Platonic Solids Turn Amorous** [Symmetry and duality in the Platonic solids]. This section presents the regular solids and discusses the idea of duality. Investigating the solids encourages students to move from the qualitative to the quantitative—how many edges, faces, and vertices does each solid have? Recording these observations and finding coincidences alert us to look for reasons and relationships that show a deeper structure. In this case, we are led to the idea of duality.

---

*General Themes*

Finding coincidences and patterns can reveal underlying structures.

---

*Mathematical Underpinnings.* (*Caveat*: These mathematical overviews are usually for the instructor only. The symbols and detailed proofs are not suitable for class presentations because mathematical symbols and terminology often present a formidable barrier for students.)

Looking at the Platonic solids is a great way for students to explore on their own. By carefully counting edges, faces, and vertices, patterns emerge that lead, among other things, to the idea of duality.

It is essential in this section to have models of all the solids for students to examine closely. The manipulative kit comes with materials to build the solids. If this task is assigned as homework, students will come to class already familiar with many key features of the solids, thus reducing the amount of class time needed for the material. Alternatively, to encourage in-class group discovery, bring your own models to class for students to examine together.

The duality relationship is most easily seen between the cube and the octahedron. A traditionally rectangular classroom is close enough to a cube to assist students in visualizing the dual octahedron. Playing with the tetrahedron allows students to discover that a solid can be its own dual.

Students will learn in Section 5.3 that there are only five Platonic solids. However, there are many semiregular polyhedra in which two or more regular polygons appear as faces. For example, if you start with an icosahedron and cut off each vertex, you get a truncated icosahedron, a shape that looks like a soccer ball. The 13 Archimedian solids are all semiregular polyhedra and include the five truncated Platonic solids.

119

Here's an alternate, very attractive approach to Sections 4.4 and 5.3. Present the Platonic solids using ideas of symmetry. Bring a bag of models to class, display the Platonic solids, but pretend there are still more in the bag. Jump to Section 5.3, which is independent of the other sections in Chapter 5. Prove that there are only five regular solids, then return to Section 4.4 to cover duality.

---

*Our Experience*

*Time spent:* One-and-a-half 50-minute sessions

*Emphasized items:* Physically handling the solids and counting sides, edges, and faces of each is a good way for students to become acquainted with them.

*Personal favorite topic or interactions:* When students really grasp the idea of duality, they often feel a satisfying "A-ha" experience, which is fun to see.

*Remarks:*

1. This section is ideal as a reading assignment not discussed in class, with students being responsible for the material for tests. It is empowering for students to realize that they can learn mathematics on their own. Homework is assigned.
2. Section 5.3 treats the Euler Characteristic and includes a proof that there are only five regular solids. You may wish to do this section on regular solids and then go immediately to Section 5.3, which has no other dependencies.

*Sample homework assignment:* Read the section and do Mindscapes I.2; II.6, 7, 8.

---

### Sample Class Activities

*Homework—make the solids.* Have students, individually or in small groups, make a complete set of the five regular solids using the materials from the kit.

*Counting.* How can the vertices, edges, and faces of the regular solids be counted in such a way that we can be confident that we have the correct answer? Ask students for suggestions. Marking numbers on the figure is one way. Counting in two different ways and getting the same answer is another. Ask how vertices and edges can be counted if you know things about the faces and how many meet at a vertex.

*Exploring the solids.* Ask students to observe features of the solids. In particular, ask them to move toward quantitative observations. For example, how many of the solids have triangular faces? How many faces meet at a vertex in each? Why isn't there a solid with four squares meeting at each vertex? A regular solid with four pentagons meeting at each vertex? A regular solid with hexagonal faces? Ask students to look for common numbers of various features.

*Duality.* Ask the students to look at the walls of the room and envision themselves in a cube. Lead them through the exercise of visualizing edges and connecting the centers of the walls, floor, and ceiling to see the two pyramids—one inverted—that make up the edges of an octahedron. Ask them why the number of faces of the cube must be the same as the number of vertices of the octahedron, and vice versa, and why the numbers of edges in the cube and in the octahedron are the same. Ask the class to formulate the idea of duality for any solid, and confirm their ideas with their chart of vertices, edges, and faces of the solids.

*Soccer ball.* Although the soccer ball is not a regular solid, ask students to explore a soccer ball and explain it in terms of the regular solids. Ask them to describe a solid that would be the dual of a soccer ball.

*Golden Rectangle.* Place an edge-model of the icosahedron on a table. Find two parallel edges, one on the top face, one opposite on the bottom. Ask students to measure the length and width of the rectangle formed, using the two parallel edges as two sides. Ask them to compute the ratio and compare that ratio to the Golden Mean.

*Extra credit homework—Golden Ratio.* Have students construct an octahedron with an icosahedron inside, with every vertex of the icosahedron touching a different edge of the octahedron. Then ask students to measure the lengths of the subsegments into which each vertex of the icosahedron divides the edge of the octahedron that it hits. Have them compute the ratio and verify that the result is the Golden Ratio. Experimental difficulties may make this problematic, but perhaps if several students do their best, the average might approach the Golden Mean.

121

### Sample Test Questions

(The Mindscapes are another good source of test questions. All test questions below are available electronically on the Test Bank CD-ROM.)

- How many faces does the dual of an octahedron have?

- Fill in the following table for the regular solids.

| Solid | Number of Vertices | Number of Edges | Number of Faces |
|---|---|---|---|
| Tetrahedron | | | |
| Cube | | | |
| Octahedron | | | |
| Dodecahedron | | | |
| Icosahedron | | | |

Indicate all the places in this table that show the concept of duality. Please briefly explain the idea of duality.

- Recall that you can take an icosahedron and cut off each vertex to get a solid that is like a soccer ball. Suppose you do a similar thing, but this time with a dodecahedron. When you cut off each vertex, you get a solid. What kinds of polygons make up the faces? How many faces does that solid have? How many vertices? How many edges?

- Draw graphs in the plane that are equivalent to the edges of the regular solids. That is, draw a graph that has the same numbers of vertices, edges, and regions as the corresponding features of each regular solid.

## Sample Lecture Notes for Section 4.5, "The Platonic Solids Turn Amorous" (2 days)

### Day One

Question of the Day (to be written on the board for students to think about and discuss before class begins): *What's the most symmetric figure you can draw in the plane? What if you could only use straight lines?*

5 minutes—Ask for answers to the first part of the Question of the Day. Eventually someone should suggest a circle. Ask the class to discuss why a circle is so symmetric, leading to the observation that it looks exactly the same when viewed from any direction. Move to the second question, where classic responses should include equilateral triangle, square, and so on. Point out that, again, each object has the property that it appears exactly the same when viewed from any vertex or any side. Introduce the idea of a *regular polygon;* note that for each polygon, all sides have the same length and all angles are the same. Note that the equal angles property of a regular polygon is basically a property of how the sides fit together. Ask students how many regular polygons there are. Note that the polygons approach the circle as the number of sides increases without bound.

10 minutes—Pose the questions in 3-space: What's the most symmetric figure they can imagine? What if they can only use flat sides? Have a bag of models handy. As students suggest the sphere, cube, or others, bring out an example of each. For each, ask for observations about the symmetries: every face looks the same, every vertex has the same number of edges meeting at the same angles, all pairs of faces come together at the same angle, and so on.

5 minutes—Introduce the idea of a *regular solid* and ask how many there are. Display and name all five Platonic solids as if these are just a few examples of regular solids. Make it appear there are still models in your bag. Ask if anyone thinks there are infinitely many, approaching the infinite symmetry of the sphere. (You could solicit several conjectures and then vote.)

15 minutes—Announce that the rest of the class will focus only on the Platonic solids. Discuss features quantitatively: shape and number of faces, total number of edges, number of edges at each vertex, and so on. Divide the class into five groups. Give each group one of the solids, and ask them to complete the table on the next page. You can have a representative from each group come to the board to fill in the data for their solid. (If you want to keep students guessing until Chapter 5, keep secret the fact that these are the only Platonic solids.)

| Solid | Number of Vertices | Number of Edges | Number of Faces | Number of Faces per Vertex | Number of Sides per Face |
|---|---|---|---|---|---|
| Tetrahedron | | | | | |
| Cube | | | | | |
| Octahedron | | | | | |
| Dodecahedron | | | | | |
| Icosahedron | | | | | |

10 minutes—Discuss various relations among the numbers in the table. Ask each group to look at the values for its solid and explain to each other why, for instance, the number of vertices times the number of edges per vertex equals twice the number of edges. Confirm these relationships with the class as a whole. (*Note:* Some students may notice the duality relations. Address this idea if time permits or suggest students think about what's going on in preparation for a discussion in the next class. (See Lecture Notes for Day Two.))

5 minutes—Finish this first class with a brief historical discussion.

**Day Two**

Question of the Day (to be written on the board for students to think about and discuss before class begins): *How do you get an octahedron from a cube?* (Copy the following table on the board to use during class.)

| Solid | Number of Vertices | Number of Edges | Number of Faces | Number of Faces per Vertex | Number of Sides per Face |
|---|---|---|---|---|---|
| Tetrahedron | | | | | |
| Cube | | | | | |
| Octahedron | | | | | |
| Dodecahedron | | | | | |
| Icosahedron | | | | | |

10–15 minutes—Discuss the Question of the Day with the goal of introducing the *dual* of a Platonic solid. Encourage students to visualize themselves inside a cube with a new vertex in the middle of each face. (See text pages 277–280.) Discuss how to join these new vertices. Make the connection between the values in the table for the cube and those for the octahedron. Do the same for the other solids, again referring to the great 3D pictures in the text.

10–15 minutes—Introduce the idea of truncating a Platonic solid to get a new kind of solid. Bring in a soccer ball to illustrate. A simpler but more abstract example is the truncated tetrahedron. Bring in a solid tetrahedron made of clay or Play-Doh. Use a fine wire to slice off each vertex. Have students work in pairs to count the number of faces, edges, or vertices in the solid that results.

**Section 4.6 The Shape of Reality?** [Non-Euclidean geometries]. This section introduces the idea of geometry in spaces other than flat ones. The section begins by looking at shortest distances between points on Earth and deducing that the great circle routes are shortest. Shortest routes on polyhedral surfaces, such as boxes, are also explored. A discussion of the sum of the angles of a triangle leads to the idea of measuring the curvature of a space. This section provides a good lesson in taking familiar ideas and applying them or interpreting them in new settings.

---

*General Themes*

Keeping our minds open to possible realities even though they may run counter to our own experience can broaden our world view.

---

*Mathematical Underpinnings.* (*Caveat:* These mathematical overviews are usually for the instructor only. The symbols and detailed proofs are not suitable for class presentations because mathematical symbols and terminology often present a formidable barrier for students.)

This section explores non-Euclidean geometry, beginning with the classic challenge: Is the shortest distance between two points always a straight line? Replacing the term "straight line" with "geodesic" helps students generalize their geometric thinking. Students see additional differences in geometric structure through triangles with angles that sum to more or less than 180° and through the idea of positive or negative curvature, as opposed to zero curvature.

In the simplest sense, spherical geometry and hyperbolic geometry arise as alternatives to Euclidean geometry by changing the parallel postulate. In spherical geometry, in which "lines" are great circles, two points still determine a unique line. However, because any two great circles on a sphere must intersect, the Euclidean parallel postulate fails: Given a line and a point not on the line, there is *no* line parallel to the first line. In fact, there are no parallel lines in spherical geometry. (This may be confusing to students unless they fully grasp that the *only* lines on a sphere are great circles. Lines of latitude may look parallel, but only the equator is a bona fide line among all latitudes.)

In hyperbolic geometry, parallel lines abound. Given a line and a point not on the line, a hyperbolic plane will offer an infinite number of lines through the point that do not intersect (are parallel) to the first line. Until quite recently, cosmologists conjectured that the universe had negative curvature consistent with a hyperbolic geometry. Current theories, however, propose that the universe has zero curvature.

While the surface of a ball provides easy access to a physical model of a 2-dimensional spherical geometry (the spherical plane), there is no easily available model of the hyperbolic plane. Models of the hyperbolic plane, such as the Poincaré disk model, exist in the plane, but they require us to redefine geodesics and distances in counterintuitive ways. Mindscape II.22, Becoming hyper, describes a way for students to build a model of the hyperbolic plane by taping triangles together.

The bug-on-a-wall example is fun to demonstrate in class using a simple cardboard box that opens up flat. A globe is also a great tool for demonstrating great circle routes. Bring in a standard flat map of the world, such as a Mercator projection, on which some great circle routes look much longer than a straight line on the map. Resolving this seeming contradiction requires students to acknowledge the distortion of the flat map. (Simply compare the size of Greenland on the globe with the size of Greenland on the flat map.)

The globe or a large ball is a good tool for displaying triangles with angles that sum to greater than 180°. Students may appreciate knowing that there is a formula for the area of a triangle on a sphere, though it is too technical for this class and requires measuring angles in radians. The analog of the Pythagorean Theorem for right triangles on a sphere is also quite technical and involves cosines.

---

*Our Experience*

*Time spent:* One-and-a-half 50-minute sessions

*Emphasized items:* Unfolding the box; sums of angles of triangles on spheres and the relation to area

*Dangers:* Building a big model of the hyperbolic plane can be fun, but it is hard to manipulate—too floppy.

*Remarks:*

1. Unfolding the box to find the shortest distance between points is a good demonstration of seeing a problem from a different vantage point.
2. One of us omits this section in class, because it is an outstanding group research project with good poster session possibilities.

*Sample homework assignment:* Read the section and do Mindscapes I.3; II.9, 10, 11, 14, 19, 20; III.26.

---

### Sample Class Activities

*Shortest paths on spheres.*   Ideally the class would be broken up into groups of three or four, and each group would have a globe or large round ball and a marker. However, an instructor demonstration will work too. Ask students to take two points on the globe, perhaps on the same latitude, and measure the distances between the points using several different paths. A string or rubber band with distances marked on it could be used. This project is perfect for a group of four—one to hold the Earth, two to hold each end of the rubber band or string, and another to move the string to various locations. Ask students to put the string on the latitude path and then lift it up above the great circle path to see how much shorter the great circle path is.

*Paths on boxes.* (Template 4.6.1) Provide each student with a shoe box, pushpins, and a long rubber band. You can also demonstrate this activity for the class. Take various pairs of points and measure the paths between them. Have students convince themselves that some paths between points on opposite walls really do go over five sides. After students have located the shortest path over five sides by measurement, ask them to cut open the box and lay it flat so that their path really is a straight line in the unfolded box.

*Triangles on spheres.* Have students draw various triangles on spheres, physically measure the angles, and add up the measurements. Ask them to measure angles in small triangles and in large triangles and to explain the differences in the sums of angles from one triangle to another. Ask different groups to use a few triangles to cover the sphere. Different groups should use different sets of triangles. Ask each group to measure the excess angles in each triangle. Then, as a whole class, deduce a theorem about the total excess over the whole sphere.

*Triangles on a saddle.* If you have access to a saddle, experiment with building a triangle on it and measuring the sum of the angles. Some classrooms, however, do not come equipped with saddles.

*Big hyperbolic plane (Thurston–Weeks).* This physical model mimics many of the properties of hyperbolic geometry. Make many copies of identical equilateral triangles—we have used triangles of about 4 inches on a side. Ask students to build a huge quilt by taping the triangles so that seven triangles come together at each vertex. Start with one circle of seven. Then add triangles to the outside so that each vertex has seven more triangles. As the quilt grows, there will be more room for more people to work. The bigger the students can make the quilt, the better. The resulting object is floppy and has folds, somewhat resembling some types of lettuce leaves.

Guide students to do several kinds of experiments. Mark a vertex as the origin, and point out the rings of triangles emerging from that point. Ask the class to pick any other point to see whether those same circles of triangles exist. Now ask them to find shortest paths. Help them notice that if they take two points on distant rings from the origin, the shortest path between them passes through rings closer to the origin. Discuss the idea that if both points are on the same ring, the path staying on that ring is longer than the path going toward the origin and back out. You might point out a rough analogy with the case of latitude paths versus great circle paths on the sphere. Ask students to draw triangles connecting three points with shortest paths. Then have them measure the angles of the triangle and take the sum. As a class, compare the sum of angles for small triangles versus triangles connecting three points equally spaced around a distant ring.

*Our universe.* Discuss experiments that might establish the curvature of the universe where the path of light is the shortest path between points. Ask what difficulties are entailed in some reasonable methods.

### *Templates for Transparencies* (located at end of chapter):

- Template 4.6.1—Box folded and unfolded showing shortest distance between points

### *Sample Test Questions*

(The Mindscapes are another good source of test questions. All test questions below are available electronically on the Test Bank CD-ROM.)

- Find the shortest path between the two points on the box.

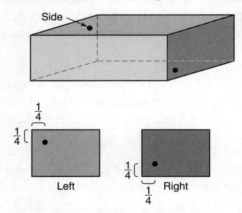

- Earth's area is approximately 200,000,000 square miles. Suppose that on Earth's surface you draw a triangle whose area is 20,000,000 square miles. Two of its angles are 70° and 80°. Assume Earth is a perfect sphere. What is the third angle of your triangle? Please show your work.

- Earth's area is approximately 200,000,000 square miles. Suppose that on Earth's surface you draw a triangle whose angles are 112°, 80°, and 60°. Assume Earth is a perfect sphere. What is the area of your triangle? Please show your work.

- If you have a quadrilateral on a sphere whose edges are great circle segments, what can you say about the sum of its angles? Can you deduce a formula relating the sum of its angles to the area it encloses?

**Sample Lecture Notes for Section 4.6, "The Shape of Reality?"**
**(1 day)**

Question of the Day (to be written on the board for students to think about and discuss before class begins): *Is the shortest distance always a straight line?*

10 minutes—Have students talk in pairs to come up with examples of the answer to the Question of the Day being no. Write suggestions on the board and lead into a discussion of what is meant by "shortest distance" and by "straight line."

5 minutes—Discuss the taxicab example (text pages 290–291), where shortest distance is not the traditional straight line. Emphasize the constraints of the given context.

10 minutes—Display a globe and a flat map of the world. Present the example of finding the shortest path from Chicago to Rome (or use two other cities at the same latitude), using string to compare the lengths of different routes. (Follow the example on text pages 291–292.) Emphasize that "straight lines" on the globe don't look straight on the flat map because the flat map distorts distances. Introduce the terms "geodesic," "great circle," and "spherical geometry." Point out that great circles as geodesics reflect the constraints of the spherical context.

10 minutes—Ask students what triangles look like on a sphere. (Distribute plastic balls and markers if you have them, otherwise use the globe or a large plain ball to demonstrate.) Remind students that all line segments on a sphere are parts of geodesics. Can anyone in class construct a triangle with two right angles? With three? Remind students that any triangle in the plane has angles summing to exactly 180°. Ask what they can say about the sum of the angles for a triangle on a sphere. Raise the idea of *curvature* and relate it to the angles of a triangle.

10 minutes—Ask students to visualize a saddle (or bring one in!) and think about what shortest paths (geodesics) look like between two points. Ask them to imagine a triangle made of segments of geodesics, then have them use their 3D glasses to view the saddle images on text pages 296–297. Discuss the apparent sum-of-angles property on the saddle surface and relate it to negative curvature. Introduce the term "hyperbolic geometry," pointing out that if the saddle surface continued infinitely in all directions, the geodesics would be actual hyperbolas.

5 minutes—Ask students how they might use properties of triangles to determine the geometry of a space. Point out the work of cosmologists trying to determine the curvature of the universe.

**Section 4.7  The Fourth Dimension** [Geometry through analogy]. This section introduces the geometry of 4-dimensional Euclidean space. The method of description builds insight and intuition by looking at the relationships between Euclidean spaces of the lower dimensions. This section is a good introduction to dealing with an abstract idea by analogy to what we already know. For many students, this section is an eye-opening experience and a highlight for the course.

---

*General Themes*

Understanding a difficult idea is easier if you think about a simpler version and use analogies to build complexity.

---

*Mathematical Underpinnings.* (*Caveat:* These mathematical overviews are usually for the instructor only. The symbols and detailed proofs are not suitable for class presentations because mathematical symbols and terminology often present a formidable barrier for students.)

This section introduces the idea of higher dimensional geometry. Getting started requires understanding the meaning of the word "dimension," a difficult term to define rigorously. Most math students see their first formal definition arising with vector spaces in linear algebra. For us, intuition serves well enough for dimensions 0, 1, 2, and 3. We informally define dimension as degrees of freedom or the minimum number of coordinate axes required to locate a point in space. A different notion of dimension is presented in Section 6.6 (Fractal Dimension).

One way to explore different dimensions is to pick one type of object and carefully examine its various manifestations in higher and higher dimensional spaces. Cubes of various dimensions can be built up using the ink-and-drag technique (text pages 314–316). It's also interesting to notice that a 2-dimensional cube (square) can be constructed by gluing together four equal-sized 1-dimensional cubes (line segments) so that they meet at right angles. Notice that the 1-dimensional cubes are glued together at specific pairs of endpoints, which are 0-dimensional cubes. Similarly, a 3-dimensional cube can be constructed by gluing together six equal-sized 2-dimensional cubes (text page 314). Notice that the 2-dimensional cubes are glued together along specific pairs of edges, which are 1-dimensional cubes. Now look at the unfolded 4-dimensional cube on text page 317. It's made up of eight equal-sized 3-dimensional cubes. To reconstruct the 4-dimensional cube, we have to glue together specific pairs of faces, which are themselves 2-dimensional cubes.

Look at the unfolded 3-dimensional cube (text page 317) and review the method for reconstructing the cube. When we perform the construction in 3-dimensional space, the 2-dimensional cubes remain perfect squares, and the angles between faces of the resulting 3-dimensional cube are all perfect right angles. If, on the other hand, we were to stay in 2-dimensional space and glue edges together to make the 3-dimensional cube, we distort many of the square faces and right angles to get the classic 2-dimensional drawing of the 3-dimensional cube:

Look again at the unfolded 4-dimensional cube (text page 317) and think about gluing together faces of the 3-dimensional cubes to reconstruct the 4-dimensional cube. The figure indicates several pairs of faces that need to be glued together. We can only work in 3-dimensional space; thus, if we actually try to carry out the reconstruction of the 4-dimensional cube, we will need to distort the 3-dimensional cubes to glue faces together. A total of 13 pairs of faces need to be glued to create the 4-dimensional cube.

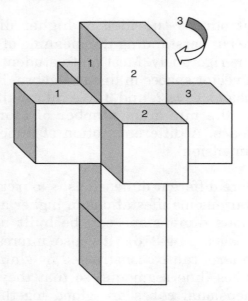

If we could actually work in 4-dimensional space, then we could glue together the 3-dimensional faces of the 4-dimensional cube without distortion.

There are other ways to explore the dimension of an object. One way is to look at its boundary. A 2-dimensional object, such as a disk, has a 1-dimensional boundary—a circle. Though that circle exists in 2-dimensional space—it curves around and could not be drawn within the confines of a 1-dimensional space—taken by itself, it is a 1-dimensional object. Once you fix an "origin" and your unit of measurement, you only need one number to specify your location on the circle. The text offers additional discussion of boundaries of 2-dimensional objects.

A 3-dimensional object, such as our planet, has a 2-dimensional boundary—its surface. Though this surface exists in 3-space and has many bumps and wrinkles, it is still 2-dimensional, requiring only two numbers, latitude and longitude, to specify a location. By analogy, the boundary of a 4-dimensional object should be 3-dimensional. This last observation is, of course, very hard to visualize, but the analogy is clear.

Yet one more way to explore a 4-dimensional object is to use level slices. You can introduce the idea of level slices of 3-dimensional objects without being as formal as in multivariable calculus. Simply use the idea of taking parallel cross sections at different heights. The cross sections of a 3-dimensional object will be 2-dimensional slices; the cross sections of a 4-dimensional object will be 3-dimensional slices.

---

*Our Experience*

*Time spent:* One-and-a-half to two-and-a-half 50-minute sessions

*Emphasized items:* The idea of understanding by analogy

*Items treated lightly:* Unfolding the cube

*Personal favorite topic or interactions:*

1. Asking students to come up with questions about the fourth dimension before actually doing the topic.
2. Untying knots and removing gold bricks from safes using the fourth dimension.

*Remarks:* Students are intrigued by the fourth dimension and enjoy the challenges of navigating in it.

*Sample homework assignment:* Read the section and do Mindscapes I.1; II.7, 12, 14; III.16, 18.

---

### Sample Class Activities

*Preliminary questions.* Ask students what questions they have about the fourth dimension before you even begin to tell them what it is. Have them work in pairs to come up with two different questions each. Emphasize that posing questions is the first important step toward understanding. Collect the questions verbally, and write them on the board. During the fourth-dimension discussions, refer to the questions as they are answered.

*Flatland.* Tell students that they are 3-dimensional creatures. Have them imagine that you, the teacher, are a 2-dimensional creature living on the board. Ask them to draw what you, a two-dimensional creature, would look like from their point of view. They will no doubt draw a smiley face. Tell them they are wrong and have them try again. Ask them where the eyes must be. Lead them to realize that the eyes couldn't see anything if they were inside the head. Also lead them to see that they could see your internal organs. Then ask them how we would look to a 4-dimensional creature. This emphasizes how different spaces look from different vantage points.

*Building up dimensions.* Describe the idea of dragging a point to generate a 1-dimensional space, dragging a 1-dimensional space to generate a 2-dimensional space, and dragging a 2-dimensional space to generate a 3-dimensional space. To help students understand this procedure, draw some segments on top of one another and label each as one of the lines that makes up 2-dimensional space. Label the 2-dimensional level of each line, perhaps $-1, -1/2, 0, 1/2, 1$ (Template 4.7.1). Now mark subsets of each one and ask students what the total set represents. For example, you might represent a circle or a letter of the alphabet. This exercise will get students used to the idea of seeing the total picture by looking at how the set intersects each level. Do a similar experiment with planes to describe 3-dimensional objects, for example, a ball, a knot, a linked pair of circles.

*The fourth dimension.* By analogy to the previous activity, ask students to identify various objects in 4-dimensional space by looking at how the objects intersect various 3-dimensional levels of 4-dimensional space. A circle placed to intersect levels in one or two points is a good example because it was used earlier in the example of the circle in the plane as it appeared at various levels. Likewise, a 3-dimensional solid ball spread out over a range of levels is good. Ask students to imagine a horizontal plane moving upward through a 2-sphere, the boundary of a ball, from the point of tangency at the south pole to the point of tangency at the north pole. Have them draw the level curves they would see (the curve of intersection between the horizontal plane and the 2-sphere) at various stages and help them discover that we see a dot, followed by growing concentric circles, followed by shrinking circles to a dot. Ask where the equator is. Finally, try an object that cannot exist in 3-dimensional space. Take a 2-sphere at level 0 with smaller 2-spheres that taper to a point at level 1 and, moving to lower levels that have smaller 2-spheres that taper to a point at level $-1$. This object is a 3-sphere, and it cannot exist in 3-dimensional space. Ask whether a point inside one of the 2-spheres can escape. Use Template 4.7.2 as necessary.

*Sealed vault.* Ask students to construct a vault in the plane that would be impregnable to a 2-dimensional being. Ask students how they, as 3-dimensional beings, would remove the contents without disturbing the vault. How would students rob a regular 3-dimensional vault if they could move in 4-space?

*Cubes warm-up.* Ask students to draw pictures of the 3-dimensional cube from as many different vantage points as possible. Invite students to come to the board and draw pictures of the different views. After you have constructed the 4-dimensional cubes, challenge students with drawing the same views but of the 4-dimensional cubes. This challenge should not be done in class.

*Cubes.* Ask students to construct cubes of dimensions 0, 1, 2, 3, 4, and 5. Instruct them to record numbers of vertices, edges, 2-dimensional faces, 3-dimensional faces, and 4-dimensional faces in a table. Ask them to draw pictures at levels of the solid cubes. Students will see that the 3-dimensional cube intersects each 2-dimensional level in a square, which is a 2-dimensional cube. They should see that a 4-dimensional cube intersects each level of 3-dimensional space in a solid cube.

*Linking.* Ask what two things link in 4-dimensional space in analogy to two circles linking in 3-dimensional space. Get the class to think about linking in lower dimensions to build up the analogy. Use Template 4.7.3 as necessary.

*Time.* Is time the fourth dimension? Ask students to consider this question. We could think of ourselves at each instant as the 3-dimensional slice of a 4-dimensional being. But does time allow the physical motion that 4-dimensional space really allows? For example, suppose you take a ball that you hold over time. What do you see at each instant? In four dimensions, you could bend that line to make a sideways **S** shape. What would that correspond to if we interpret the levels as different moments of time? How many balls would we see at an intermediate moment? Would we see some strange disappearances? Thus, time carries too much restrictive baggage to be the best analogy of the geometric fourth dimension.

**Templates for Transparencies** (*located at end of chapter*):

- Template 4.7.1—Looking at space by slicing

- Template 4.7.2—A slice view of 4-dimensional space

- Template 4.7.3—Linking in dimensions 1, 2, and 3

**Sample Test Questions**

(The Mindscapes are another good source of test questions. All test questions below are available electronically on the Test Bank CD-ROM.)

- Explain how you could unlink a chain without cutting the links using the fourth dimension.

- Even using the fifth dimension, it would be impossible to remove an object from a sealed 4-dimensional cube. True or False?

- Give a careful sketch of a 4-dimensional cube.

- How many edges emanate from a vertex of a 4-dimensional cube? Explain your answer.

- How many edges emanate from a vertex of an $n$-dimensional cube? Explain your answer.

- A 1-dimensional cube is a line segment; a 2-dimensional cube is a square; a 3-dimensional cube is a cube; and so on. Complete the following table:

| Dimension of the Cube | Number of Vertices | Number of Edges | Number of 2D Faces | Number of 3D "Faces" |
|---|---|---|---|---|
| 1 | | | | |
| 2 | | | | |
| 3 | | | | |
| 4 | | | | |

- A 0-dimensional triangle is just a point. A 1-dimensional triangle is a line segment; you know what a 2-dimensional triangle looks like; a 3-dimensional triangle is a tetrahedron. The pattern involves adding a vertex in the next dimension and joining it to all the vertices before with edges. Complete the following table:

| Dimension of the Triangle | Number of Vertices | Number of Edges | Number of 2D Faces | Number of 3D "Faces" |
|---|---|---|---|---|
| 1 | | | | |
| 2 | | | | |
| 3 | | | | |
| 4 | | | | |

- In regular 3-dimensional space, two circles link up as pictured. You cannot pull one circle free of the other in 3-space without pulling one through the other. In the fourth dimension, a sphere can be linked with a circle. Please draw a sphere and a circle in 4-space that are linked in the sense that you cannot pull the circle away from the sphere without pulling it through the sphere.

## Sample Lecture Notes for Section 4.7, "The Fourth Dimension" (2 days)

### Day One

Question of the Day (to be written on the board for students to think about and discuss before class begins): *What is dimension? What question(s) do you have about the fourth dimension?*

5 minutes—Have groups of students discuss the questions. Ask each group to tell you two of their responses; write the questions on a side board for reference throughout the class.

10 minutes—Discuss the idea of dimension, using the point, line, and plane as examples. In helping students grasp the idea of a fourth dimension, ask them to think about how they might build up from dimension 0 to dimension 1, from 1 to 2, and from 2 to 3. Present the ink-and-drag construction (text pages 308–310).

5 minutes—Discuss how, in theory, it is possible to ink and drag 3-dimensional space in a "new" direction to create 4-dimensional space. Acknowledge the abstract quality of this exercise. Ask students whether they think the resulting space is real. Then ask if they think a line is real, or a plane is real!

10 minutes—Focus on the plane as a 2-dimensional universe. Ask students to suggest objects that can exist in the plane: disk, square, anything flat with no thickness, anything you can draw on a piece of paper. Draw some examples on the board. Ask students to describe the boundaries of the objects, leading to the observation that all the boundaries are 1-dimensional.

10 minutes—Have students imagine what 2-dimensional creatures would look like. (See pages 133–134 of this *Instructor Resources and Adjunct Guide*). Have them draw the face of a 2-dimensional math teacher (Ed). Be sure they get the eyes, nose, mouth, and so on, on the boundary. Have them draw a second person (Mike), and ask how Ed and Mike see each other. Now have them draw a body for Ed with a few vital organs. Can Mike see Ed's heart? Looking from a 3-dimensional viewpoint, can students see Ed's heart?

10 minutes—Now ask them to look at each other. What do they see? Can they see any inner organs? Hold up an apple, ask if they can see anything other than the boundary, the 2-dimensional skin. Now ask them to imagine that our world is part of a larger 4-dimensional world in which 4-dimensional creatures are watching us. How will the 4-dimensional creatures see us? How would they see an apple? Discuss the value, and limitations, of this exercise in trying to understand the fourth dimension.

**Day Two**

Question of the Day (to be written on the board for students to think about and discuss before class begins): *What do you get when you slice a block of cheddar cheese? What do you get when you slice a 4-dimensional block of cheddar cheese?*

10 minutes—Announce that to better understand the fourth dimension, the class will now construct cubes in as many dimensions as they can. Remind students of the ink-and-drag technique. Have pairs of students try it, but this time have them drag only a short, prescribed distance. (Suggest about 2 inches.) Starting with a point, they create a line segment, then a square, and then a cube. Comment that each of these objects might be called a "cube" (see text page 314). Focus on the construction of the 3-dimensional cube: they start with a square (2D cube), which outlines one of the faces of the 3-dimensional cube. Once they've dragged this square, it outlines the opposite face of the 3-dimensional cube. Corresponding vertices of these faces are joined by edges created in the dragging process. Observe that the boundary (faces) of the 3-dimensional cube consists of a bunch (six) of 2-dimensional cubes glued together in a special way.

10 minutes—Ask how students could use this technique to create a 4-dimensional cube. They start with a 3-dimensional cube, which outlines one "face" of the 4-dimensional cube. Once they've dragged this cube, it will outline the opposite face of the 4-dimensional cube. Ask students how vertices of these two faces are joined. Acknowledge the difficulty of drawing the result in two dimensions. Point out that the boundary of the 4-dimensional cube consists of a bunch (how many?) of 3-dimensional cubes.

5 minutes—Ask how students might create a 5-dimensional cube. Do they think these objects, or higher dimensional spaces in general, really exist? Point out the need for more than three dimensions in understanding and solving problems with more than three variables (text page 316).

5 minutes—Bring up the topic of cheese. Show the class a thin square slice of cheese without revealing the block of cheese from whence it came (a paper facsimile of a cheese slice will also work here). Ask what this slice tells them about the shape of the orginal block of cheese. Point out how nicely the slice fits on a square cracker. Observe also that the knife blade is basically 2-dimensional. Now show them a bunch of square slices, saying these were cut in succession from the same chunk of cheese. Can they describe the original chunk with confidence?

138

5 minutes—Observe that slicing a cubic block of cheese is analogous to slicing an abstract 3-dimensional cube with a plane (a 2-dimensional space). The resulting cross section is a square, just like a square cracker (or a 2D cube). Be clear that your slice is parallel to one of the original cube faces—a *level slice*. (For extra fun, bring in several cubical chunks of cheese, a box of square crackers, and have a party.)

5 minutes—Introduce the idea of level slices as a way of exploring the shape of a 3-dimensional object. What shape crackers should they use for a cheese log? A cheese ball?

10 minutes—Recall the Question of the Day. What kind of crackers will match the shape of a slice from a 4-dimensional block of cheese? Point out that now the knife and the slices are 3-dimensional. Parallel slices of a 4-dimensional cube will be 3-dimensional cubes, as long as you start out with the knife parallel to one of the faces of the 4-dimensional cube. Discuss level slices of 4-dimensional objects as yet one more method of exploration.

# Template 4.3.1  Architecture with the Golden Rectangle

# Template 4.3.2   Golden Rectangle with a square inside

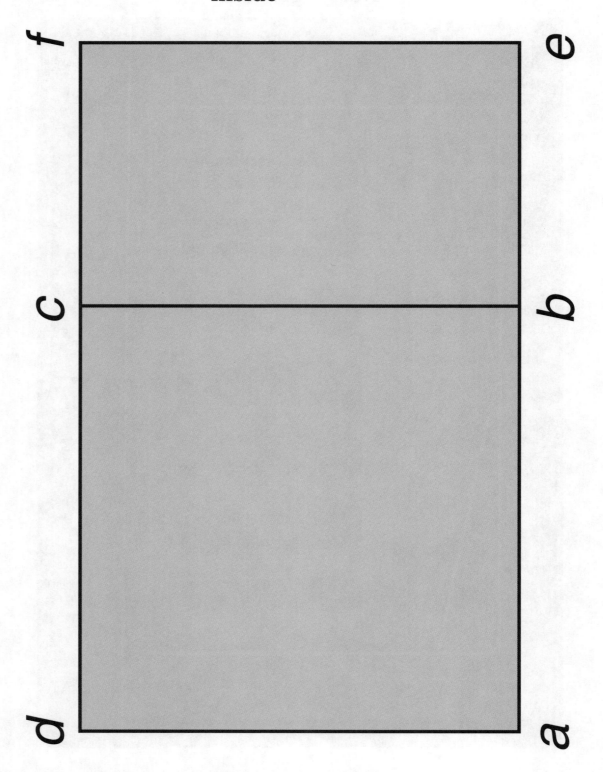

# Template 4.3.3   The logarithmic spiral in a
## Golden Rectangle

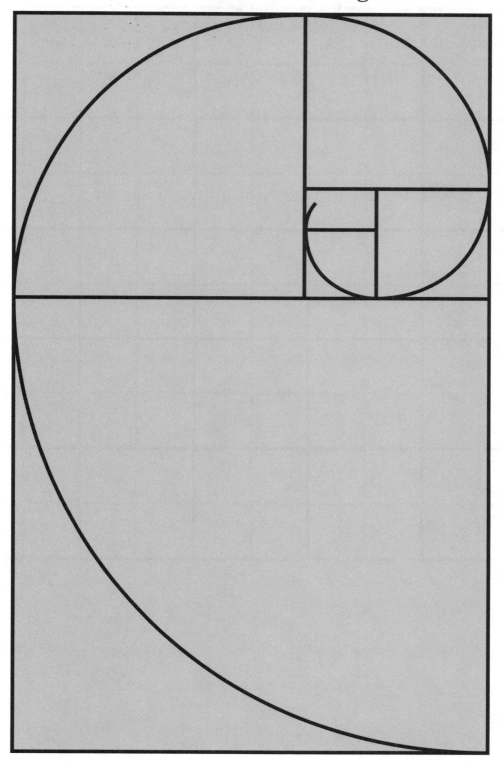

# Template 4.4.1  Square tiling

# Template 4.4.2   Square tiling with grouping showing symmetry of scale

# Template 4.4.3   Equilateral triangle tiling

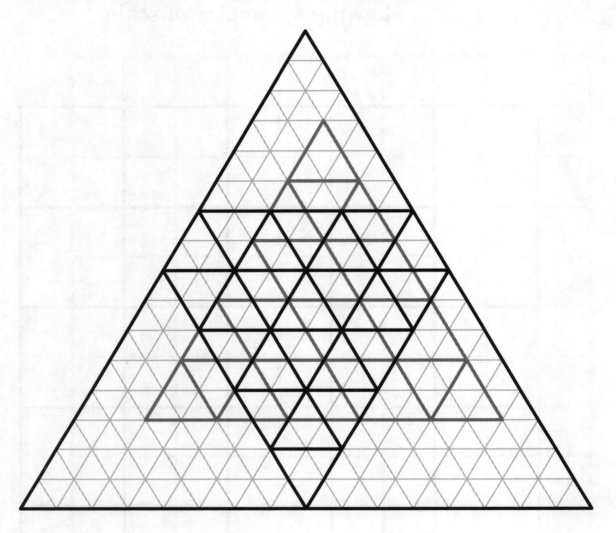

Template 4.4.4   Five identical 1, 2, $\sqrt{5}$ right
triangles, creating a larger one

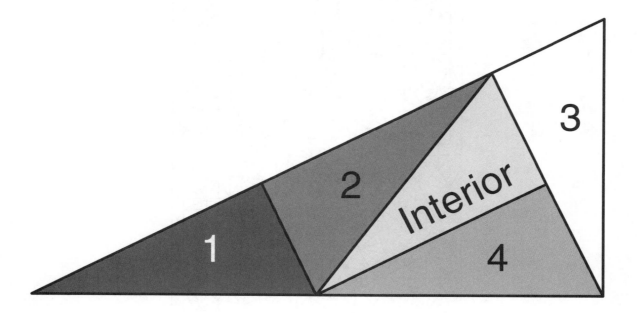

# Template 4.4.5   The Pinwheel Pattern

# Template 4.4.6   Five-unit super-tiles, mismatching

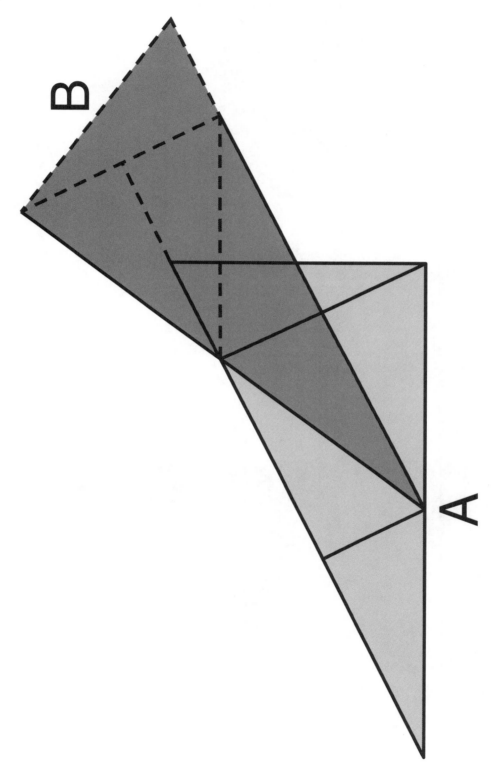

# Template 4.6.1  Box folded and unfolded showing shortest distance between points

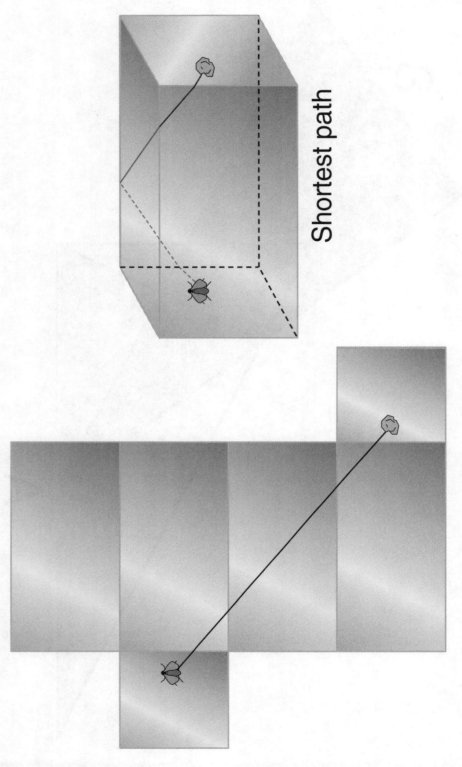

Shortest path

# Template 4.7.1  Looking at space by slicing

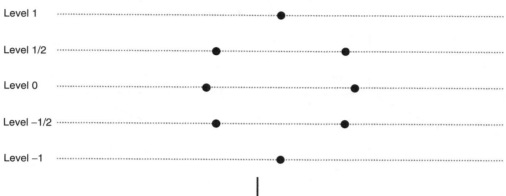

Level 1

Level 1/2

Level 0

Level −1/2

Level −1

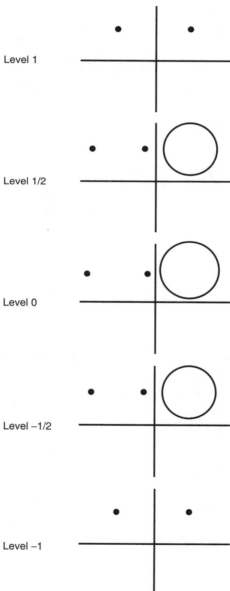

Level 1

Level 1/2

Level 0

Level −1/2

Level −1

# Template 4.7.2  A slice view of 4-dimensional space

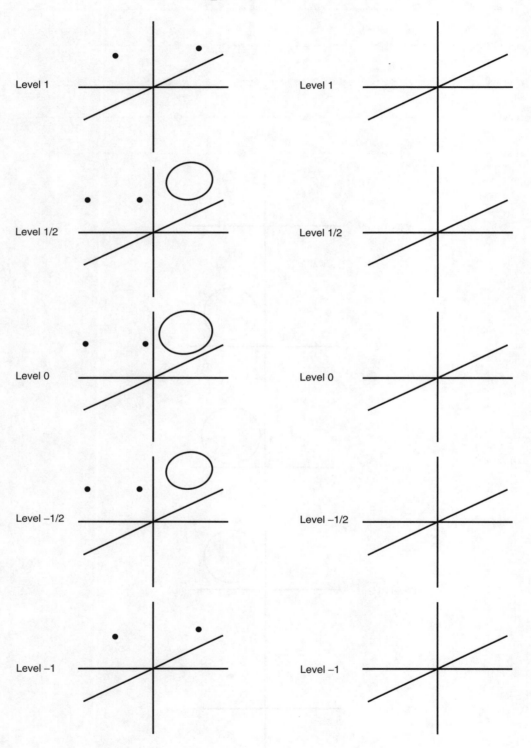

Level 1

Level 1/2

Level 0

Level −1/2

Level −1

Level 1

Level 1/2

Level 0

Level −1/2

Level −1

# Template 4.7.3   Linking in dimensions 1, 2, and 3

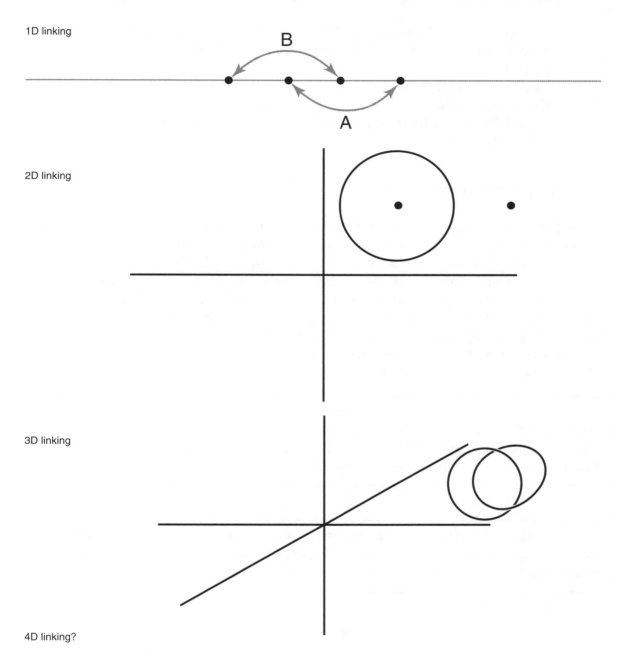

1D linking

B

A

2D linking

3D linking

4D linking?

# Chapter 5   Contortions of Space

## I.   Chapter Overview

Chapter 5 introduces the general field of topology. Basically, topologists study those aspects of a structure or object that remain unchanged when the object is stretched or shrunk without tearing or cutting. This chapter presents a collection of examples and ideas from geometric topology, including some perennial favorites, such as removing a vest without removing one's jacket, experiments with the Möbius band, and a bit of knot theory. One of this chapter's philosophical goals is to demonstrate the idea that we can choose notions of equivalence to meet our needs or interests. This chapter is a mind-stretching exercise for students.

---

### Dependencies
Section 5.3 (The Euler Characteristic) involves the Platonic solids, which were introduced in Section 4.5. The sections in this chapter are largely independent, so any may be used or omitted.

---

### Sections to Assign as a Reading Assignment or Term Project
We often assign Sections 5.4 and 5.5 as term projects. Any sections that you choose not to do in class are good paper or extra credit topics.

---

### *Reminders*
1. Class activities are suggested for each section. These can be done with students working in groups of two or three and then gathering their responses or as one global class discussion.
2. Before introducing a new topic, engage students with enticing questions, which they quickly discuss with neighbors; solicit responses within a minute. Use the responses as the springboard for the introduction.

3. We have included more sample class activities than you will probably want to use. Choose only the ones that are right for you and your class.
4. Each section can be treated in more or less depth. It is frequently a good idea to omit difficult technicalities in order to treat the main ideas well and then move on.

## II. Section-by-Section Instructional Suggestions

**Section 5.1 Rubber Sheet Geometry** [Topological equivalence by distortion]. This section presents several rubber sheet puzzles and solutions designed to illustrate some geometric distortions that open students' minds to what is possible if stretching, bending, shrinking, and twisting are allowed. We include in this section the idea of spaces (technically, 1-manifolds and 2-manifolds) in which every point has a neighborhood equivalent to Euclidean space. In particular, we talk about the circle, sphere, torus, and two-holed torus. The eye-opening lesson for students is that they can think about many features of the world using unusual ideas of sameness to discover interesting insights.

---

*General Themes*

Changing, modifying, or tweaking one aspect of reality can reveal hidden structure in the world.

---

*Mathematical Underpinnings* (*Caveat:* These mathematical overviews are usually for the instructor only. The symbols and detailed proofs are not suitable for class presentations because mathematical symbols and terminology often present a formidable barrier for students.)

This first section presents three topological challenges and then describes what it means for two objects to be *equivalent by distortion:* Each can be obtained from the other by stretching, shrinking, bending, or twisting, without cutting or gluing. Topologists call two such objects *topologically equivalent.* More precisely, objects $S$ and $T$ are topologically equivalent if there is a function called a *continuous deformation* from $S$ to $T$. However, to define such a function requires more basic topology than is reasonable to present in this text.

155

This section's informal, intuitive presentation on equivalence by distortion is motivated by three challenges. Doing one or more of these challenges in class can be very effective, especially if students are directly involved. The vest-removal challenge is the easiest to demonstrate. Ask for a student volunteer ahead of time; encourage the student to practice with an oversized tank top. (One of us has successfully had a student demonstrate the Dropping Trou puzzle from Chapter 1—dressed discreetly, of course.)

Turning an inner tube inside out is more difficult to demonstrate. You can display the figures on text page 331 using Template 5.1.1. Illustrate steps 4, 5, and 6, which are the hardest for students to visualize, by creating a model with two strips of paper and a rectangle. Of course, putting a big enough hole in a real inner tube allows for a true demonstration, but this is very messy.

The ring challenge is the most difficult of the three. Using Template 5.1.2, outline the linked circles in red and the outer boundary of the rubber sheet in blue. After the distortion, emphasize that the ring is still linked with the red circles but not with the blue one.

The text continues with a discussion of *topological invariants;* or those characteristics of an object that do not change when the object is distorted. Emphasize the technique of removing a point, circle, or disk from an object to see what happens. Underlying this discussion is the idea of a *manifold*. An object is a *1-manifold* if in a small neighborhood around each point, the object looks like a line segment. Examples include the real line ($\mathbb{R}$) or the curves below.

An object is a *2-manifold* if in a small neighborhood around each point, the object looks like a disk. Examples include the plane ($\mathbb{R}^2$), the surface of a sphere, and the surface of a torus. A *3-manifold* looks locally like $\mathbb{R}^3$, and an *n-manifold* looks like $\mathbb{R}^n$. Although the text does not introduce this formal terminology, it does suggest ways to determine whether two 2-manifolds, such as the torus and the sphere, are equivalent by distortion.

### Sample Class Activities

*Pre-chapter homework.* A good way to get students involved before discussing the chapter in class is to have them try the puzzles physically or pictorially, working in groups. Suggest that they try the puzzles before looking ahead and reading the solutions. Of course, if they want to look ahead, that's perfectly fine, too. You could also have them revisit Dropping Trou from Chapter 1, "Fun and Games."

*Vest removal.* The first puzzle is to remove a vest without taking off one's jacket. Physically demonstrating this trick is a good introduction to the chapter; better yet, get a student to demonstrate the trick.

*Inverting an inner tube.* Turning a torn inner tube inside out is entertaining, but extremely messy with a real inner tube. Perhaps a plastic swim ring would be neater. We supply pictures (see Template 5.1.1) from which transparencies can be made to demonstrate the deformation just as well and without the mess. After showing the move, repeat the pictures, but this time, place a transparency over the first transparency to show the meridianal curve and the longitudinal curve. Have students discuss for a few minutes where the meridianal and longitudinal curves will end up after the distortion.

During the discussion, you may want to point out that during distortions, any curve on a surface (as well as any subset) must also be moved during the distortion, and that following the progress of the curves during the distortion can be important. Such is the case in the third challenge of this section.

*The ring.* The third challenge is extremely difficult to do physically. We supply Template 5.1.2, which presents a sequence of pictures demonstrating the distortion. One of the key points to make is that the ring and the two circles that are linked in the first picture remain linked in the final picture.

*The ring revisited.* Draw a ring around the unlinked circle in the final picture. Ask students to follow it backward through the distortion to show that it remains unlinked throughout. This example is so counterintuitive that students need to be reassured that genuinely impossible things, such as unlinking linked curves or linking unlinked curves, are in fact not happening.

*Equivalence by distortion.* Show examples illustrating the idea of equivalence by distortion. Point out various features that such equivalences allow or do not allow. Template 5.1.3 contains several drawings of pairs of objects. For each pair, ask students to decide whether the first object can be distorted to look like the second. During the feedback, point out the idea that some features are preserved during distortions and some are not.

*Locally Euclidean spaces.* In distinguishing between the circle and the circle with sticker, explain that the circle with sticker has a point with a neighborhood different from every point's neighborhood on the circle. Discuss the idea of an object, every one of whose points has a neighborhood equivalent to a Euclidean space—the line or the plane. Note that we do not use the terminology of $n$-manifold in the text.

*The idea of invariants.* Draw the sphere and the torus. Ask students to discuss whether it would be possible to distort one to look like the other. During the discussion, interrupt students to ask them where a meridianal curve would end up if there were a distortion taking the torus to the sphere. Point out that the meridianal curve would have to be taken to some possibly wiggly curve on the sphere.

Using that curve as a hint, discuss the idea that a subset (in this case a curve) that separates the object upon its removal in one position would have to separate the object after the distortion. Thus, that kind of a property is useful in distinguishing objects that cannot be manipulated to look like one another by distortion.

*Unlinking the double torus.* Display the transparency (see Templates 5.1.4a and 5.1.4b) that shows the apparently linked double torus and the apparently unlinked double torus. We often ask for a vote about the equivalence or inequivalence of these two objects. Even though this example is illustrated in the book, there are points you can emphasize by going through it in class—of course, there are also the occasional students who did thoroughly read the section before class. Use the transparency overlay to show the pair of linked curves. Ask students why the existence of these curves does not conclusively demonstrate the impossibility of contorting the linked figure to the unlinked one.

Raise the following two points during the discussion: (1) Can the linked curves actually be unlinked by distortion? The answer is no. Of course, we do not actually prove this fact, and it is probably best not to embark on such a discourse. Simply explain that such curves cannot be unlinked by distorting them. Armed with this correct conviction, students should realize that somehow the two curves must remain linked. So, (2) ask students to discuss and draw pictures to illustrate how the curves look during the contortion process that takes the apparently linked two-holed torus to the apparently unlinked one. Students can draw the migration of the linked curves during the distortion on the illustrations in their books.

*Knotted Jell-O puzzle.* Transparencies can be made from Templates 5.1.5a and 5.1.5b to indicate a sequence of moves that changes the cube with one straight and one knotted hole into a cube with two straight holes. The geometric idea for students to understand is that one hole can slide up the side of the other hole. Again, following the future position of some interesting curves is valuable. To illustrate this idea, draw a knotted curve that goes along the originally knotted hole and then around the boundary. Ask students to follow the future positions of that curve during the distortion.

***Templates for Transparencies*** *(located at end of chapter):*

- Template 5.1.1—Inverting a torus

- Template 5.1.2—Unlinking a two-holed sheet from a ring

- Template 5.1.3—Equivalent by distortion?

- Template 5.1.4a,b—Double torus, linked and unlinked

- Template 5.1.5a,b—Cube with straight and knotted hole

## Sample Test Questions

(The Mindscapes are another good source of test questions. All test questions below are available electronically on the Test Bank CD-ROM.)

- If the letters R and P are made from lines, then they are equivalent by distortion. True or False?

- Consider the following letters: A B C D E F G H I J K L M N O P Q R S T U V W X Y Z. Arrange the letters into groups that are equivalent to one another by distortion.

- **a.** Follow the loop. In the series of pictures that follows, a rubbery sheet with two holes originally seems to link the ring twice, but at the end links it only once. In the first picture, a dotted loop has been drawn. Show where the dotted loop should appear in each of the subsequent pictures.

- **b.** Is it possible to separate the ring from the two-holed sheet? Why or why not?

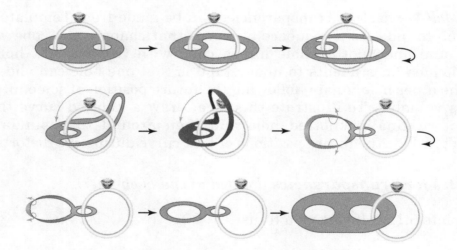

- **a.** The left and right squares in the series of pictures below represent a square rubbery sheet with fixed sides. The small squares labeled A and C are little areas of the rubber. Draw on the small squares to show how you can stretch the rubbery sheet, while leaving the boundary fixed, to start at the left picture and get to the right picture.

**b.** Is it possible to draw a curve connecting A to A, another curve connecting B to B, and a third curve connecting C to C within the large square so that none of the three curves intersects another? If so, draw the three connecting curves. If not, explain why it can't be done.

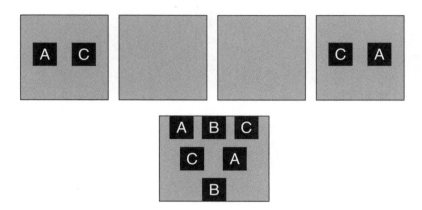

- Let $S$ be the surface of a torn inner tube. Is $S$ equivalent by distortion to a disk with two holes that has a tube glued from one hole to the other, as shown in the diagram? Explain your answer.

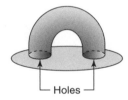

Holes

## Sample Lecture Notes for Section 5.1, "Rubber Sheet Geometry" (1 day)

Question of the Day (to be written on the board for students to think about and discuss before class begins): *Is Earth a ball or a donut?*

20 minutes—Pose several challenges for the class, such as the vest removal, the inner tube, or the ring. (Also consider Mindscape II.10.) If you asked for volunteers in advance, have those students demonstrate the challenges they prepared. Illustrate the inner tube challenge with Template 5.1.1 and strips of paper, as previously discussed. Illustrate the ring challenge with Template 5.1.2.

5 minutes—Introduce topology as *rubber sheet geometry,* followed by the idea of *equivalence by distortion* (objects can be stretched, shrunk, or twisted, but not cut or glued). Explain how the starting and ending configurations of all the challenges are equivalent by distortion. Give additional examples: coffee cup and donut, solid ball and cube, circle and square, letter C and letter S, and so on.

5 minutes—Ask if everything is really the same if we allow unlimited stretching. When are two objects topologically different? Raise the Question of the Day. Point out that locally, Earth looks flat; if we lived on a really big donut, then locally, it would look flat as well.

10 minutes—Introduce the following technique for checking whether two objects are equivalent by distortion: Remove a point from each object and see what happens. Illustrate with simple figures such as block letters as on page 333. Have students work in pairs on Mindscape II.7. Ask how the problem would change if a different font were used (serif rather than sans serif).

5 minutes—Extend the technique to surfaces: Remove a circle to see what happens. Illustrate with the torus and the sphere, as on text page 336.

5 minutes—Finish with another example, such as Mindscape III.26. Give students a minute to talk about the question in pairs, then discuss as a class.

**Section 5.2  The Band That Wouldn't Stop Playing** [Möbius band and Klein bottle]. This section features several tactile experiments, including cutting the Möbius band in various ways. It also introduces the idea of representing the twisted Möbius band by a rectangular strip with oppositely pointing arrows to indicate how the Möbius band is constructed. This idea reflects the general lesson that looking at issues or objects in various representations is an effective way to deepen one's understanding and discover why certain phenomena occur.

---

*General Themes*

Surprising discoveries can often be understood by looking at them in a new way. Taking an action often leads to new insight into mysterious ideas.

---

*Mathematical Underpinnings.  (Caveat:* These mathematical overviews are usually for the instructor only. The symbols and detailed proofs are not suitable for class presentations because mathematical symbols and terminology often present a formidable barrier for students.)

This section's primary focus is the Möbius band, an ideal topic for hands-on experimenting in mathematics. Many students will be familiar with Möbius bands, but few may have undertaken a detailed analysis of the curious, one-sided strip. Provide scissors, paper strips, and tape for students to experiment in class, either individually, in groups, or as a class demonstration. The life lesson here is to *do it!*

The Klein bottle is a subtler construct than the Möbius band. A physical model is of only a little help, because we can't study the bottle in its true, 4-dimensional habitat. Template 5.2.1 presents a 2-dimensional drawing of a Klein bottle.

Creating a Möbius band and verifying its basic properties is fairly straightforward. Cutting the strip in half down the middle is particularly fun. Cutting a band in thirds will likely be new for most students. Most students should be successful with this experiment if given careful instructions; understanding the results, however, is another matter. This is a great opportunity to introduce schematic models as a way to better understand a structure with confusing properties. Topologists call these schematic drawings *identification diagrams* because the arrows indicate the manner in which two sides of a polygon are glued together or *identified*.

The identification diagram of the Möbius band greatly clarifies the results of both cutting experiments, as explained on text pages 349–350. The same idea offers an elegant model for the Klein bottle—an object that exists only in 4-dimensional space clearly represented by a 2-dimensional diagram.

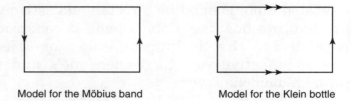

Model for the Möbius band     Model for the Klein bottle

Of course, simpler objects can be modeled in the same way:

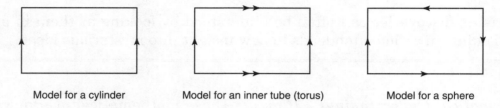

Model for a cylinder     Model for an inner tube (torus)     Model for a sphere

These last three objects differ from the first two in many ways, but one way is not mentioned in the text. The cylinder, inner tube, and sphere are what topologists call *orientable,* whereas the Möbius band and Klein bottle are *nonorientable.* Informally, a surface is orientable if moving around in the surface does not change your orientation. As you move across a boundary in the rectangle models above, the arrows maintain the same direction. In the Möbius band and Klein bottle, however, crossing a boundary flips your orientation as you follow the arrows.

Another idea touched on only briefly in the text is that of a connected sum. On text page 351, we describe an attempt to create a one-sided surface with no edge by "sewing" a disk along the boundary of a Möbius band. Though the result in this example is the somewhat obtuse projective plane, a surface that truly exists only in four dimensions, we can use the method to create many different surfaces. Sewing or gluing two surfaces together along boundaries results in what topologists call a *connected sum.* As an example, start with two tori and remove a small disk from each, as in Figure 1. Then glue the two tori together along the boundaries of the removed disks to get a two-holed torus, as in Figure 2 (multiholed tori were introduced in Section 5.1, text pages 336–337).

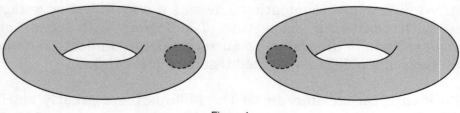

Figure 1

166

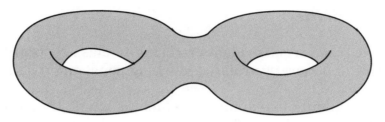

Figure 2

The two-holed torus can also be modeled with the following identification diagram; however, this diagram is too complex for most students to grasp.

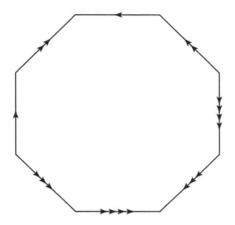

*Our Experience*

*Time spent:* One to one-and-a-half 50-minute sessions

*Emphasized items:* The physical Möbius band experiments

*Items omitted or treated lightly:* The Klein bottle

*Dangers:* The self-intersection of the Klein bottle is a difficult point. Students can confuse this situation with the rules against cutting and amalgamating points during distortions.

*Personal favorite topic or interactions:* Having students cut Möbius bands in the classroom—give students scissors and strips or ask them to bring scissors. Peel-off stickers are handy for taping together Möbius bands.

*Remarks:* The cutting activity is a huge hit.

*Sample homework assignment:* Read the section and do Mindscapes I.3; II.8, 9, 14, 25; III.33; IV.36.

## Sample Class Activities

*Pre-section homework.*  Ask students to bring to class several Möbius bands made from strips of paper, two colored marking pens, and a pair of scissors.

*Counting.*  Ask students to examine the Möbius bands they brought to class and to mark the edge without lifting their pens. Likewise, have them draw a line down the center to see that it has only one side. Discuss why the Möbius band has only one edge and one side. Suggest that a strip could be put together in many ways with several twists. Ask how they could tell whether a strip with its ends glued has one side or two. One edge or two. Help students discover that the issue comes down to the oddness or evenness of the number of twists.

*Cutting.*  Have students do the two cutting experiments from the section—cutting the Möbius band down the middle and cutting the Möbius band while hugging the right side. Discuss why cutting the band in thirds can yield a Möbius band of the same length as the original Möbius band and a new object, a twice-twisted band, with edges twice as long as the center line of the Möbius band.

*The rectangle representation.*  Explain the rectangle with identifications. Ask students to explain the results of the cutting experiments by referring to the rectangle representation.

*Exploring the Klein bottle.*  The Klein bottle is difficult to understand at first. If you have a model of the Klein bottle, definitely bring it to class. If not, we provide Template 5.2.1 to help students see the bottle's features. First ask students to take a rectangular piece of paper, perhaps 11 inches long and 3 inches wide. Ask them to place the two long edges next to one another before actually taping the pieces together, then ask them to draw arrows on the two shorter sides—one up and one down, as in a Möbius band. Then have them tape the long edges to one another to make a tube. Tell them to bring the two tube ends around toward each other, as if they were making a torus. Have students note that the arrows are going in opposite directions from one another. Encourage them to try to arrange the tube ends so the arrows do match up. Of course, students will not be able to do so. Then ask them to cut a hole in the side of the tube near one end and stick the other end through it. Help students see that now the arrows on the tube ends are going in the same direction. Instruct them to tape the ends together. Explain how the disk that was cut out of the side of the tube is actually part of the Klein bottle and could physically be put back in 4-dimensional space without self-intersections.

Now discuss whether it is possible to cut lengthwise along the tube direction (avoiding the hole) in such a way that they get two Möbius bands. This process will probably not work in class.

*Homework.* Assign as homework the exercise of first building a Klein bottle and then cutting the bottle in two along the correct path to show that it is actually the union of two Möbius bands glued along their boundaries.

*Class discussion—novel representations.* Finding new representations for objects is a good exercise. Ask students to choose some object or idea outside of mathematics and create a representation for it that helps them see it in a different way.

***Templates for Transparencies*** *(located at end of chapter):*

- Template 5.2.1—Klein bottle

### Sample Test Questions

(The Mindscapes are another good source of test questions. All test questions below are available electronically on the Test Bank CD-ROM.)

- Suppose you want to create a decorated, transparent Möbius band. You wish to decorate it with little circles with arrows going around each circle, as shown. You want the directional arrows for nearby circles to be going in the same direction. What goes wrong with your plan?

- On a transparent Möbius band, draw two circles that cross exactly once. By "transparent" we mean that when you draw on the Möbius band, it shows through on both "sides"—that is, it stains all the way through.

- If you took the rectangle below and joined the right and left edges with a twist, you would obtain a Möbius band. Suppose you cut the Möbius band on the lines drawn. What would you get?

- Draw a Klein bottle and shade a Möbius band that is part of it.

## Sample Lecture Notes for Section 5.2, "The Band That Wouldn't Stop Playing" (1 day)

Question of the Day (to be written on the board for students to think about and discuss before class begins): *Take a strip of paper and tape the short ends together to make a loop. How many pieces do you get if you cut the loop down the middle?*

5 minutes—Distribute precut strips of paper, scissors, and tape to groups of students. Have students create at least two Möbius bands per student. (If your class is too large, have students bring prepared Möbius bands and a pair of scissors to class.)

10 minutes—Have students draw along the edge of one of their Möbius bands without lifting their pens. Then have them draw a line down the center of the band. Encourage groups to discuss what these two drawing exercises show about the number of sides and edges of the Möbius band. Ask what would happen if they created a band using two twists instead of just one. Ask for conjectures about how the number of twists relates to the number of sides (edges) of the resulting band.

5 minutes—Instruct students to look at the band they marked and think about what would happen if they cut the band along the center line. Have each student write their prediction in their notes. Do not allow discussion. (It's much more fun this way!) Now tell them to cut along the center line. Make sure they cut only along the line, crimping the band if necessary to start the cut. Now discuss the results.

5 minutes—Explain how students can cut their second Möbius band into thirds, "hugging" the right side as described on page 348. Before they start cutting, ask them to conjecture what will happen. Have them note their predictions privately; again do not allow discussion. Be sure to give very clear instructions—stay 1/4 inch from the right side when cutting. When they're done, discuss the results.

15 minutes—Introduce the identification diagram model for the Möbius band, and use it to explain the side and edge characteristics of the Möbius band itself and the results of the first cutting experiment. Have students discuss among themselves how the diagram can be used to understand the results of the second cutting experiment. Confirm and clarify their thinking with a general class discussion.

10 minutes—Ask if the class can think of a way to create an object with only one side and no edges. (If they seem confused, point out that a sphere has no edges, though it does have two sides.) Suggest they start with the rectangle diagram. Lead into the orientation diagram for the Klein bottle. Ask students to discuss among themselves the actual construction of such a structure, then confirm that it exists only in 4-dimensional space.

**Section 5.3 Feeling Edgy?** [The Euler Characteristic]. This section describes how to draw a connected graph edge by edge and that $V - E + F$ (vertices minus edges plus faces) continues to equal 2 whenever you add a new edge. This inductive argument is presented, but not in the formal style of a proof by induction. If you do formal inductive proofs in class, this is a good proof to use. However, we do not suggest a rigorous or formal discussion of induction. The Euler Characteristic provides a good example of a feature of a situation that remains constant during a building process, a thought-provoking idea as applied to the real world. In this section, we apply the Euler Characteristic to prove there are only five Platonic solids.

---

*General Themes*

Looking carefully can reveal structure even within seemingly structureless objects.

---

*Mathematical Underpinnings.* (*Caveat:* These mathematical overviews are usually for the instructor only. The symbols and detailed proofs are not suitable for class presentations because mathematical symbols and terminology often present a formidable barrier for students.)

This section presents the Euler Characteristic of the plane and uses it to prove that there are only five Platonic solids. The Euler Characteristic arises in relation to plane graphs. The text presents an informal induction proof; what follows is a more rigorous version. Rather than defining graphs from scratch, we present the following summary in a manner tailored to our needs. Please note that the text uses the term "graph" to mean "plane graph."

For our purposes, a *plane graph* is an object drawn in the plane using a finite set of points, or *vertices,* with line segments or curves, called *edges,* connecting selected pairs of vertices. The edges must be drawn so that they do not intersect each other or themselves except at vertices. Some examples:

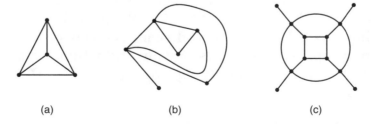

(a)          (b)          (c)

Each example shown is actually a *connected plane graph,* because in each case it is possible to get from any vertex to any other vertex by traversing edges in the graph. Notice that Figures (b) and (c) have at least one vertex joined to only one edge. Such an edge is called a *pendant edge*.

The edges and vertices of a plane graph divide the plane into *regions,* or *faces*. Including the exterior, infinite region, Figures (a), (b), and (c) divide the plane into four, four, and six regions, respectively. Note that an edge is a pendant edge if and only if it does not border on two faces. The Euler Characteristic is stated as follows:

**Claim:** *If $G$ is a connected plane graph with $V$ vertices, $E$ edges, and $F$ faces, then $V - E + F = 2$.*

**Proof:** We use mathematical induction on the number of edges in $G$. If $E = 0$, then $G$ has one vertex and one face. So, $1 - 0 + 1 = 2$, as desired.

Now suppose the Euler Characteristic holds for all connected plane graphs with $E$ edges, for some $E \geq 0$. Let $G'$ be a connected plane graph with $E' = E + 1$ edges, $V'$ vertices, and $F'$ faces.

If $G'$ has a pendant edge, create the graph $G$ by removing the pendant edge and one vertex at its tip. Now, $G$ has $E' - 1 = E$ edges, $V' - 1$ vertices, and $F'$ faces. Note also that removing the pendant edge will not disconnect the graph. Thus, $G$ is a connected plane graph with $E$ edges. By our induction hypothesis, we have $(V' - 1) - (E' - 1) + F' = 2$. Simplifying, we get $V' - E' + F' = 2$, so our result holds for $G'$.

If $G'$ has no pendant edge, then every edge of $G'$ must border on two faces, as shown. Deleting the edge will collapse two faces into one. (In the figure, the dotted lines represent arbitrary chains of edges.)

Once again, we create the graph $G$ by removing an edge, but this time we leave the vertices. Now, $G$ has $E' - 1 = E$ edges, $V'$ vertices, and $F' - 1$ faces. Removing the edge does not disconnect the graph. Thus, $G$ is a connected plane graph with $E$ edges. By our induction hypothesis, we have $(V') - (E' - 1) + (F' - 1) = 2$. Simplifying, we get $V' - E' + F' = 2$, so again our result holds for $G'$.

Therefore, by the Principle of Mathematical Induction, the Euler Characteristic holds for all connected plane graphs.

To relate this result to the Platonic solids, the text presents the method of stereographic projection to transform any regular solid into a connected plane graph. Because each such graph must satisfy the Euler Characteristic, some elementary algebra and case-by-case analyses reveal that only five possibilities exist. This proof is explained in detail in the text (pages 364–366) and is quite accessible.

The Euler Characteristic for plane graphs that are not connected is still a constant for all graphs with the same number of connected components. This is a great result for students to discover themselves.

More sophisticated students can play with graphs drawn on the torus. Be sure they have edges going through the hole (longitudinal edges) and around the hole (meridian edges). If they play around with multiholed tori, they'll find the Euler Characteristic for graphs with at least one each of all longitudinal and meridian edges is a constant that depends only on the number of holes in the torus.

---

*Our Experience*

*Time spent:* Two 50-minute sessions if the proof of five Platonic solids is done; one session otherwise

*Emphasized items:* Seeing how the Euler Characteristic remains constant as a connected graph is built

*Items omitted or treated lightly:* One of us omits the proof that there are five regular solids.

*Personal favorite topic or interactions:* Having students doodle randomly and then count vertices, edges, and faces

*Remarks:* Bring in a balloon and a magic marker. Draw a Platonic solid on the balloon, then stretch it out to see stereographic projection in action.

*Sample homework assignment:* Read the section and do Mindscapes I.2; II.7, 9, 13; III.26; IV.40.

---

### Sample Class Activities

*Random drawing.* Ask students to take out a piece of paper, put their pens at a random place, draw a sweeping curve that crosses randomly over itself, and then stop. Have them put dots at the beginning point and the stopping point and at every crossing. Have each student count vertices, edges, and faces for his or her drawing to confirm that $V - E + F = 2$. Now, ask them to retrace very darkly the same squiggle—edge by edge—and count $V - E + F$ at each stage. They should see that each additional edge either adds one more vertex or one more region; thus, the $V - E + F = 2$ formula continues to hold.

*Euler Characteristic.* After the squiggle activity, point out that adding a new edge to a connected graph, even if it is not one continuous drawing, still leaves $V - E + F$ the same. Then provide the statement and proof of the Euler Characteristic Theorem.

*Extraneous class activity—two coloring.* For this auxiliary exercise, ask students to color the faces of the squiggle graph with two colors by shading one region, polka-dotting all its adjacent regions, shading adjacent regions adjacent to polka-dotted ones, and so. Discuss why it is possible to color the faces of such a graph with two colors. Although the text does not explore coloring issues, it is a wonderful topic. If you wish, it would fit in here quite naturally.

*Regular solids.* Apply the Euler Characteristic to prove that there can be only five regular solids. The first step is to reteach the idea of stereographic projection. Although this topic was introduced in Chapter 3, "Infinity," it probably needs to be reintroduced. We supply Template 5.3.1 to show how a cube can be blown up on a balloon so that its edges, faces, and vertices lie on the balloon. Simply by stretching the balloon out on the plane, the stereographic projection puts the vertices, edges, and faces on the plane. Discuss how to use the regularity of the surfaces to deduce relationships among the numbers of sides of each face, and the numbers of faces, vertices, and edges. The arithmetic and algebra need to be handled with care. Even though this proof is given in detail in the book, students will certainly need to see the steps presented and discussed in class.

*Euler Characteristic of nonconnected graphs.* Ask students to explore and make a conjecture about the Euler Characteristic of two-component, three-component, or $n$-component graphs. This is a good activity for getting students to see patterns and analyze related situations.

*Euler Characteristic of other surfaces.* (This activity is included in the Mindscapes for Section 5.3.) Template 5.3.2 is a transparency of a polyhedral torus. Ask the class to compute $V - E + F$ for this torus. Then ask how that number changes as more edges are added. Mention that a small connected graph on a little spot on the torus would have $V - E + F = 2$, like the sphere.

However, if you start by adding two curves, one meridianal and one longitudinal, then $V - E + F = 0$. It will continue to equal 0 with further subdivisions or with any bigger connected graph that contains the figure eight shape. Reinforce the proof of the Euler Characteristic on the plane by asking students how to prove these facts about the torus.

*Euler Characteristic of connected sums.* (This activity is included in the Mindscapes for Section 5.3.) Template 5.3.3 illustrates two tori that create a triangulated two-holed torus by connected sum. Ask students to compute the Euler Characteristic. Then, view the torus as two tori with a triangle removed from each and glued together on the boundary (i.e., the connected sum). Confirm that if you know the Euler Characteristic of each piece, then the connected sum has Euler Characteristic equal to the sum of the two Euler Characteristics minus 2—the two removed triangular faces.

*Auxiliary question on maps.* To show another application of the Euler Characteristic, ask why any map has a country with five or fewer neighbors.

### Templates for Transparencies *(located at end of chapter):*

- Template 5.3.1—Square projected onto balloon

- Template 5.3.2—Polyhedral torus

- Template 5.3.3—Connected sum of two polyhedral surfaces

### Sample Test Questions

(The Mindscapes are another good source of test questions. All test questions below are available electronically on the Test Bank CD-ROM.)

- What is the Euler Characteristic for the plane? Illustrate with an example.

- **a.** Is it possible to draw a connected graph with 14 vertices and 12 edges? Explain.

  **b.** Is it possible to draw a connected graph in the plane with 8 vertices, 10 edges, and 5 faces?

- **a.** Using the solid lines of the connected graph below, compute

  (number of vertices) – (number of edges) + (number of faces).

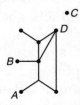

  (The outside region counts as a region.)

- **b.** If you added an edge from point *A* to point *B*, how would the number you just calculated change? Explain very briefly.

- **c.** If you added an edge from point *C* to point *D*, how would the number you just calculated change? Explain very briefly.

- **d.** Prove that $V - E + F = 2$ for any connected graph in the plane.

- In considering the Euler Characteristic in class, we looked at connected graphs drawn in the plane, that is, doodles drawn without lifting the pen. Suppose we now consider a graph made up of *two* connected pieces that do not touch each other. For any such graph, what would $V - E + F$ equal? Illustrate your answer with an example.

## Sample Lecture Notes for Section 5.3, "Feeling Edgy?" (2 days)

### Day One

Question of the Day (to be written on the board for students to think about and discuss before class begins): *Can I read into your psyche?*

1 minute—Announce that you will demonstrate your ability to reveal something about each student's subconscious by using an artistic creation by each student. (Follow text on pages 359–361.)

4 minutes—Have each student place his or her pen on a piece of paper. Ask them to create a sweeping, random doodle without lifting their pens. Demonstrate on the board, suggesting that they not make anything too complex because they will have to count pieces. Once they are done, have them draw a big dot at each crossing point and at the start and finish points.

5 minutes—Close your eyes and claim to see that all the dots, except the start and finish dots, have four lines coming out. If any student's drawing has a dot with six or more lines, suggest they weren't really drawing randomly and ask them to draw another figure. Close your eyes again and proclaim that each drawing has exactly two fewer edges than dots plus regions.

10 minutes—Demonstrate how to count edges and regions using your sample drawing. Have students count dots, regions, and edges in their drawings. Caution them to include the infinite region around the outside. Claim your powers of clairvoyance; take a dramatic bow.

15 minutes—Sheepishly admit mathematical chicanery. Confess that mathematicians have known this result for more than 250 years, thanks to Euler. Introduce terminology and notation: *graph, connected graph, vertex, edge, face, V, E, F.* State the Euler Characteristic for the plane: $V - E + F = 2$.

10 minutes—Work through the proof following the presentation on page 361. Have students build up connected graphs one edge at a time, computing $V$, $E$, and $F$ at each stage. Point out how the result fails if a new edge and two new vertices are added in such a way that the new graph is not connected. Acknowledge that although this method is not quite rigorous, it is very similar to the formal way mathematicians would prove the theorem.

5 minutes—Suggest that students play with the expression $V - E + F$ using graphs that aren't connected to see if they find any patterns. Finish with a salute to one more life lesson: Start simple and build up.

**Day Two**

Question of the Day (to be written on the board for students to think about and discuss before class begins): *How many Platonic solids are there? What do Euler and Plato have in common?*

2 minutes—Remind students about the Platonic solids. Write the following table on a side board for reference during class.

| | Number of Vertices | Number of Edges | Number of Faces | Number of Edges per Vertex | Number of Edges per Face |
|---|---|---|---|---|---|
| Tetrahedron | | | | | |
| Cube | | | | | |
| Octahedron | | | | | |
| Dodecahedron | | | | | |
| Icosahedron | | | | | |

5 minutes—Ask students if they think there are any other regular solids. Can anyone relate regular solids to connected graphs and the Euler Characteristic? Claim that the answer to this question is the key to one proof that there are only five regular solids.

15 minutes—Bring a balloon to class, blow it up, and draw a small, connected graph on it. Ask students to check that the Euler Characteristic holds for the graph on the surface of the balloon. Point out that by stretching the balloon's air hole, any connected graph on the balloon can be transformed into a connected graph in the plane with the same number of vertices, edges, and faces (stereographic projection). Now blow up another balloon and draw a cube on it. Draw the stereographic projection on the board. Help students acknowledge that the vertex-edge-face values and relations are identical for the balloon graph, the plane graph, and the actual Platonic cube. Look at a projection of the tetrahedron. (The other solids are a little harder, but fun for students to try.)

5 minutes—Because any regular solid can be represented with a connected graph in the plane, point out that the Euler Characteristic must hold for the vertices, edges, and faces of a regular solid. State the theorem: *There are only five regular solids.* Acknowledge that to prove this claim requires some additional observations about regular solids.

180

10 minutes—Derive the results on text page 365 for an arbitrary regular polygon: $F = 2E/s$ and $V = 2E/c$, where $s$ is the number of edges per face and $c$ is the number of edges per vertex. (Refer to the table on the board for examples to help students see that $Fs = 2E$ and $Vc = 2E$.) Derive the formula $E(2/c + 2/s - 1) = 2$.

10 minutes—Analyze the formula to yield broad constraints on $c$ and $s$. Eliminate all possible cases except those that correspond to the Platonic solids (text page 366).

5 minutes—Finish with a hats off to counting. The proofs for both the Euler Characteristic and the Platonic solids rest on the simple idea of counting.

**Section 5.4  Knots and Links** [A little knot theory]. This section presents the idea of a knot (a closed loop) and a link (a collection of closed loops). The unknot is described, and various common knots are pictured. The Borromean rings are also presented as an interesting counterintuitive example of a link. The problem of untangling an arc attached to the ceiling at one end and to an unknotted simple closed curve at the other is as an example of attacking a difficult problem by establishing a measure of complexity and then systematically and incrementally reducing that complexity.

*General Themes*

Making a tiny bit of progress over and over can result in dramatic progress toward answering profound questions.

*Mathematical Underpinnings.* (*Caveat:* These mathematical overviews are usually for the instructor only. The symbols and detailed proofs are not suitable for class presentations because mathematical symbols and terminology often present a formidable barrier for students.)

This section introduces the branch of topology known as knot theory. The study of knots and links is both playful and profound. In addition to their applications in biology, knots offer students hands-on experience in understanding what it means for two mathematical objects to be equivalent. The broader life lesson of reaching a goal through incremental steps is also valuable.

One of the fundamental questions for topologists is to determine when two knots are really the same. Given two knots, when can one be rearranged until it looks exactly like the other? In particular, when is an apparent tangle really the same as the unknot? The text explores this last idea in the example with the rings at the gym (text pages 378–379), illustrating the approach of rearranging the knot to reduce the number of crossings by one. If one can do this repeatedly, eventually the tangle will be reduced to the unknot. Here we see the power of a slight reduction in the complexity of a challenge.

The ring-on-a-rope example (text page 379) is revealing enough to have students try it in class. Construct a model like the one in the text. (If needed, take a piece of rope and tie a loop at one end in place of the ring.) Have a student successively reduce the number of crossings by one until the rope is completely untangled. Alternatively, use Template 5.4.2 and draw successive steps, though this is less dramatic.

For creating knots in class, a short extension cord works well: Tie a simple knot and then plug the ends together to create a bona fide mathematical knot. String works pretty well on an overhead projector, and it's easy to create many examples. If you want to demonstrate a lengthy crossing reduction, use a lightweight chain, as it stays in place more reliably.

Knot theory offers an excellent example of a fundamental unanswered problem in mathematics: Given pictures of two knots, there is no known way to determine if they are really the same knot simply by looking at the crossings.

Despite this, crossing patterns are studied closely. A knot is called *alternating* if, when following the rope around, the pattern of crossings alternates between over and under. The trefoil knot on the left is alternating, but the figure-eight knot on the right is not.

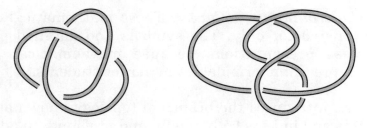

There are actually two trefoil knots.

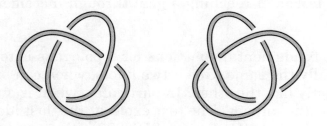

These two knots are considered topologically equivalent because they are mirror images of each other. But physically, they are different knots. If you had a string tied like the trefoil knot on the left, no matter how you manipulated it, you would never get it to look exactly like the other one. (Note that flipping it over will give you a knot like the one on the right but with all the crossings reversed!)

The section ends with a look at the Borromean rings, an appealing example of a link. This link is named for a family of the Italian Renaissance who used it on their coat of arms.

> *Our Experience*
>
> *Time spent:* One 50-minute session
>
> *Personal favorite topic or interactions:*
>
> 1. Having students build a human knot
> 2. Having students untangle examples of the rope and ring
>
> *Danger:* Knots and links allow for many fascinating pictures, but real proofs about knots usually require an explanation of the Reidemeister moves, which we chose not to do.
>
> *Remarks:* We usually omit this section in class, because it is an outstanding group research project with good poster session possibilities.
>
> *Sample homework assignment:* Read the section and do Mindscapes I.1; II.6, 13, 14, 17; III.27; IV.37.

### Sample Class Activities

*Human knots.*   Bring students to the front of the class to enact Mindscape II.6.

*Knots.*   Take two pictures of knots, or, preferably, use real rope. Ask students if they can maneuver one so that it is the same as the other. Template 5.4.1 shows a complicated knot that is really the unknot. Ask students if they can maneuver the knot in the template into the unknot. Discuss whether there is any systematic method by which students could determine how to get from knot to knot. Then point out that no such method is known yet.

*The rope with ring.*   Template 5.4.2 shows the rope hanging from the ceiling with an unknotted curve attached. Demonstrate how counting the crossings is an effective measure of complexity that can be reduced. If you have presented a more formal idea of inductive proofs, this is a good example of induction applied in a nonnumerical setting.

Ask students why such a method would not work to untangle any unknot. Point out that it would work if you could find an effective way of reducing the crossings by one, but no one knows such a method.

*Borromean rings.* If you have a model of the Borromean rings, direct students to try to undo them. An alternative is to make the rings out of three hula hoops. If you make the rings of string, you or students could manipulate them to a position where two of the rings are separate and round while the third is tangled between them. The configuration of the Borromean rings leads to the question of how to find a set of four or more unknotted simple closed curves such that, upon the removal of one, all the rest come apart. That is, looking at the Borromean rings in the nonsymmetrical way, with two rings nested one within the other and the third entangled, is actually helpful in figuring out the four-loop case. The general lesson is that looking at situations, even known ones, from different vantage points may help solve unknown problems. See Template 5.4.3.

**Templates for Transparencies** *(located at end of chapter):*

- Template 5.4.1—Complicated unknot

- Template 5.4.2—Rope with ring attached

- Template 5.4.3—Two views of the Borromean rings

**Sample Test Questions**

(The Mindscapes are another good source of test questions. All test questions below are available electronically on the Test Bank CD-ROM.)

- Draw intermediate steps to show how to untangle this knot to become the unknot.

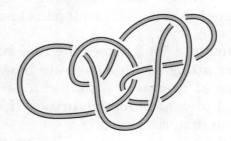

- The picture shows a rope attached to the ceiling at one end and to a ring at the other end. What move would you make to reduce the crossings by one?

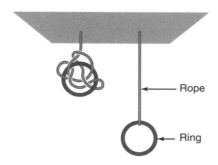

Two gym rings: one tangled, one not

- Here is a picture of the Borromean rings with two of the circles round and disjoint and the third one looped and connecting them. Notice that if you remove any one of these rings, the other two come apart. Can you draw a set of four Borromean rings so that if you remove any one ring, the other three come apart to be three unknotted, disconnected rings?

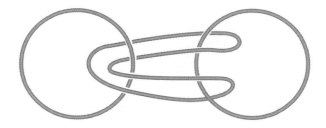

## Sample Lecture Notes for Section 5.4, "Knots and Links" (1 day)

Question of the Day (to be written on the board for students to think about and discuss before class begins): *When is a knot not a knot?*

10 minutes—Introduce the concept of a mathematical knot. Display examples in several ways: string or an extension cord, transparencies, drawings on the board. Discuss the DNA example (text pages 375–376), pointing out that DNA strands exist in knotted tangles. When the strands split apart, the tangles sometimes interfere, forcing one strand to break apart and pass to the other side of another strand. This example inspires the idea of changing crossings in a knot.

10 minutes—Analyze several simple examples by looking at crossings on an overhead projector. Have students try to reduce the number of crossings. Include a knot like the one in Template 5.4.1 to illustrate equivalence to the unknot. Include a trefoil knot to illustrate that sometimes the number of crossings cannot be reduced.

5 minutes—Provide a physical model of the ring-on-a-rope challenge (text pages 378–380). Have each student come up and reduce the number of crossings one by one until the rope is untangled.

10 minutes—Ask for five volunteers and perform Mindscape II.6. Point out the significance of the changed crossing: changing the crossing definitely changed the knot.

10 minutes—Introduce the concept of a link. Display several examples. Display a physical model or drawing of the Borromean rings (see Template 5.4.3). Have students work in pairs to see if they can draw a set of four Borromean rings. If someone succeeds quickly, have them work on five!

5 minutes—Finish with a look at the Borromean rings that exist in the icosahedron as a link of three Golden Rectangles (text page 381). Urge students to use their 3D glasses to look at the picture in the text.

**Section 5.5  Fixed Points, Hot Loops, and Rainy Days** [The Brouwer Fixed Point Theorem]. This section contains the statements and proofs of cases of the Brouwer Fixed Point Theorem and the Borsuk-Ulam Theorem. As usual, we present the underlying ideas without formal definitions. The concept of a fixed point during some transformation is geometrically and philosophically intriguing and has applications in the real world, such as to the weather. Students can look for analogies to fixed points in other settings.

---

*General Themes*

Building a simple model can bring truths about our world into focus.

---

*Mathematical Underpinnings.* (*Caveat:* These mathematical overviews are usually for the instructor only. The symbols and detailed proofs are not suitable for class presentations because mathematical symbols and terminology often present a formidable barrier for students.)

This section discusses the ideas underlying the Brouwer Fixed Point Theorem, a subtle and profound result from topology. The theorem is difficult to prove, but the text offers an intuitive discussion about why the result makes sense. We recommend *not* discussing this part of the section in class. Winding numbers are difficult to understand, though a bright, ambitious student could explore Brouwer's Theorem and proof as part of an independent project.

The Brouwer Fixed Point Theorem can also be examined in one dimension. Think of two pieces of string, each the same length, lined up together.

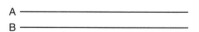

Now fold string B so that every point is still opposite some point of string A.

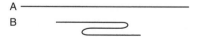

The Brouwer Fixed Point Theorem establishes that there will always be a point in string B that is directly opposite the same point on string A that it was lined up with before it was folded.

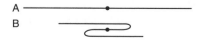

191

This result can be illustrated in class using a "string" of students. (See the Sample Lecture Notes.)

A much simpler version of Brouwer's Theorem, the Hot Loop Theorem, is proven in detail in the text. The proof uses the Intermediate Value Theorem and offers a fine example of function notation. Here's a more compact version of the argument that emphasizes the importance of continuity and the use of the Intermediate Value Theorem.

**Hot Loop Theorem:** *If we have a circle of variably heated wire, then there is a pair of opposite points at which the temperatures are exactly the same.*

**Proof:** Consider an arbitrary circle of heated wire. Label 12 points on the circle as a clock is labeled: 1, 2, ..., 12. Label, with appropriate decimal numbers, all points in between these integer points. For each point $x$ on the circle, let $x'$ denote the point opposite $x$, and let $temp(x)$ denote the temperature at the point $x$. We want to find a point $x$ for which $temp(x) = temp(x')$.

Note that because heat varies *continuously,* the function *temp* varies continuously as we move around the circle. Thus, the difference function $D(x) = temp(x) - temp(x')$ is also continuous. Notice that $D(12) = temp(12) - temp(6)$. If $D(12) = 0$, then $temp(12) = temp(6)$, and we have our point. If $D(12) \neq 0$, then it must be positive or negative. Suppose it is positive. (The argument is analogous if $D(12)$ is negative.) Then, $D(6) = temp(6) - temp(12)$ is negative. But $D$ is a continuous function, so by the Intermediate Value Theorem, there is a point $x$ between 6 and 12 for which $D(x) = 0$. Thus, $x$ is a point for which $temp(x) = temp(x')$.

This result offers an excellent example of a function. You may want to convey the idea of continuity informally, along with the Intermediate Value Theorem.

The Hot Loop Theorem answers the question about whether there exist antipodal points on Earth with the same temperature: Simply consider the equator as a "hot loop." In fact, because temperature varies continuously, any circle on Earth will contain two antipodal points with the same temperature. A stronger result is true as well. At any moment, there are two antipodal points on Earth with exactly the same temperature and barometric pressure. The Hot Loop Theorem guarantees two points with same temperature and two points with the same pressure, but to get both values equal at the same pair of points requires a much stronger theorem: the Borsuk–Ulam Theorem.

**Borsuk–Ulam Theorem:** *Any map $f$ from $S^n$ to $\mathbb{R}^n$ must identify a pair of antipodal points of $S^n$. (That is, any function from the n-dimensional sphere to n-space must map two antipodal points to the same function value.)*

For $n = 2$, $S^2$ is simply the surface of a traditional sphere, and $\mathbb{R}^2$ is simply the set of all ordered pairs of real numbers. If we define a function $f$ to map each point on Earth's surface to the pair of numbers giving the temperature and pressure at that point, then the Borsuk–Ulam Theorem guarantees that two antipodal points have the same function value—that is, the same temperature and pressure. Students might enjoy hearing about this result, which is stated in the text as the Meteorology Theorem (text page 395).

---

*Our Experience*

*Time spent:* One 50-minute session

*Emphasized items:* The Hot Loop Theorem and how it relates to the Meteorology Theorem

*Items omitted:* The Brouwer Fixed Point Theorem

*Remarks:* We usually omit this section in class, because it is an outstanding group research project with good poster session possibilities.

*Sample homework assignment:* Read the section and do Mindscapes I.3; II.6, 7, 8; III.16, 17.

---

### Sample Class Activities

*The one-dimensional Brouwer Fixed Point Theorem.* Have students line up across the front of the room, standing shoulder to shoulder and holding hands. Number the students and, if there is a board, put each student's number directly behind him or her. Challenge students to move and reassemble themselves along the front without letting go of the hands they are holding, doubling back if they want, in such a way that no person is lined up with their own blackboard number. Follow with a discussion of the 1-dimensional fixed point theorem, and point out the relationship with the line of students.

*Demonstrating the Brouwer Fixed Point Theorem.* Ask each student to tear a page from a notebook, crumple it up, and smash it back down on the next page. Ask them to find a point that lies directly above where it started.

*Proof of the Brouwer Fixed Point Theorem.* Indicate a large, round circle on the floor. Ask a student to walk around that circle always facing the class, that is, always facing the same direction the whole time. The student will sometimes be walking sideways and backward. Tie a rope at the student's left hip and let another student stand somewhere in the circle and hold the other end of the rope. As the circling student circles, the other student is free to move anywhere within the circle but must end up back where he or she started when the circling student gets back to the start. Point out that the rope has wrapped around the waist of the circling student exactly once. Explain winding number. Then have the two students repeat the activity, but this time have the circling student traverse only a small circle near the center of the original big circle and have the other student at least 5 feet away and constrained to move only 1 foot at most. In this case, they just shuffle a bit. The winding number is 0. Ask students to explain what happens to the winding number as the circling student moves in from the outside and traverses slightly smaller circles. Explain the necessity for a discontinuity of winding number and its implications for the proof of the Brouwer Fixed Point Theorem.

*Hot Loop Theorem.* Draw a circle and ask students to think of it as Earth's equator. Ask them to give you temperatures at different places around the equator, trying to choose them so no two diametrically opposed points have the same temperature. As they try futilely to defeat the Hot Loop Theorem, point out the issues of continuity and intermediate value.

### Sample Test Questions

(The Mindscapes are another good source of test questions. All test questions below are available electronically on the Test Bank CD-ROM.)

- Here is a picture of the equator with temperatures marked. Where would you expect to find antipodal points on the equator where the temperatures are the same?

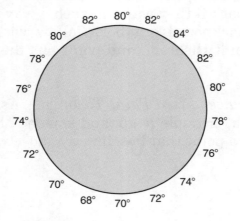

- Here is a loop in a circle with points indicating how the loop corresponds to the circle. What is the winding number obtained if you march around the circle, constantly pointing at the corresponding point on the loop?

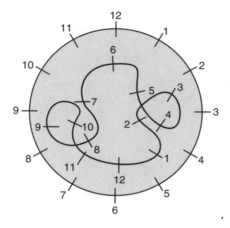

- If the winding number for the boundary of the disk is 1, why do you know that the winding number for a nearby circle is also 1?

- A long line of people encircles the track in a stadium. They hold hands so that each person next to any person has a last name with a letter at most one removed. That is, if your name is Smith, the people on both sides of you must have last names beginning with the letter R, S, or T. Why is it that someone in the circle must be directly across from a person whose last name is within one letter of his or her own?

## Sample Lecture Notes for Section 5.5, "Fixed Points, Hot Loops, and Rainy Days" (1 day)

Question of the Day (to be written on the board for students to think about and discuss before class begins): *What's the temperature on the other side of the world?*

2 minutes—Draw a circle on the board. Ask students to think of it as the equator and to imagine they know the temperature at each point. Ask if they think there will always be two diametrically opposed points with the same temperature. Suggest that this couldn't possibly be true for every conceivable set of temperatures. Convince them such a claim can't be true.

10 minutes—Have several students come to the front of the room and stand holding hands in a line in front of the board. Number the students and write each student's number on the board behind him or her. Also give students a piece of paper to hold in front of them with their original number on it. (Work with at least five students; if you don't have a board, tape numbered sheets of paper to the wall.) Ask students to shift positions while holding hands, so that as many as possible are standing in new positions. Point out that one or more students can share a position if they wish. Have them try several arrangements. Ask the rest of the class to check whether any student is still in front of his or her original number.

5 minutes—Introduce the idea of a fixed point and relate it to the human chain activity. Have the volunteers sit down. Illustrate further with two pieces of string on an overhead projector, or simply draw line segments on the board. Make the claim that a fixed point will always exist. Acknowledge that there is a more general result, the Brouwer Fixed Point Theorem.

5 minutes—Return to the circle/equator on the board. Mark a number of points around the circle and ask for possible temperatures at these points; write a temperature beside each point. Define what it means for two points on the equator to be diametrically opposed (antipodal). Ask how many students think there will always be two antipodal points with the same temperature. Find possible pairs given the suggested temperatures.

3 minutes—Now ask students to think again whether it is possible to label every point of the circle so that no pair of antipodal points has the same temperature. Acknowledge the importance that temperature varies continuously. Mention the Intermediate Value Theorem, if appropriate. Finish by stating the Hot Loop Theorem.

# Template 5.1.1  Inverting a torus

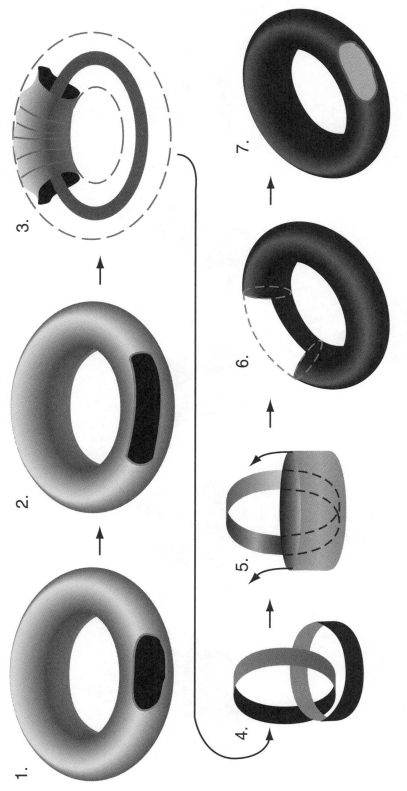

# Template 5.1.2  Unlinking a two-holed sheet from a ring

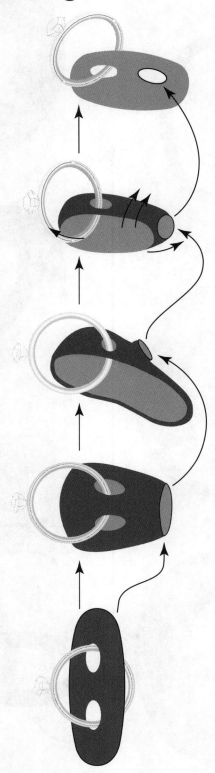

# Template 5.1.3  Equivalent by distortion?

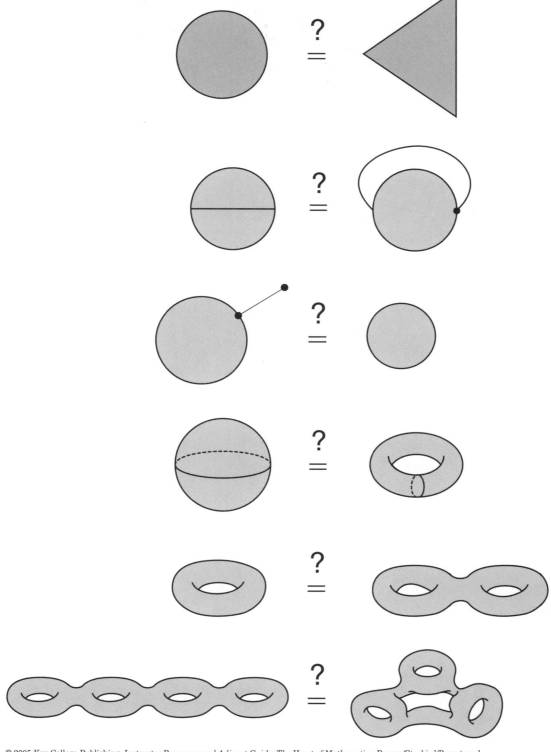

# Template 5.1.4a  Double torus, linked

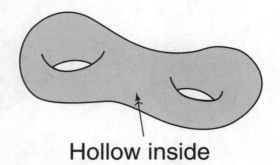

Hollow inside

# Template 5.1.4b   Double torus, unlinked

# Template 5.1.5a  Cube with straight and knotted hole

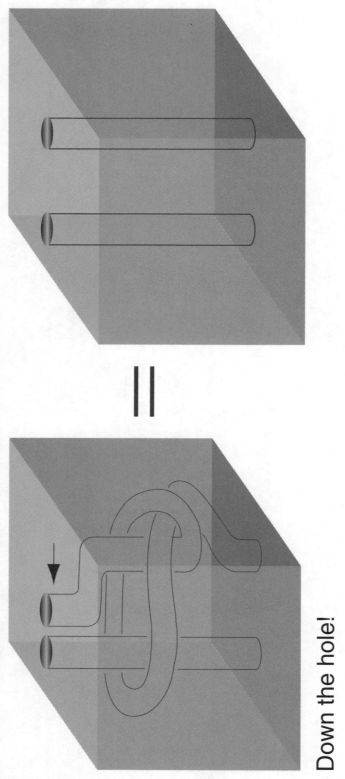

Down the hole!

# Template 5.1.5b    Cube with straight and knotted hole

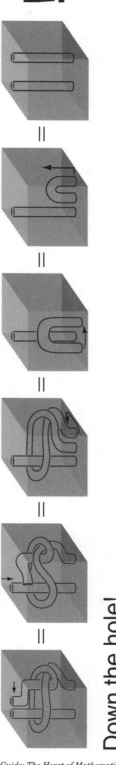

Down the hole!

# Template 5.2.1　Klein bottle

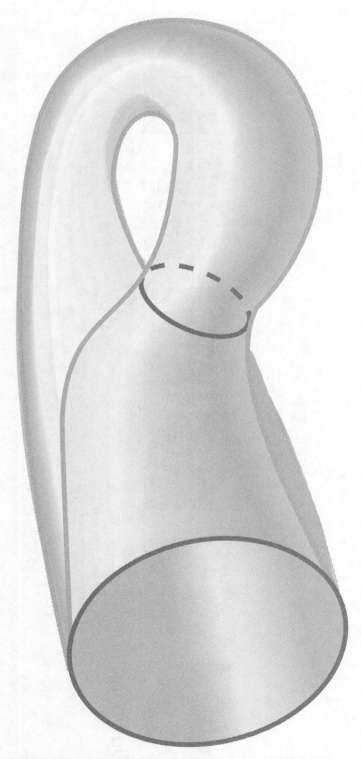

# Template 5.3.1    Square projected onto balloon

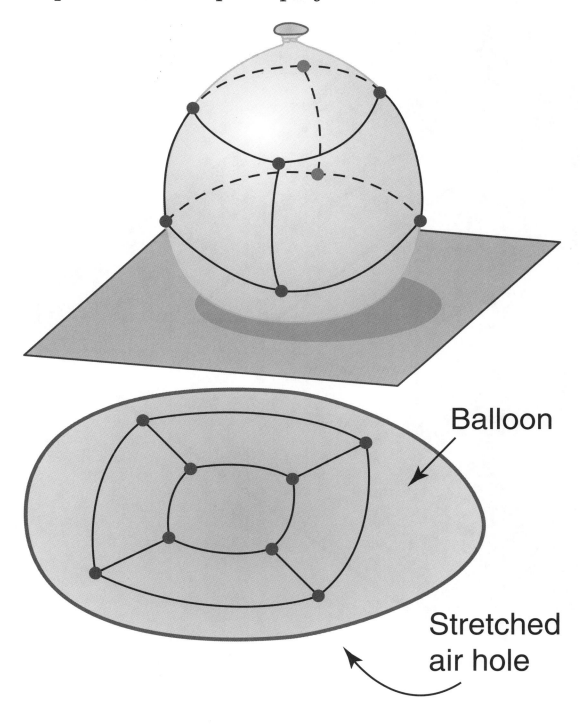

Balloon

Stretched
air hole

# Template 5.3.2  Polyhedral torus

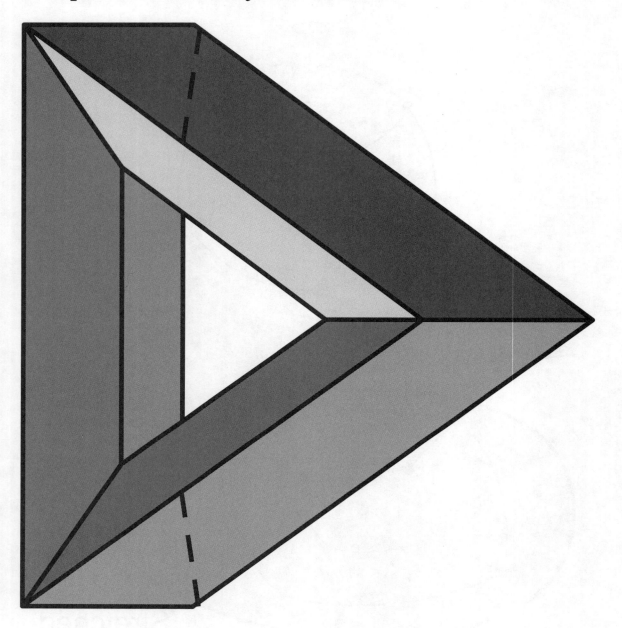

# Template 5.3.3   Connected sum of two polyhedral figures

Remove both faces, and glue edges together.

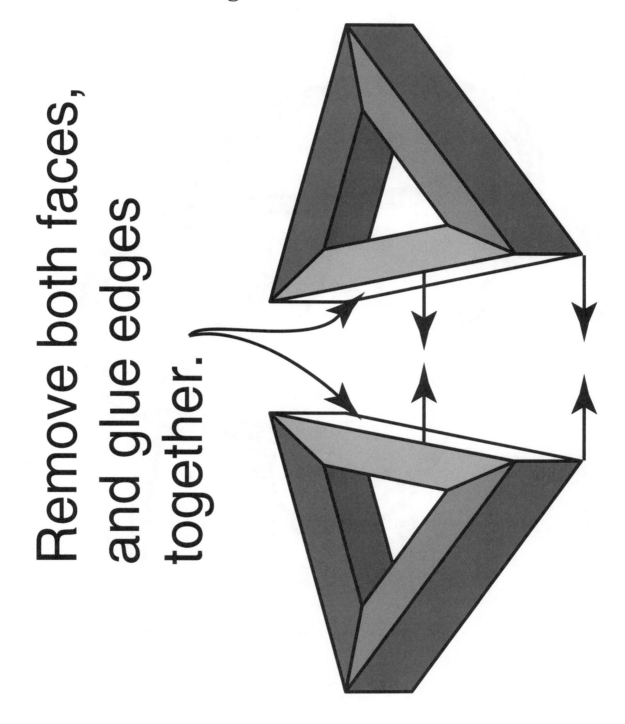

# Template 5.4.1   Complicated unknot

# Template 5.4.2   Rope with ring attached

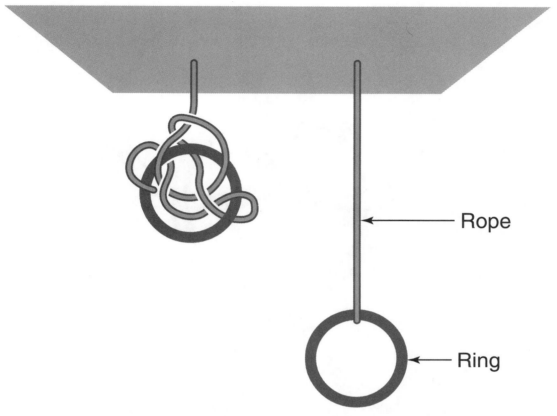

Rope

Ring

Two gym rings: one tangled, one not

# Template 5.4.3   Two views of the Borromean rings

# Chapter 6   Chaos and Fractals

**6.1**   **Images** [A gallery of fractals]
**6.2**   **The Dynamics of Change** [Repeated applications of simple processes]
**6.3**   **The Infinitely Detailed Beauty of Fractals** [Creating fractals through repeated processes]
**6.4**   **The Mysterious Art of Imaginary Fractals** [Julia and Mandelbrot Sets]
**6.5**   **Predetermined Chaos** [Deterministic chaos]
**6.6**   **Between Dimensions** [Fractal dimension]


## I.   Chapter Overview

This chapter is an introduction to fractals and deterministic chaos. Fractal images are now household objects; one of the goals of this chapter is to make fractals meaningful as well as intriguing. Most of the ideas in this chapter are subordinate to the notion of a repeated or iterative process. Repeating a simple rule for replacement of a part of a picture, repeating a simple rule for shrinking and duplicating a picture, or repeating a simple quadratic function all lie at the core of the infinite complexity that is the hallmark of fractals and chaos.

---

**Dependencies**

Section 6.5 (Deterministic chaos) is partially based on themes described in Section 6.2 (Simple processes). Section 6.6 (Fractal dimension) involves objects defined in Section 6.3 (Creating fractals).

---

**Sections to Assign as a Reading Assignment or Term Project**

It would not be unreasonable to do Sections 6.1 through 6.3 and then omit the rest of the chapter, perhaps assigning Section 6.5 as a reading assignment. Section 6.4 on the Mandelbrot and Julia sets, Section 6.5 on deterministic chaos, and Section 6.6, on fractal dimensions, are ideal sources for a research project or extra credit, if time does not permit their full treatment in class. Any omitted sections are good paper or extra credit topics.

---

*Reminders*

1. Class activities are suggested for each section. These can be done with students working in groups of two or three and then gathering their responses or as one global class discussion.
2. Before introducing a new topic, engage students with enticing questions, which they quickly discuss with neighbors; solicit responses within a minute. Use the responses as the springboard for the introduction.
3. We have included more sample class activities than you will probably want to use. Choose only the ones that are right for you and your class.
4. Each section can be treated in more or less depth. It is frequently a good idea to omit difficult technicalities in order to treat the main ideas well and then move on.

## II. Section-by-Section Instructional Suggestions

**Section 6.1  Images** [A gallery of fractals]. This section displays a collection of images that have become icons for the study of mathematical chaos and fractals. Each image is explained in the subsequent sections. The meta-lesson here is to observe carefully and look for patterns.

---

*General Themes*

Looking at objects closely reveals surprising structure.

---

*Mathematical Underpinnings.* (*Caveat:* These mathematical overviews are usually for the instructor only. The symbols and detailed proofs are not suitable for class presentations because mathematical symbols and terminology often present a formidable barrier for students.)

This opening section is intentionally presented without formal text. It emphasizes the value of just looking at things. Have students study the images and discuss what they see as broadly or as specifically as they wish.

There is no technical, mathematical definition of a *fractal,* the term coined by Benoit Mandelbrot. Informally, a fractal is a structure with infinite detail. Students may enjoy hearing that mathematicians do occasionally use language in a vague and somewhat imprecise way. If students seem unnerved by this notion, reassure them that in the discussion of dimension in Section 6.6, the idea of a *fractal dimension* does arise more formally.

### Sample Class Activities

*Looking at the images.*   All the images in this section suggest infinite detail. Many show self-similarity of one kind or another. Ask students to describe the implied detail that does not appear in the images themselves. Ask them to point out self-similarity at different scales. Have students guess at processes or techniques by which such infinitely detailed images might be conceived, generated, and rendered.

### Sample Test Questions

(The Mindscapes are another good source of test questions. All test questions below are available electronically on the Test Bank CD-ROM.)

- Circle two parts of the picture that are identical to the whole picture but at different scales.

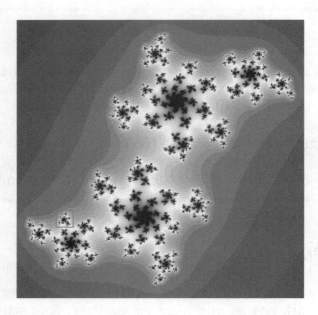

- The upside down, missing triangle in the middle of the Sierpinski Triangle has area exactly one-quarter of that of the original big triangle in which the Sierpinski Triangle sits. What is the total area of all the missing triangles of the Sierpinski Triangle?

## Sample Lecture Notes for Section 6.1, "Images" (1/2 day)

Question of the Day (to be written on the board for students to think about and discuss before class begins): *How can you convince your parents that you're eating enough broccoli?*

5 minutes—Have students look at the images in the section and discuss their impressions among themselves. Encourage them to look for self-similarity and the potential for infinite detail.

5 minutes—Ask students which images suggest truly infinite detail. Acknowledge that representing such detail in an actual image is impossible.

10 minutes—Ask for other examples of fractal-like structures: trees, branches, twigs; clouds; a complex river system. Suggest some common objects, such as a book or person or pen, and ask for careful explanations as to why they are not fractals.

5 minutes—End with some speculation about how one might create a fractal image.

**Section 6.2  The Dynamics of Change** [Repeated applications of simple processes]. Compound interest and population growth are two examples of repeated applications of simple processes. Many models of phenomena have an iterative character, and most of this chapter is devoted to the study of these dynamical systems. The idea of seeing what happens when the same process is repeated over and over provides a way of looking at the world that students can apply to many life situations.

---

*General Themes*

Using simple models for complex systems can help illuminate important features.

---

*Mathematical Underpinnings.* (*Caveat:* These mathematical overviews are usually for the instructor only. The symbols and detailed proofs are not suitable for class presentations because mathematical symbols and terminology often present a formidable barrier for students.)

This section offers several examples of iterative dynamical systems; it's worth pointing out that iteration is a fundamental idea in many areas of mathematics. The Fibonacci numbers introduced in Chapter 2 were defined using an iterative process. Mathematical induction is a method of proof that can be thought of as iterative in an informal sense. One could even think of constructing the positive integers by starting with 0 and then adding 1 repeatedly as an iterative process. In each case, the output of some process is filtered back to be the input at the next stage.

Conway's Game of Life, a highly simplified model of real life, offers an excellent example of surprising complexity that arises from an iterative process. Conway created the rules of the game to maximize the possibility that different starting configurations would yield many generations of varied results before either dying out or settling into periodicity. The Game of Life, which is an example of what computer scientists call a *cellular automata,* has been widely studied, with many starting configurations falling into classifications with names like blinker, still life, and Methuselah. Though very abstract, the Game of Life is a fun way to explore iteration, especially if done using a computer. Encourage students to play Conway's Game of Life using the CD-ROM packaged with their text. There are also many implementations and discussions of the game on the Web, offering interesting examples of initial populations giving rise to surprising behavior.

If your classroom is equipped with computer capabilities, the strongest activity is to use Conway's Game of Life provided on the CD-ROM. Instructions are included in the program. Otherwise, you can hand out graph paper with starting configurations as listed in the text. To help students keep track of their cells, mark a central cell in the grid spaces for successive generations.

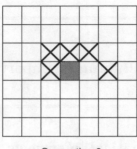

Generation 0

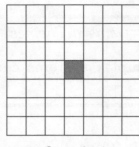

Generation 1

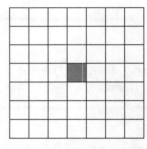

Generation 2

The Verhulst Model of population growth offers a more algebraic and concrete example of an iterative process. The idea of population density is abstract, but the model is accessible.

Mindscape III.30 introduces the idea of a *cobweb plot,* which records the results of repeatedly applying a function $f(x)$. Start with $x_0$. If we let $y_1 = f(x_0)$, then $y_2 = f(y_1)$, $y_3 = f(y_2)$, and so on. Each starting value generates a sequence $y_1$, $y_2$, $y_3$, . . . . The figures in the text (page 425) show that these values are simply the $y$-coordinates of the points on the graph of $f(x)$ that are reached by following the path of the cobweb plot.

All these examples of iterative dynamical systems foreshadow the idea of chaos, or how the behavior of a system can be radically different, even with only small changes in the initial conditions. This topic is explored in more detail in Section 6.5.

_Our Experience_

_Time spent:_ One to one-and-a-half 50-minute sessions, depending on the coverage and depth

_Emphasized items:_ For one of us, Conway's Game of Life

_Items omitted or treated lightly:_ The Verhulst Model

_Dangers:_ The Verhulst Model can be difficult and boring for some classes.

_Remarks:_

1. One of us treats this section as a reading assignment.
2. You can spend more or less time on this section depending on how much time you spend on Conway's Game of Life and the Verhulst Model. You could spend quite a while on each, or, alternatively, you could omit one or both entirely.
3. The Verhulst Model is a good example of chaos, which is discussed in Section 6.5. If you intend to emphasize it later, laying a foundation here is a good idea. Otherwise, you might just skip it.

_Sample homework assignment:_ Read the section and do Mindscapes I.2; II.6, 8, 12, 13; III.27, 28.

### Sample Class Activities

_Compound debt._   Ask students how much they would owe on a credit card every year if they maxed it out at $5000, were charged 18% interest per year, and paid nothing for 10 years. They can use calculators, to see how this process is repeated.

_Conway's Game of Life._   Hand out sheets of grid paper with various initial positions drawn. Ask students to mark with numbers or colors which squares will be alive after one, two, three, and so on generations. Choose initial conditions that lead to total death, a population explosion, a repeating pattern, and a migrating pattern. Give an example where a small change in the initial condition leads in one case to a population explosion and in another case to total death. Programming the Game of Life provides a good computer homework exercise, if some members of the class are inclined that way.

*Verhulst Model.* After explaining the reasonableness of the Verhulst Model, ask students to compute the population futures for various starting populations and constants. Programmable calculators or computers are ideal for such experiments. In Section 6.5, students will learn how sensitive the future populations are to the starting populations and to the choice of constants. Congratulate students who discover these sensitivities here.

### Sample Test Questions

(The Mindscapes are another good source of test questions. All test questions below are available electronically on the Test Bank CD-ROM.)

- Given the following initial populations in the Game of Life, determine the next three generations. [Note to Instructor: Choose one or two of these initial populations for an in-class test. It would take too long for students to do them all.]

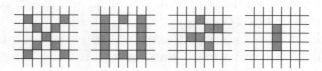

- For each of the following initial populations in the Game of Life, determine whether the configuration is periodic, explosive, or doomed to extinction. [Note to Instructor: Choose one or two of these initial populations for an in-class test. It would take too long for students to do them all.]

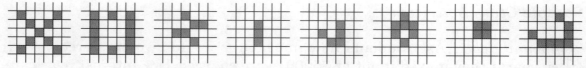

Choices: (1) population explosion; (2) extinction; (3) a stable pattern; (4) a periodic pattern; (5) a migratory pattern

- Suppose the density of next year's population of fish in a pond is given by $P_{n+1} = P_n + 2.5P_n(1 - P_n)$. If the population today is 1000 and the maximum sustainable population of the pond is 5000, what does this model predict the population to be next year?

## Sample Lecture Notes for Section 6.2, "The Dynamics of Change" (1 day)

Question of the Day (to be written on the board for students to think about and discuss before class begins): *What do you need to know to predict the world's population in the future?*

10 minutes—Ask students for examples from the natural world or their lives of systems that have a repeating quality. (If they seem stumped, offer the example of the seasons. Other examples to add if their list is short: Earth's daily rotation, tides or annual floods, commuter rush hour, new population of mosquitos born every year, and so on.) Introduce the term "dynamical system" to describe appropriate examples: Earth and the solar system, populations in an ecosystem, and so on. Discuss how complex systems often involve one or more simple processes repeated over and over.

30 minutes—Introduce John Conway's Game of Life. (If possible, bring in a computer to display the text's CD-ROM implementation.) Describe the rules and present several examples. Hand out graph paper with several initial populations drawn, and ask students to determine a few iterations. Have them work in pairs to compare their results and catch errors. Present more examples of interesting behavior. Discuss the five outcomes described in the text (page 416). Give more examples.

10 minutes—Discuss the limitations of the Game of Life as a model of real-world population growth. Ask students to suggest different ways to model population change, focusing only on population size. Ask if there are two or three numbers corresponding to a population that might help predict the population size the following year. (Size, birthrate, and death rate are likely suggestions. Carrying capacity is more abstract but could be suggested in the form of "amount of food.") Discuss the possibility that a reasonable model might be designed to take a few numerical input values and then predict future population size. Urge interested students to read about the Verhulst Model in the text (pages 417–420).

## Section 6.3 The Infinitely Detailed Beauty of Fractals [Creating fractals through repeated processes].

Replacing a part of a picture with smaller copies of itself, using collage-making instruction sets, and choosing points randomly by playing the Chaos Game all lead to fractal pictures.

---

*General Themes*

Repeating a simple process can result in a complex structure.

---

*Mathematical Underpinnings.* (*Caveat:* These mathematical overviews are usually for the instructor only. The symbols and detailed proofs are not suitable for class presentations because mathematical symbols and terminology often present a formidable barrier for students.)

Building on the idea of iteration, this section presents several techniques for producing fractal images. The primary method starts with a simple figure, reduces it, duplicates it, and then creates a new image in a prescribed way. This *collage method* clarifies one source of infinite detail in a true fractal and provides an excellent opportunity for students to imagine what happens when they repeat something an infinite number of times.

Although most examples of the collage method require a computer to render many iterations, students can draw the Koch Curve fairly successfully for at least two or three iterations. Template 6.3.1 shows the curve through five iterations. (Note that applying the Koch Curve iteration to all three sides of an equilateral triangle will yield what is called the Koch Snowflake.)

Mindscape II.14 asks students to determine the number of bends at the $n$th stage of the Koch Curve construction. It's also fun to consider finding the length of the curve at each step. If the starting shape has length 4/3, then the curve at the next stage has length 16/9. Because at each stage the total length increases by a factor of 4/3, the length after $n$ iterations is $(4/3)^{n+1}$. Invite students to consider the length after an infinite number of iterations. The idea of a curve of infinite length confined in a finite area is challenging.

The collage method can be introduced as a way to construct the Sierpinski Carpet from the Tight Weave story in Chapter 1 or the Sierpinski Triangle. Doing these examples carefully is particularly useful if you plan to cover Section 6.6 on dimension. If your class found computing the length of

225

the Koch Curve to be fun, you can also compute the area of the Sierpinski Carpet. In this case, start with a square of area 1, and then remove the center square of area 1/9, so the initial image in the collage process has area 8/9. At each step, you again remove one-ninth of the area, leaving eight-ninths of what you started with. So, after $n$ interations, the area of the carpet is $(8/9)^{n+1}$. Thus, after an infinite number of iterations, the area of the completed carpet is $\lim_{n \to \infty}(8/9)^{n+1} = 0$. In Section 6.6, however, students will learn that even though this object has no area, its dimension is greater than 1.

In general, the collage method allows arbitrary size reductions of the original image and more random placements to create the next image. The text presents a detailed example (text pages 437–440). Encourage students to create repeated-image collages by exploiting the size-reduction features of a photocopy machine. Emphasize the fact that the initial image is irrelevant— Because each iteration requires some size reduction(s) of the image from the previous step, in the limit, the image is a point.

The Chaos Game offers an alternate approach—using a random process rather than an iterative one—to create a fractal. The idea may be fascinating to students, but the process itself requires precision to be successful. A computer model of the process is more reliable than plotting points by hand. The CD-ROM that accompanies the text contains many fractal-based activities.

---

*Our Experience*

*Time spent:* One-and-a-half to two 50-minute sessions

*Items omitted:* None

*Dangers:* The Chaos Game requires some care.

*Personal favorite topic or interactions:* That the collage instructions, rather than the starting picture, determine the fractal

*Remarks:* It is useful to have software connected to a display device, but it is not necessary.

*Sample homework assignment:* Read the section and do Mindscapes I.3, 5; II.21, 22, 23, 24, 25; III.26, 27.

---

## Sample Class Activities

*A Tight Weave.* Revisit the Tight Weave story from Chapter 1, "Fun and Games." Demonstrate how the repeated application of a process leads to the resulting Sierpinski Carpet.

*Replacement parts.* Use Template 6.3.1. Start with an interval and demonstrate how repeatedly replacing each segment with four smaller segments generates the Koch Snowflake. Give students some initial pictures and a replacement scheme, and ask them to draw the resulting image.

*Collage-making.* After explaining the idea of a collage-making instruction set, ask students to follow sets of instructions to generate different images (see Template 6.3.2). Have them start with two totally different images and do perhaps four repetitions to convince themselves that following the same instruction set results in the same final collage.

*Approximations with collages.* Collage-making instruction sets can be chosen in such a manner that the final collage will be an infinitely detailed approximation of some target image. Give students a picture, and ask them what collage-making instruction set would generate a final collage that approximates that picture. If you can use a computer program that demonstrates consecutive images, you could enter student suggestions and show the result. Choosing a collage-making instruction set in which every copy is small, as opposed to one in which some size reductions are modest, will demonstrate that instruction sets that reduce each copy dramatically may lead to better approximations.

*The Chaos Game.* Use Template 6.3.3. The Chaos Game creates fractals, such as the Sierpinski Triangle, by a random process rather than a replacement process or a collage-making scheme. Give to groups of students a die and identical copies of a triangle on a transparency. Have them each roll their die to decide which corner to go halfway toward. Ask them to repeatedly and carefully put down small dots. After each has marked 20 or 30 dots, collect some transparencies and overlay them. You should see a good approximation of the Sierpinski Triangle. Alternatively, of course, a computer program could plot many iterations of dots.

***Templates for Transparencies*** *(located at end of chapter):*

- Template 6.3.1—Koch Curve

- Template 6.3.2—Collage method

- Template 6.3.3—Chaos Game

### *Sample Test Questions*

(The Mindscapes are another good source of test questions. All test questions below are available electronically on the Test Bank CD-ROM.)

- **a.** In the Chaos Game for the Sierpinski Triangle, what sequence of die rolls would you need to throw to land in the shaded sub-triangle?

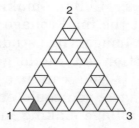

  **b.** Indicate in which sub-triangle the sequence of rolls 2, 3, 1 puts you.

- Here is the first step in a collage process. What is the next step? Roughly what would the final fractal look like?

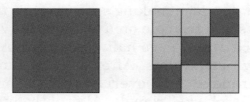

- Draw the first three steps for constructing the Koch Curve.

- Draw the first three steps for constructing the Sierpinski Triangle.

228

## Sample Lecture Notes for Section 6.3, "The Infinitely Detailed Beauty of Fractals" (1 day)

Question of the Day (to be written on the board for students to think about and discuss before class begins): *How can you turn a triangle into a snowflake?*

10 minutes—Have students create the Koch Curve. Encourage them to use a pencil. Suggest that they turn their paper sideways and start with quite a large initial image. Be clear that each of the four segments should have the same length. Have them draw on a second sheet of paper a copy of the image one-third of its original size. Ask them to make three more, for a total of four smaller copies. Explain that each of these copies will be placed in the same relative configuration as the original four line segments.

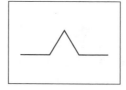

Initial image

Draw the smaller copies in the prescribed configuration, one by one.

 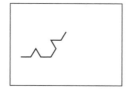  

Now have students repeat the process. Have them draw four 1/3-size copies of the final image above in the prescribed configuration. (You might suggest that it's easier just to take their last image and draw a triangular bump on each of the 16 line segments.) Display Template 6.3.1 on an overhead projector so they can see how the process creates an image with greater and greater detail.

5 minutes—Ask students what would happen if they applied the same procedure to each side of an equilateral triangle. Discuss the idea of repeating the process an infinite number of times. Give the names Koch Curve and Koch Snowflake.

5 minutes—Return to the Koch Curve, with Template 6.3.1 on display. Ask students to imagine that the last curve is the true Koch Curve—the image obtained after an infinite number of steps. Clearly, the curve has infinite detail. Ask students to describe any features of self-similarity to reinforce the fractal quality of the curve. Now ask how they could generalize this process to create other fractals.

15 minutes—Introduce the collage method. Present the instructions on text pages 438–439. Accompany the discussion with an example, such as the Sierpinski Triangle. If possible, follow with a less regular example in which an image is reduced to several different smaller sizes and arranged asymmetrically. (Barnsley's Fern is a great example. There are many implementations on the Web if you have access to an online computer display in your classroom.) Discuss the properties of infinite detail and self-similarity that an object constructed with the collage method will have.

10 minutes—Present Template 6.3.2 with only the three initial images showing. Announce that you plan to apply the Sierpinski Triangle collage process to each image. Ask students to predict how the three cases will differ after three iterations. Show the first iteration and ask for new predictions. Do the same for each successive iteration. Ask for explanations about why all three look the same.

5 minutes—Ask students if the result of any collage process will be independent of the starting image. Discuss why the answer is yes.

**Section 6.4 The Mysterious Art of Imaginary Fractals** [Julia and Mandelbrot Sets]. This section gives a brief introduction to the geometry of the addition and multiplication of imaginary numbers; however, it also points out that the geometric procedure is all that is actually required to understand these sets. This section gives a down-to-earth explanation for exotic-looking objects, perhaps encouraging students to seek reasons for phenomena they see around them.

---

*General Themes*

Understanding the intricate nuances of a phenomenon may not be possible until we can physically see it.

---

*Mathematical Underpinnings.* (*Caveat:* These mathematical overviews are usually for the instructor only. The symbols and detailed proofs are not suitable for class presentations because mathematical symbols and terminology often present a formidable barrier for students.)

The Mandelbrot Set and Julia Sets give rise to some of the most fascinating and familiar fractal images. Understanding how these sets and their beautiful visualizations arise can be undertaken on several levels. At the simplest level, each point in the plane corresponds to its own Julia Set. Each Julia Set consists of either one "piece" or more than one piece. The Mandelbrot Set is the set of points in the plane for which the corresponding Julia Set has just one piece.

Such a simple explanation, however, is highly unsatisfying. This section presents true explanations, bringing students to a complete and actual understanding of Julia Sets and the Mandelbrot Set. The most thorough explanation includes an introduction to arithmetic with complex numbers. Alternatively, students can view Julia Sets by focusing more on geometric definitions of the sets in the plane.

After introducing complex numbers and complex arithmetic, the text presents a thorough explanation of how each complex number $a + bi$ yields a Julia Set (text pages 466–471). First, given a fixed complex number $a + bi$, define a function for all complex numbers $z$ to be $f(z) = z^2 + a + bi$. Write $f^n(z)$ to denote $f(f(f(\ldots z)))$, where $f$ is applied iteratively $n$ times. Now, look at the set $S$ of all points $z$ in the complex plane such that iterations of the function $f$ starting at $z$ remain bounded. So, $S = \{z$ in complex plane for which there is a fixed constant $C$ such that $|f^n(z)| < C$ for all $n = 1, 2, 3 \ldots\}$. The Julia Set for $a + bi$ is the *boundary* of this set, that is, it is the set of points on the edge between $S$ and the points of the complex plane not in $S$. It's often easier to focus on $S$ as the "filled-in" Julia Set for $a + bi$.

231

The Mandelbrot Set is the set of all points, $a + bi$, in the complex plane for which the corresponding Julia Set forms a single piece in the complex plane. That is, the Mandelbrot set = {$a + bi$ | ($a + bi$)-Julia Set, is connected}. It is extremely difficult for students to comprehend the Mandelbrot Set, because each point in it gives rise to an entire Julia Set, the structure of which determined that the original point should be in the Mandelbrot Set in the first place.

In the case where $a = b = 0$, the ($a + bi$)-Julia Set isn't very interesting— it is just the unit disk. However, this example is described geometrically in the text in a way that avoids complex arithmetic altogether. We can show that the geometric and algebraic descriptions are equivalent. First, recall that a complex number $a + bi$ has modulus $r = \sqrt{a^2 + b^2}$ and argument $\theta$, as the diagram shows. Note that the modulus $r$ is the distance from $a + bi$ to the origin.

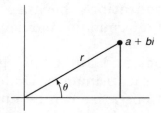

**Claim:** *Let $z = a + bi$ be a nonzero complex number with modulus $r$ and argument $\theta$. Then, $z^2$ has modulus $r^2$ and argument $2\theta$.*

**Proof:** First notice that $a = r\cos\theta$ and $b = r\sin\theta$. Notice also that the result is trivial if $b = 0$. If $a = 0$, then $z^2 = (bi)^2 = -b^2$, which clearly has modulus equal to $r^2$. It's also easy to check that whether $b$ is positive or negative, the argument of $(bi)^2$ is $2\theta$.

Now, $z^2 = (a^2 - b^2) + (2ab)i$, as shown on page 463. The modulus of this number is $\sqrt{(a^2 - b^2)^2 + (2ab)^2}$. But $(a^2 - b^2)^2 + (2ab)^2 = a^4 - 2a^2b^2 + b^4 + (2ab)^2 = a^4 + 2a^2b^2 + b^4 = (a^2 + b^2)^2$. So, $z^2$ has modulus $a^2 + b^2$, which equals $r^2$.

The argument of $z^2$ is the angle $\beta$ such that $\sin\beta = 2ab/(a^2 + b^2)$ and $\cos\beta = (a^2 - b^2)/(a^2 + b^2)$. If we substitute $a = r\cos\theta$ and $b = r\sin\theta$ and simplify, we find $\sin\beta = 2\cos\theta\sin\theta$ and $\cos\beta = \cos^2\theta - \sin^2\theta$. Recognizing the double-angle formulas, we find $\sin\beta = \sin 2\theta$ and $\cos\beta = \cos 2\theta$. The only way to satisfy both these conditions is if $\beta = 2\theta$, and our results hold.

Students can appreciate the ideas that lead to the Julia Sets without all these technicalities. The Sample Lecture Notes suggest a presentation that avoids complex numbers altogether, but that still conveys a sense of how the sets are defined. Discussing these ideas in the abstract can lead to an understanding of where the beautiful pictures come from. Plan to use a computer display in class with the CD-ROM for dramatic illustrations of various Julia Sets and the Mandelbrot Set.

The CD-ROM fractal images show the actual points in the fractal as black, surrounded by colorful regions. For the Julia Sets, the different colors indicate how fast the iterates fly off to infinity. For the Mandelbrot Set, the colors are related to how disconnected the corresponding Julia Sets are.

To add to the human drama of fractals, you may want to relate that Gaston Julia was severely injured during World War I and lost his nose. He did much of his research between painful surgeries, and he published his groundbreaking paper in 1918. He wore a nose patch for the rest of his life, as pictured in the text. Despite his injury, Julia made significant contributions to the field of dynamical systems.

---

*Our Experience*

*Time spent:* One-and-a-half or two 50-minute sessions

*Dangers:* The challenge is to string several steps together—any one of which is not too difficult. The idea of an iterative process, the method of complex arithmetic, determining whether a point goes off to infinity, the idea that every complex number leads to a different Julia Set, and deciding whether a point is in the Mandelbrot Set based on some feature of the associated Julia Set is a difficult sequence of steps.

*Remarks:* We sometimes omit this section in class, because it is an outstanding group research project with good poster session possibilities.

*Sample homework assignment:* Read the section and do Mindscapes I.4; II.6, 8, 10, 12, 13; III.26–28, 29–30, 31.

---

### Sample Class Activities

*0-Julia Set.* After explaining the squaring function for complex numbers, ask students to determine the 0-Julia Set.

*$(1 + i)$-Julia Set.* Ask students to use programmable calculators. Have them program their calculators to follow the first 25 steps of taking an imaginary number, squaring it, and adding $1 + i$ to the result. As an alternative, you can

demonstrate this procedure on a computer. In any case, ask students to start with random numbers in the imaginary plane and determine whether the number is in the $(1 + i)$-Julia set. Take a big number like $4 + 5i$, and get the class (without calculators) to see that squaring quickly makes the point very far from the origin. Let them start with a fixed point or periodic point and see the pattern.

*Julia Set.* Show a picture of a Julia Set. Ask students to confirm its properties by using a computer or calculator to take appropriate iterates of points near or on the Julia Set.

*Mandelbrot Set.* For each Julia Set in the text, ask students to verify whether the imaginary number associated with that Julia Set is or is not in the Mandelbrot Set. Check student answers by looking at the Mandelbrot Set.

### Sample Test Questions

(The Mindscapes are another good source of test questions. All test questions below are available electronically on the Test Bank CD-ROM.)

- The pictures show the $(0.3 + 0.6i)$-Julia Set and the $(0.101 + 0.6i)$-Julia Set.

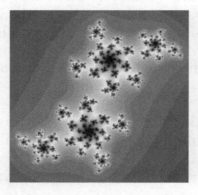 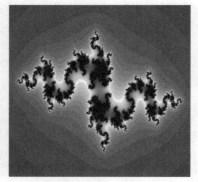

  **a.** Is $0.3 + 0.6i$ a point in the Mandelbrot Set? Why or why not?

  **b.** Is $0.101 + 0.6i$ a point in the Mandelbrot Set? Why or why not?

- Compute $(10 + 12i)^2 + (1 + i)$.

- Is the point $10 + 12i$ in the $(1 + i)$-Julia Set? Why or why not?

- Extra credit: Is the Mandelbrot Set locally connected? Why or why not?

## Sample Lecture Notes for Section 6.4, "The Mysterious Art of Imaginary Fractals" (2 days)

*(NOTE: This lecture requires a computer display.)*

### Day One

Question of the Day (to be written on the board for students to think about and discuss before class begins): *What's the story behind this picture?* [Note to Instructors: Display the Mandelbrot Set.]

1 minute—Announce that today everyone will be an artist, using the entire plane as a canvas. The goal is to paint a fractal. Remind students that until now you've used only the collage method to create fractals. Suggest that there are other ways and that all the ways share a key ingredient: a repeating process. To paint today's fractal, tell students that for each point in the plane, they have to decide whether to include that point in the picture. That is, should they include it in their fractal? Explain that they will make this decision by applying an infinite process to each point.

5 minutes—Draw the $x$- and $y$-axes on the board, mark the origin $O$, and choose a point $P$ in the first quadrant. For now, the goal is to decide whether $P$ will be in the fractal. The repeating process has two steps. First, draw a line from the origin to $P$, mark the angle from the positive $x$-axis to this line, and then double the angle. (Draw a long line from the origin at this new angle. Stress that all angles are measured counterclockwise from the positive $x$-axis.) Second, take the length of the segment $OP$, square it, and then mark the point that is this distance from the origin on the new line. (Don't really compute a square, just say that's what you're doing and mark a point at a reasonable distance.) The process will then be repeated with the new point and iterated to create a bunch of points (call them "iterates").

10 minutes—Do several easy examples. Start with the point $(0, 2)$; the next points are $(-4, 0)$, $(16, 0)$, $(256, 0)$, and so on. Note that the iterates get farther and farther from the origin. Start with the point $(-1/2, 0)$; the next points are $(1/4, 0)$, $(1/16, 0)$, $(1/256, 0)$. Start with a point in the first quadrant farther from the origin than $(0, 2)$. Ask students if they think the bunch of points created by this starting point will get arbitrarily far away from $O$ or can they "lasso" the points. Start with yet another point quite close to $O$ in some other quadrant. Ask the same question.

10 minutes—Announce that you should include (or put *in* the fractal) any starting point whose iterates can be lassoed. Each starting point whose iterates move arbitrarily far from $O$ will *not* be included. Have students experiment using different starting points to see if they can determine which points will be included. Give them enough time (and hints, if necessary) to discover that the fractal is just the unit disk. Acknowledge that this may be a rather bland painting, but it's an example of a Julia Set, named after mathematician Gaston Julia, who defined these sets just before 1920.

5 minutes—Add an extra wrinkle: Before starting the painting, choose an arrow. (Draw an arrow starting at $O$.) Now begin: Pick a starting point, double the angle, and square the distance, *then* move the arrow so it starts at this new point. The point where the arrow ends will be the first iterate of the starting point. Continue to iterate, always using the arrow chosen at the very beginning. As before, if the iterates can be lassoed, then the starting point is in the fractal; otherwise, it's not.

10 minutes—Do many examples with the arrow pointing 1 unit to the left. Comment that the fractal resulting from this arrow is the Julia Set for $(-1, 0)$—when drawn from $O$, the arrow terminates at the point $(-1, 0)$. Note that the first Julia Set simply reflects this process using the zero arrow; thus, it is the Julia Set for $(0, 0)$.

5 minutes—Show the computer display using the CD-ROM. Show students the Julia Set for $(0, 0)$. Ask what they think the Julia Set looks like for $(-1, 0)$. Bring up the display for $(-1, 0)$, and let students gaze with wonder and amazement. Remind them that points inside the set stay bounded when iterated, whereas points outside the set blow up. Zoom in on the boundary to show the infinite detail and beauty of the image.

5 minutes—Show many different Julia Sets, noting that each one corresponds to the particular point (arrow) chosen at the very beginning of the process. Explain the significance of the different-colored regions outside each Julia Set. Admit that the fractal from the Question of the Day is *not* a Julia Set, so this class has a cliff-hanger ending.

**Day Two**

Question of the Day (to be written on the board for students to think about and discuss before class begins): *We still don't know the story behind this picture. What's going on?* [Note to Instructors: Display the Mandelbrot Set.]

10 minutes—Bring up the Julia Set display. Remind students that each Julia Set corresponds to an arrow, and each arrow (when drawn from the origin) determines a point. Give many examples, both connected and disconnected. Ask students to separate the Julia Sets into two very simple categories. (Aim for something like "one piece" or "more than one piece.") As students make other suggestions, respond, "Great idea. What else?" until they hit upon the right categories.

10 minutes—Once the two categories are established, announce that it's time for a new painting, er . . . , fractal. For each point, include that point if its Julia Set has one piece. If its Julia Set has more than one piece, the point is not included. Then click on the Mandelbrot Set. Demonstrate the feature that displays the Julia Set when you click on a point in the Mandelbrot Set. Go back and forth over the boundary to show the one piece–two piece contrast. Point out that this time the different-colored regions outside the fractal are related to the disconnectedness of the corresponding Julia Sets. Acknowledge how amazing it is that such simple categories as "one piece" and "more than one piece" could yield such a complex structure.

5 minutes—Finish by zooming in on the boundary of the Mandelbrot Set to emphasize the infinite detail and pattern repetition. Comment that Gaston Julia defined and studied his sets as abstract structures in 1918, to great mathematical acclaim. Benoit Mandelbrot did not discover his set until near the time of Gaston Julia's death in 1978. The computer imaging of these sets first occurred in 1980. Sadly, Julia never saw this aspect of his beautiful work.

**Section 6.5 Predetermined Chaos** [Deterministic chaos]. This section presents the idea of sensitivity to initial conditions. Repeatedly hitting the sine key of a calculator or computing predictions of populations given by the Verhulst Model are both examples of deterministic systems that are sensitive to initial conditions—that is, they exhibit mathematical or deterministic chaos. The topic of chaos helps students appreciate the difficulties inherent in trying to predict the future using iterative schemes.

---

*General Themes*

Ignoring the minor discrepancies within our midst can lead to chaos.

---

*Mathematical Underpinnings.* (*Caveat:* These mathematical overviews are usually for the instructor only. The symbols and detailed proofs are not suitable for class presentations because mathematical symbols and terminology often present a formidable barrier for students.)

An iterative process is surprisingly dependent upon the initial starting value. Seemingly insignificant deviations result in wildly different outcomes. In the case of a numerical process repeated many times, round-off error alone can yield two very different results, even if the starting points were extremely close together. The text presents a compelling example, using results from two calculators that have slightly different round-off precision (text page 485). Even when starting with the same value and iterating the same process, the two calculators end up diverging wildly. This example can be the core of an engaging class.

Students may not appreciate a particular subtlety of this example. They may mistakenly presume that one of the calculators is somehow correct and the other, being less precise, is wrong. In fact, except for the initial value of 0.7391, all of the numbers produced by both calculators are actually wrong. That is, they are only approximations to the true values being requested, which are all irrational numbers. It's important to stress that correct answers do exist. However, no calculator or computer can ever hold an exact irrational number in decimal form, because doing so requires an infinite number of digits. If we represent such numbers in some other form, such as $\sqrt{2}$ or $\pi$, then we have the number exactly. Of course, it's hard to do calculations with these symbolic presentations of irrational numbers, but they do allow us to state formulas precisely and to do symbolic computations to some extent.

The calculator illustration is an example of *chaos,* a term that describes any system in which very small differences in initial conditions result in vastly different outcomes. Chaos precludes prediction. In the calculator example, rounding to seven digits instead of eight gave similar results for a few steps, but completely different answers after 35 iterations.

Another familiar chaotic system is weather. Given the weather today and some knowledge of past weather patterns, one could predict tomorrow's weather fairly accurately. No one will ever predict the weather 60 days from now, however. It's not that we're not smart enough or that our computers aren't powerful enough, it's that weather is too chaotic. The classic butterfly effect is real: A tiny change in air currents in one part of the world could have, after many days, a significant affect on the weather far away.

The Mindscapes for this section offer a version of cobweb plots similar to those in Section 6.2. But here they are called *cobweb tents* and are plotted on a grid, as shown in the figure. The implementation on the CD-ROM provides a nice class presentation or student exploration. Start at any point on the diagonal, move up or down vertically until you reach the tent, then move left or right horizontally until you reach the diagonal, and repeat.

On the figure, a sample path is traced out using the starting point (0.2, 0.2). The process stabilizes at an intersection of the tent and the diagonal.

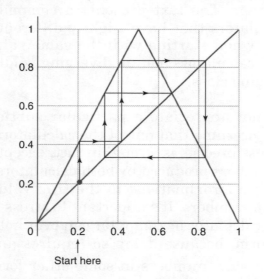

These plots can be described using an iterated function, as in Section 6.2, but the function is more awkward to describe. Using absolute value, the formula is $f(x) = 1 - |2x - 1|$. Defined piecewise,

$$f(x) = \begin{cases} 2 - 2x, & \text{for } x \geq 1/2 \\ 2x, & \text{for } x < 1/2 \end{cases}.$$

240

The cobweb plot records the results of repeatedly applying $f(x)$. Start with $x_0$. If we let $y_1 = f(x_0)$, then $y_2 = f(y_1)$, then $y_3 = f(y_2)$, and so on. Each starting value generates a sequence $y_1, y_2, y_3, \ldots$. These values are the $y$-coordinates of the points on the graph of $f(x)$ that one reaches by following the path of the cobweb.

Using these equations, it's easy to find the terminal point of the path to be (2/3, 2/3). As with the original cobweb plots, cobweb tents illustrate many aspects of iteration and chaos. The starting point (0.2, 0.2) yields a path that terminates. Making a very small change in the starting point to (0.2001, 0.2) yields a path that will probably never terminate.

---

*Our Experience*

*Time spent:* One or one-and-a-half 50-minute sessions

*Emphasized items:* Calculator iterations

*Items omitted or treated lightly:* Sometimes we treat the Verhulst Model and cobweb plots lightly. Other times we do quite a bit with them.

*Dangers:* The Verhulst or tent functions and cobweb models can be difficult and might be omitted. On the other hand, they allow students to see chaos in action. Deterministic chaos is an interesting idea, with or without the cobweb plots. Sometimes we choose to present the calculator and population examples and let it go at that.

*Remarks:* We often omit this section in class, because it is an outstanding group research project with good poster session possibilities.

*Sample homework assignment:* Read the section and do Mindscapes I.3; II.6, 7, 9, 12, 13; III.28.

---

### Sample Class Activities

*Calculator experiments.* Give students a decimal number and ask them to enter it and repeatedly strike the sine key on their calculators, multiplying the result by 180° each time. Ask them to record perhaps every fifth result. If two students have calculators with different rounding algorithms, then instead of recording the answers, you can have one student read his or her answer until the answers start to diverge from the other student's. Then have both students read. Soon the diverging answers will become clear.

A second calculator experiment is to have students program the Verhulst formula and compute the 50th generations. Students should start with initial populations that differ only in the thousandths place. Point out the significance of such a result for predictions about population growth.

*Tent function.*   Present students with the tent function on the interval [0, 1]. Ask them to follow the futures of several points. Point out the cobweb plot for this process—going from a point on the diagonal in the unit square up or down to the graph of the tent function, then going horizontally back to the diagonal and repeating. Ask students to follow the futures of some periodic points. Then ask them to follow the futures of some nearby points whose futures are spread over the whole interval.

*Attractors and repellers.*   Ask students to consider the squaring operation on the real numbers. Discuss what happens under repeated applications of that operation to points near the fixed point 0 (they get closer to 0) and to points near 1 (they get farther from 1). Give them an example of an attracting periodic orbit in the Verhulst Model. Ask them to follow the futures of nearby points to see that the points get closer to being periodic. Then ask them to follow the futures of points near periodic points of the chaotic tent function or Verhulst function $4x(1 - x)$.

## Sample Test Questions

(The Mindscapes are another good source of test questions. All test questions below are available electronically on the Test Bank CD-ROM.)

- Sketch the first three steps of the cobweb plot of the tent function $f(x) = \{2x,$ if $x \leq 1/2; -2x + 2$ if $x \geq 1/2\}$, starting from the point indicated.

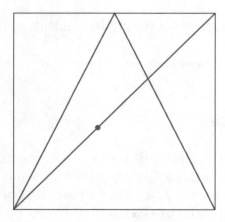

- For the tent function $f(x) = \{2x,$ if $x \le 1/2; -2x + 2$ if $x \ge 1/2\}$,

  **a.** Find a point of period 3.

  **b.** Sketch its cobweb plot.

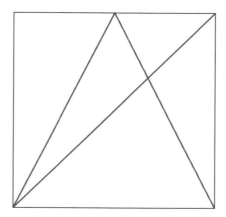

- Suppose you took a calculator and computed $180\sin(x)$ repeatedly. Suppose you started with 0.236 and made a list of your first 50 results. Then you started with 0.237 and made a list of your first 50 results. How close would you expect the 50th results to be to one another?

- Do you think that 100 years from now, weather prediction will become so exact that meteorologists will be able to predict three months in advance whether it will rain at a given place on a given day?

## Sample Lecture Notes for Section 6.5, "Predetermined Chaos" (1 day)

Question of the Day (to be written on the board for students to think about and discuss before class begins): *How close is your calculator's answer to the correct answer?*

10 minutes—Discuss the Question of the Day. Ask several students how many decimal places their calculator displays. (Bring in two of your own calculators to be sure you have examples with different precision.) Compute $\sqrt{2}$ using calculators with at least two different displays. Write the answers on the board, and discuss the differences. Point out that most calculator results are only approximations. The number displayed has been rounded off, creating round-off error.

10 minutes—Ask whether round-off error matters, especially if it's in the seventh or eighth decimal place. Ask under what circumstances such an error might propagate. (Cheer if someone suggests "iteration.") Describe the iterative sine computation (text pages 483–485). Display the results from the original time in Table 1 (text page 485). (Keep the other results hidden.) Then follow the story on text pages 484–485 that produced the second set of values in the table. Starting with the fifth row, reveal the rows one-by-one. Discuss the results. Emphasize that a tiny change in the input yielded vastly diverging results.

5 minutes—Now present the dueling calculators example (text pages 485–487). Describe the scenario, and display the rows of Table 2 (text page 486), one by one. For each row, have students note the first decimal place at which the values differ. By row 29, the divergence has reached the first decimal place. Row 30 brings the values closer; but after that, they diverge quickly.

5 minutes—Point out that each calculator was functioning perfectly and each used exactly the same procedure. At some point in the iterations, however, very small differences in input ultimately led to very large differences in output.

5 minutes—Introduce the term "chaos." Ask students for other examples of dynamical systems that exhibit chaotic behavior. (Be sure students mention weather.)

15 minutes—Discuss weather prediction. Ask students how well they think we predict the weather now: one day in advance, two days, a week, a month. Ask for examples from their experience when the weather prediction was wrong. Ask if they've heard of the butterfly effect. Finish with a discussion of the impossibility of accurate long-range weather forecasting. Even students' great-great-great-grandchildren will not be making accurate 60-day weather forecasts.

**Section 6.6 Between Dimensions** [Fractal dimension]. This section presents the idea of fractal dimension for self-similar fractals. We develop the idea of fractal dimension by looking at dimensional relationships in familiar objects, such as squares and cubes, and defining fractal dimension by analogy. This section provides a good example of the technique of exploring and clarifying one's understanding of the familiar in order to develop tools for understanding the bizarre.

---

*General Themes*

Searching for patterns leads to new ideas.

---

*Mathematical Underpinnings.* (*Caveat:* These mathematical overviews are usually for the instructor only. The symbols and detailed proofs are not suitable for class presentations because mathematical symbols and terminology often present a formidable barrier for students.)

The idea of a 1-, 2-, or 3-dimensional object is fairly intuitive. Some of the fractals constructed earlier in this chapter turn out to have dimensions that lie between integral dimensions. By carefully analyzing the dimensions of some simple objects, this section develops the idea of a *fractal dimension,* or a dimension that is not a whole number. The process offers great potential for students to discover ideas on their own.

An elegant definition of dimension arises by asking how many copies of an object are needed to create a larger copy of the object. To make an object twice as large as the original, one needs two copies of a line segment, four copies of a square, and eight copies of a cube, as summarized in this table (see also page 505 in the text).

| Original Object | Dimension of the Object | Scaling Factor to Make a Larger Copy | Number of Copies Needed to Build the Larger Copy |
|---|---|---|---|
| Line | 1 | 2 | $2 = 2^1$ |
| Square | 2 | 2 | $4 = 2^2$ |
| Cube | 3 | 2 | $8 = 2^3$ |

If we require $N$ copies to create a larger version scaled by a factor of $S$, then it makes sense to define the dimension of the object to be the number $d$ satisfying the equation $S^d = N$. The text demonstrates this relation with triangles and rectangles, as well as with different scaling factors.

247

This definition extends naturally to the many self-similar fractals constructed with the collage method in Section 6.3. Because three copies ($N = 3$) of the Sierpinski Triangle are needed to construct a new one with side length twice as long as the original ($S = 2$), the dimension of the Sierpinski Triangle will be the number $d$ satisfying $2^d = 3$.

To find $d$, we take the natural logarithm of both sides of $S^d = N$ to obtain $\ln(S^d) = \ln(N)$. So $d\ln(S) = \ln(N)$ and $d = \ln(N)/\ln(S)$. So, the dimension of the Sierpinski Triangle is $\ln(3)/\ln(2)$, which is approximately 1.26185. This material offers an interesting review of logarithms, though students can get by simply by pushing buttons on their calculators.

Notice that with $d$ defined in terms of natural logs, fractal dimensions are often irrational. An interesting challenge would be to use the collage method to construct a self-similar fractal that has a rational dimension.

---

*Our Experience*

*Time spent:* One 50-minute session

*Emphasized items:* The relationship between the collage method of constructing a fractal and the fractal dimension

*Remarks:*

1. This topic is good for getting students to develop a new idea on their own (fractal dimension), with the proper guidance.
2. We often omit this section in class, because it is an outstanding group research project with good poster session possibilities.

*Sample homework assignment:* Read the section and do Mindscapes I.3; II.7, 9, 10, 11; III.17; IV.21.

---

### Sample Class Activities

*Familiar objects.* Have students draw a filled-in square or triangle. Ask them to record how many squares or triangles of that size are required to make a square or triangle that is $n$ times as large on each side—in other words, a square or triangle that has been blown up by a factor of $n$. Have them express the numbers in terms of $n$ to a power. Do a similar exercise with a cube. Again, students should note that the exponent corresponds to our idea of dimension.

*Developing new ideas.* Ask students to state how the exponent is involved in counting the number of small objects needed to cover an $n$-fold blowup of a $k$-dimensional object. Ask them to guess how many boxes are needed to cover an object $n$-fold larger if that object has dimension 1-1/2.

*Fractal dimension.* Take a fractal object that was constructed by a replacement process. For example, consider the Sierpinski Carpet obtained by starting with a unit square and replacing the square by eight of the nine $1/3 \times 1/3$ subsquares, leaving out the central subsquare. Ask students to compute how many Sierpinski Carpets are necessary to make a Sierpinski Carpet that is $n$-fold larger. Ask them to deduce an appropriate measure of the fractal dimension of the Sierpinski Carpet.

*Build a fractal.* Ask students to describe a construction for a fractal of dimension ln(5)/ln(3). Help them by starting with a square, dividing it into a ticktacktoe board pattern, and asking what collage-making process would result in a fractal of dimension ln(5)/ln(3) and why. Seeing that they could choose any five of the nine repeated squares will give them the insight.

### Sample Test Questions

(The Mindscapes are another good source of test questions. All test questions below are available electronically on the Test Bank CD-ROM.)

- Below are the first two steps in the construction of a fractal. What is the fractal dimension of the resulting fractal?

- Give a method for the construction of a fractal of fractal dimension $\dfrac{\ln 7}{\ln 4}$?

- What is the fractal dimension of the Sierpinski Triangle? Why?

- What is the fractal dimension of the Koch Curve? Why?

## Sample Lecture Notes for Section 6.6, "Between Dimensions" (1 day)

Question of the Day (to be written on the board for students to think about and discuss before class begins): *What is the dimension of a cloud?*

5 minutes—Discuss the Question of the Day. Clouds seem to have height, width, and length, but they have a cotton candy–like quality; they're not really solid. Where do they really begin and end? Help students admit that a cloud might have dimension between 2 and 3.

5 minutes—Before exploring objects that might have fractional dimension, ask students to define dimension. (Typical answers: number of degrees of freedom, length and width, etc.) Acknowledge the limitations of these "definitions" when applied to objects like the Koch Curve, which clearly has length but may or may not have width.

10 minutes—Admit the need for a clearer definition of dimension. Proclaim the value of looking at simple things deeply. Have students start with a line segment, a square, and a cube. Ask them how many copies they need of each to create a line (square, cube) that is twice as long (on a side). Create a table as on text page 505. Have them repeat the exercise with a scaling factor of 3, and expand the table as on text page 506.

5 minutes—Ask students if they see a pattern in the data. Can they suggest a definition of dimension? Lead them to the $S^d = N$ definition. Then point out that this definition does not require $d$ to be a whole number.

10 minutes—Work with the class to discover the dimension of the Koch Curve (text pages 507–508). Once you reach the equation $2^d = 3$, acknowledge that this equation does not solve easily for $d$. Announce that $d = \ln(3)/\ln(2)$. Use appropriate reference to calculator keys. (Logarithmically literate students can solve for $d$ themselves, of course.) Compute $d$ to reveal a true fractal dimension. Discuss whether a dimension somewhat larger than 1 makes sense.

10 minutes—Have students discover the dimension of the Sierpinski Triangle themselves. Analyze the answer.

5 minutes—Finish with the general dimension formula: $d = \ln(N)/\ln(S)$.

# Template 6.3.1   Koch curve

## Creation of the Koch Curve

# Template 6.3.2   Collage method

5th copy

4th copy

3rd copy

2nd copy

1st copy

Initial image

# Template 6.3.3 Chaos Game

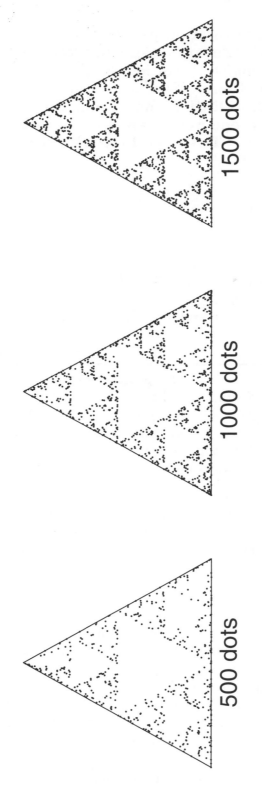

1500 dots

1000 dots

500 dots

# Chapter 7   Taming Uncertainty

## I.   Chapter Overview

This chapter presents an introduction to the study of probability and statistics. It focuses on the question of how we can deal effectively with the uncertain and the unknown. The chapter's overall goal is to encourage students to grapple with the idea of how to give a reasoned, quantitative analysis to situations that involve random processes and likelihoods.

---

**Dependencies**

Sections 7.1 and 7.2 contain the basic idea of probability, which is used throughout the chapter. All subsequent sections depend on Sections 7.1 and 7.2. Section 7.7 refers to some material from Section 7.6, but otherwise Sections 7.3–7.7 are independent of one another.

---

**Sections to Assign as a Reading Assignment or Term Project**

Skipping Section 7.4 (Systematic counting) is one way to shorten the chapter. In the past, we have assigned Sections 7.3 and 7.4 as project topics. Any omitted sections are good paper or extra credit topics.

---

*Reminders*

1. Class activities are suggested for each section. These can be done with students working in groups of two or three and then gathering their responses or as one global class discussion.
2. Before introducing a new topic, engage students with enticing questions, which they quickly discuss with neighbors; solicit responses within a minute. Use the responses as the springboard for the introduction.
3. We have included more sample class activities than you will probably want to use. Choose only the ones that are right for you and your class.

**4.** Each section can be treated in more or less depth. It is frequently a good idea to omit difficult technicalities in order to treat the main ideas well and then move on.

## II. Section-by-Section Instructional Suggestions

**Section 7.1 Chance Surprises** [Unexpected scenarios involving chance]. This section entices students to think about the idea of probability. Several intriguing scenarios show students that there are many probabilistic issues in life and that being able to give a quantitative evaluation of likelihood is a valuable refinement to their thinking about uncertainty.

---

*General Themes*

Thinking about situations that run counter to our intuition can lead to new and important insights.

---

*Mathematical Underpinnings.* (*Caveat:* These mathematical overviews are usually for the instructor only. The symbols and detailed proofs are not designed for class presentations because mathematical symbols and terminology often present a formidable barrier for students.)

The scenarios presented in this section all involve probability. Though later sections provide more formal background and definitions, this introduction helps students appreciate that clear definitions and techniques are needed before probabilities can be determined. In the process, students begin to develop an intuitive idea of probability. Stress details and approaches that will become important later on: Probability is always measured in a context, simulations help make valid predictions, an event may be much more likely than it seems at first. All of these concepts underscore the need for precise definitions.

The two penny experiments are great fun to do in class. Part of the fun is that initial intuition is often wrong. Be prepared for the possibility that the outcome of your spinning or balancing experiment may be different from the text examples. Whatever the outcome, emphasize that the data come from two entirely different experiments—balancing and spinning—thus reinforcing the idea that probabilities are measured in a particular context, the *sample space.* Though the chapter never introduces this terminology, see the Mathematical Underpinnings of Section 7.2 in this *Instructor Resources and Adjunct Guide* for further discussion.

The Let's Make a Deal and the two reunion scenarios are easily simulated with playing cards, as described in the text. Simulations help students look for probabilities based on actual data, rather than jumping to some conclusion based on faulty reasoning. Doing an experiment over and over will also help them better understand the relation between *probability* and *relative frequency,* terms discussed in Section 7.2.

The Birthday Question is a great one for class discussion. The highly counterintuitive answer can motivate students to test it on groups of friends in their dorms, on sports teams, in the dining hall, and so on, before they derive the solution more formally in a later class.

---

*Our Experience*

*Time spent:* One 50-minute session

*Items omitted:* None

*Dangers:* Be prepared for the following argument: In Reunion Scene—Take Two, some students will say there are four possibilities; Jonathan-Boy, Jonathan-Girl, Boy-Jonathan, Girl-Jonathan. Reply with the hypothetical that all boys are named Jonathan.

*Personal favorite topic or interactions:* All of them

*Remarks:* All the scenarios are surprising and provoke controversy. Making bets about the outcomes is a good way to keep attention high. Have students vote on the outcomes before the experiment is performed and record the votes.

*Sample homework assignment:* Read the section and perform experiments.

---

### Sample Class Activities

*Pennies-on-edge.* Bring new pennies to class, hand each student one or more pennies, and describe the Lincoln on Edge scenario. Before performing the experiment, take a vote of students' expectations. After tabulating the votes, instruct each student to perform the experiment five times. Make a table to show the outcomes. Emphasize that when seemingly equally likely events do not arise roughly equally after many trials, you must become suspicious and seek an underlying cause for the disagreement between intuition and reality.

*Other penny experiments.*   Repeat the same process of voting, experimenting, and tabulating with the Dizzy Lincoln experiment. Finally, repeat the same process but with students flipping their pennies.

*Let's Make a Deal.*   Remind students of the Let's Make a Deal story from Chapter 1, "Fun and Games." If you haven't done so before, have students play the game with their neighbor, first using the stick method, then using the switch method. Record the results.

*Reunion scenarios.*   Have students pair up. Ask each person to flip a coin. If both people in the pair get tails, the trial is thrown out. Otherwise, they record whether they got two heads or just one. Have each pair record at least five trials. Before gathering the data, ask students to vote on what percentage of the outcome will be heads-tails and what percentage will be heads-heads. Then gather the data.

*Birthday problem.*   Select about 50 students in your class or have them choose themselves and a relative to get up to a pool of 50. Make a bet that two people will have the same birthday. Keep the money.

### Sample Test Questions

Because the scenarios presented in this section are analyzed in Section 7.2, the sample test questions about them occur in Section 7.2.

## Sample Lecture Notes for Section 7.1, "Chance Surprises" (1 day)

Question of the Day (to be written on the board for students to think about and discuss before class begins): *How many people are needed in a room so that the probability of two people having the same birthday is roughly 0.9?*

10 minutes—Remind students that flipping a fair coin is a standard act in which each outcome, heads or tails, is clearly understood to have a probability of 1/2. Now suggest another penny experiment. Distribute several pennies (the newer the better) to each student or pair of students. Explain that students will balance the pennies on edge, bang the desktop to topple the pennies, and then record the number of heads and tails facing up. Subtly get students to agree that the outcome will be about equal numbers of heads and tails. Proceed with the experiment, have students repeat it several times, and record the results on the board. Surprise! Briefly discuss the results.

10 minutes—Do the penny-spinning experiment, again getting students to predict 50-50. Record class results on the board and discuss briefly. Ask students to look for further explanations before the next class.

15 minutes—Present the class reunion scenarios. Ask students to briefly discuss their predictions. Record their predictions on the board. Reveal the two different probabilities. In response to protests, suggest a simulation to reinforce your claim. Distribute several decks of playing cards, and have students simulate the two scenarios as described in the text (pages 518–519). Gather data and discuss. Acknowledge that a simulation alone does not validate the claimed probabilities; indicate that the text gives a more rigorous explanation in Section 7.2.

10 minutes—End with the Let's Make a Deal scenario that was initially presented in Chapter 1. Ask for predicted probabilities, and then reveal the true answer (or ask if anyone remembers the answer from Chapter 1). Distribute sets of three appropriate cards to groups of students. Have some groups simulate switching while others simulate not switching. Tabulate data and discuss.

5 minutes—Discuss the Question of the Day, and then write student answers on the board. Reveal the answer as 41. For emphasis, announce that with 23 people in a room, the probability that two share a birthday is 0.5, and with 70 people, the probability is more than 0.999. As students react, assure them that this seemingly counterintuitive result will be derived in a subsequent class (cliff-hanger!). Use this example to reinforce the idea that probability can be very tricky.

1 minute—Acknowledge the pitfalls of using intuition to determine probabilities and the need for clearer definitions and techniques.

## Section 7.2 Predicting the Future in an Uncertain World

[Probability]. This section presents a basic introduction to probability. It begins with rolling a die, leading students to consider the idea of equally likely outcomes and that the fraction of successes divided by total possible outcomes is a good measure of likelihood. This ratio becomes the definition of probability. This section explains the surprising scenarios introduced in Section 7.1 and illustrates how simple examples can help us develop methods for analyzing more complicated settings.

---

*General Themes*

Quantifying a situation often leads to better understanding.

---

*Mathematical Underpinnings.* (*Caveat:* These mathematical overviews are usually for the instructor only. The symbols and detailed proofs are not designed for class presentations because mathematical symbols and terminology often present a formidable barrier for students.)

The definition of probability provided in this section is a simplified version of a standard finite probability space in which all outcomes are equally likely. A more formal definition would include the following: A *sample space* is the set of all possible outcomes of an experiment (or random process). An *event* is a subset of a sample space. If $S$ is a finite sample space in which all outcomes are equally likely, and $E$ is an event in $S$, then the probability of $E$ (denoted $P(E)$) equals $|E|/|S|$. This formula can actually be obtained from a more fundamental set of axioms, but it is clearly motivated by the intuition that the probability of an event occurring should be approximately the proportion of times it occurs if an experiment is repeated many times, the *relative frequency* of $E$. This idea is formalized as the *Law of Large Numbers:* If $n$ is the number of times an experiment is repeated, then as $n$ approaches infinity, the relative frequency of $E$ will approach $P(E)$.

The section discusses neither the more abstract definition of a *probability space* nor the idea of an infinite sample space. Some students might be intrigued, however, by the idea that if a sample space is infinite, then there may be events with probability 0 that can actually occur, as well as events with probability 1 that may not always occur. The groundwork for a classic illustration of this is given in Section 2.7. If the sample space is the set of all real numbers under the experiment of picking a number $x$ at random, then the probability of the event "$x$ is rational" is 0, even though it is technically possible to pick a rational number. Similarly, the probability of the event "$x$ is irrational" is 1, even though it is technically possible to pick a number that is not irrational. Understanding how these probabilities arise is a challenge, but the example underscores the general understanding that to say an event has probability 1 really means that it will *almost* certainly occur.

In presenting the solution to the Birthday Question, the text uses a fundamental property of any probability space: Given an event $E$, the probability that $E$ occurs is 1 – the probability that $E$ does *not* occur; that is, $P(E) = 1 - P(\text{not } E)$. This result is very intuitive and should be readily accepted by students.

Students may also be tempted to use another important idea in probability that is much more subtle, namely, the idea of *independent events*. The Birthday Question is answered very carefully in the text in a way that does not invoke independence. However, when throwing dice and tossing coins, students often multiply probabilities without realizing what they are doing. In fact, $P(E \text{ and } F \text{ occur}) = P(E)\ P(F)$ if and only if $E$ and $F$ are *independent events*. Independence is usually defined using conditional probability, though the equality itself, $P(E \text{ and } F) = P(E)\ P(F)$, can also be used as a defining condition for independence. This is quite an abstract definition. Students can grasp the idea of independent events such as coin tosses, but sometimes their thinking gets clouded when asked, "After 10 tosses of a fair coin turning up heads, what's the probability that the 11th toss will be heads?" The text does not formally address the idea of independent events (though the concept is alluded to in Section 7.4); therefore, it may be best to emphasize the definition of probability as a proportion that requires counting desired outcomes and total number of outcomes rather than multiplying probabilities.

Here's a rather crude but reality-based example that might help recalcitrant students realize the risks of multiplying probabilities. Suppose that in the month of April, you kept track of rainy days and found that it rained 15 out of 30 days. Thus, in the abstract, the probability of rain on a random April day was 1/2. Suppose you also noticed that on exactly 15 out of 30 days you carried an umbrella, giving the probability of an umbrella on a random day as 1/2. Now, what's the probability that on a random day it rained *and* you carried an umbrella? One hopes the answer would not be $(1/2)(1/2) = 1/4$ if looking out the window or listening to the weather forecast were at all involved in the decision about carrying an umbrella. Clearly a rainy day and carrying an umbrella are not independent events.

The Mindscapes offer many interesting probability questions to challenge students. Mindscape III.30 is particularly fun to try in class. Bring in three cards, one blue on both sides, one red on both sides, and the third red on one side and blue on the other (plain cardboard squares marked with X and O work just as well). Display the three cards so students clearly see how the cards are marked. Put the cards behind your back, draw one at random, and show students one side. Ask for the probability that the other side of the card is the same color (has the same symbol). Many students will say 1/2. To lead them to the true answer of 2/3, refer to the definition of probability and help students determine what experiment is being done and what the outcomes are. It helps to break the experiment into two parts: First choose a card, then choose a side to display. There are three choices of card and two choices of side for each card, so there are six outcomes to the experiment. Diagramming the six possibilities on the board helps clarify the result. If the displayed card is red, for example, then two out of three such outcomes come from the card with the other side red as well.

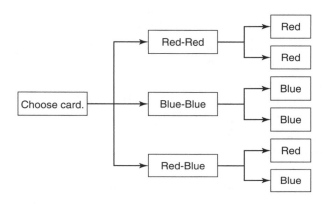

---

*Our Experience*

*Time spent:* One to one-and-a-half 50-minute sessions

*Emphasized items:* Equally likely outcomes; probability of something not happening is $1 - P$.

*Items omitted or treated lightly:* Gombauld's dice games

*Personal favorite topic or interactions:* Reunion scenarios and birthday question

*Remarks:* Constantly focusing on listing the equally likely possibilities helps students see probability as natural and straightforward.

*Sample homework assignment:* Read the section and do Mindscapes I.2, 4; II.7, 8, 12, 18, 19, 20; III.28, 30; IV.40.

---

## Sample Class Activities

*Dice.* Show or hand out a table of the 36 equally likely outcomes of rolling a pair of dice (see text page 529). Ask students to compute various probabilities from the table. For example, ask for the probability of rolling a 9 or the probability of getting doubles (both dice the same). Then display a table with the numbers 2 through 12 on it. Ask what is the probability of rolling an 8, a 9, or a 12. Ask students to explain the situation, guiding them to crystallize in their minds the importance of considering equally likely outcomes.

*Conditional probability.* One of the examples in this section is to flip two coins and cover up the result. The flipper then looks at the coins and announces, "At least one coin is a heads." What is the probability that there are two heads? This example requires explanation and discussion. A persuasive approach is to have students work in pairs to do the experiment several times. Then gather the results. If, when students flip the coins, both coins are tails, they ignore the flip. If at least one is a heads, they mark it down as either both heads or one head. Gathering the data will soon show the 1/3 probability of two heads. Showing the chart of four equally likely outcomes with the tails-tails outcome crossed out gives a good means of showing the 1/3 probability.

This activity introduces the idea of conditional probability; however, this term is not mentioned nor is the notion highlighted in the text. Instead, the activity is simply presented as finding the total number of possible outcomes (3) and dividing that into the number of successful outcomes (1). You may wish to emphasize the idea of conditional probability more prominently than the text does.

*Let's Make a Deal.* You might want to revisit the Monty Hall question from Chapter 1, "Fun and Games." If you did not actually do the experiment at that time or in Section 7.1, you might want to have students do the experiment in pairs, with some pairs doing the switch method and the other pairs doing the stick method. Taking the data from several dozen such experiments is usually quite persuasive. Looking at this question from the perspective of equally likely outcomes provides a good discussion.

*Birthday problem.* The section concludes with a discussion of the Birthday Question. The surprising result is definitely worth actually doing the activity in class. After doing the experiment, it will still require some effort on your part to explain the result. After you have proved your high probability of success, you will probably want to return your winnings, but you don't have to.

*Homework.* Certainly, students need to do many examples of computing the probability of some outcome using coins, dice, or cards followed by actually

266

trying the experiment. Point out the correspondence between the probability computed by analyzing equally likely outcomes and the data they collect concerning relative frequencies of outcomes. This correspondence can provide many examples of the Law of Large Numbers.

## Sample Test Questions

(The Mindscapes are another good source of test questions. All test questions below are available electronically on the Test Bank CD-ROM.)

- Theoretically, it's possible to build a weighted die so that the probability of rolling a 1 is 1/3, a 2 is 1/6, a 3 is 1/6, a 4 is 1/12, a 5 is 1/12, and a 6 is 1/12. True or False?

- It is more likely than not that in an arbitrary group of 30 people, 2 people will share the same birthday. True or False?

- Suppose you are playing a game with three cards. One of them is red on both sides, another is blue on both sides, and the third is red on one side and blue on the other. You are shown one side of one card, and it's red. What is the probability that the other side is blue?

- Assume that the probability of having a boy or a girl is the same.

  **a.** What's the probability of having all boys in a family of three? A family of four? A family of five?

  **b.** How many children must a couple have in order to have more than a 0.95 probability that at least one of the children will be a girl? The following equations may be helpful: $1/8 = 0.125$; $1/16 = 0.0625$; $1/32 = 0.03125$.

- Suppose you are on a game show like *Let's Make a Deal,* except this time there are four doors. After you choose one, Monty Hall opens another door, behind which is an empty bucket. What is your probability of winning the one grand prize if you stick to your original choice? What is your probability of winning the grand prize if you switch to one of the other unopened doors? Please explain your reasoning.

- Complete the probability of the following scenes:

  **a.** Rolling a 6 three times in a row with a fair die

  **b.** Selecting an ace from a regular deck of cards, given that a king was previously removed from the deck

**c.** Selecting an ace from a regular deck of cards, given that an ace was previously removed from the deck

**d.** Randomly selecting a real number for which the digit 3 *never* occurs in its decimal expansion

**e.** Flipping a half-dollar coin eight times and seeing at least one head

**f.** Suppose I have rolled two dice, and I announce that the sum of the two dice is 4 or less. Compute the probability that I rolled a total of 3.

**g.** You have a penny, a nickel, and a die. Compute the probability of flipping a head or rolling a 2.

**h.** Compute the probability of rolling two fair dice and having the sum exceed 4.

**i.** Suppose I have rolled two dice, and I announce that the sum of the two dice is 8 or more. Compute the probability that I rolled a total of exactly 9.

## Sample Lecture Notes for Section 7.2, "Predicting the Future in an Uncertain World" (1 day)

Question of the Day (to be written on the board for students to think about and discuss before class begins): *When you roll a pair of dice, are you more likely to roll snake eyes (two 1's) or a total of 7? Why?*

5 minutes—Distribute pennies to students. Ask them each to flip one penny three times and record the sequence of heads and tails that result. Have them repeat the experiment two more times, and record the results. Record all possible outcomes of such an experiment on the board. For each outcome, record the number of times it occurred during the student flipping experiments. Assuming a penny is fairly balanced, ask, "What's the probability of getting exactly two heads? Of getting at least one tail?"

5 minutes—Acknowledge that probability hasn't yet been defined, but intuition is a reasonable start. Use the penny-flipping example to lead to the definition of *probability*. Point out that probability is always a value between 0 and 1 and that an event with probability 1 is considered certain to occur. Emphasize the assumption that all outcomes are assumed to be equally likely.

5 minutes—Introduce *relative frequency*. Compare the relative frequencies of various outcomes from the penny-flipping example, leading into a discussion of the Law of Large Numbers.

10 minutes—Distribute pairs of dice among groups of students. Have them toss the pairs five times and record the sum of the dots (*pips*) showing. Write the outcomes 2, 3, . . . , 12 on the board, and record the relative frequencies. Pose the Question of the Day, leading to the observation that the listed outcomes are *not* equally likely. Ask students what the real outcomes are, and create a table on the board (or display one on a transparency). Revisit the Question of the Day. Reinforce the difference between an event and an outcome. Do more examples.

5 minutes—Looking at the table of dice outcomes, ask for $P$(even sum). Then ask for $P$(odd sum). Ask students for a general rule and state it formally: $P(\text{not } E) = 1 - P(E)$. Do more examples with dice and coins.

10 minutes—Discuss the birthday problem. Present the solution described in the text (pages 530–534). Emphasize that at each stage, the probability of no shared birthdays is simply the proportion of desired outcomes over the total possible outcomes.

10 minutes—Finish with Mindscape III.30, as described in this section's Mathematical Underpinnings.

**Section 7.3 Random Thoughts** [Coincidences]. This section discusses situations in the world, particularly coincidences, that involve randomness. The philosophical point is to see that random behavior is one of the governing forces in our experience of the world. The section contains discussions of daily-life coincidences, stock prices, monkeys typing, Buffon's needle, and random walks. The mathematical content shows that outcomes with a low probability are almost certain to occur if the experiment is repeated often enough.

```
┌────────────────────────────────────────────────────────────────────┐
│                                                                      │
│                          General Themes                              │
│                                                                      │
│  Applying careful quantitative analysis can change our whole world   │
│  view.                                                               │
│                                                                      │
└────────────────────────────────────────────────────────────────────┘
```

*Mathematical Underpinnings.* (*Caveat:* These mathematical overviews are usually for the instructor only. The symbols and detailed proofs are not designed for class presentations because mathematical symbols and terminology often present a formidable barrier for students.)

The discussion of coincidence and randomness in this section offers a lively and fun opportunity to reinforce clear thinking about probability that students will remember for a long time. When considering a seemingly unlikely event, students often do not recognize that any probability judgment takes place in the context of a particular sample space. The idea of randomness is also key.

A classic example for the mathematician involves answering the following question, known as Bertrand's Paradox: What's the probability that a random chord chosen in a circle of radius 1 will have length greater than $\sqrt{3}$ (the side length of an inscribed equilateral triangle)? What makes this question so tricky is that you get different answers depending on how you interpret the meaning of "a random chord." There are different ways to "choose" a chord at random, several of which give rise to different probabilities. If you do a search on the Web for Bertrand's Paradox, you'll find many discussions of this problem.

The idea of randomness and the impact of repeating an experiment a very large number of times are key to understanding probability. The amusing scenario of 1,000,000 monkeys typing randomly on keyboards (text pages 544–546) reinforces this idea. We want to estimate how much time will pass until a monkey almost certainly types out "To be or not to be: that is the question."—a 41-character sequence (not counting the shift key for capital letters). Because there are just more than 31,536,000 seconds in a

271

year, each monkey can type about 769,171 41-character sequences in a year (no bathroom breaks!). With a million monkeys, this gives a total of $7.69171 \times 10^{11}$ 41-character sequences generated per year. Because there are $48^{41}$, or about $8.5289 \times 10^{68}$ such sequences, we can estimate that each sequence would be typed once every $(8.5289 \times 10^{68})/(7.69171 \times 10^{11})$ years, or in about $1.1088 \times 10^{57}$ years. Thus, as stated in the text, we would expect to wait more than $10^{57}$ years for that classic line from *Hamlet* to appear.

Even though the sample space here is enormous, and the probability of any particular outcome is miniscule, if the experiment is repeated a sufficient number of times, every possible outcome should occur. If we think of our lives as a long series of events strung together, we would expect to see at least a few very unlikely events in our lifetime. The moral of the story: Expect the unexpected.

---

## Our Experience

*Time spent:* One 50-minute session

*Emphasized items:* Personal coincidences, stock investments, monkeys typing

*Items omitted or treated lightly:* Buffon's needle, random walks

*Personal favorite topic or interactions:* Personal coincidences, monkeys typing

*Remarks:* This whole section is entertaining and lively. Students enjoy the discussion of coincidence. They come to understand how common coincidences must be, which is sometimes a shock to them. This section emphasizes the difference between a single event being rare and the likelihood that some rare event will occur sometime.

*Sample homework assignment:* Read the section and do Mindscapes I.2; II.23; III.26, 29, 30, 32.

---

### Sample Class Activities

*Coincidences.* The most potent way to show that coincidences are common is to uncover some coincidences that students have experienced or some coincidence in the recent news. Students often recall startling coincidences that may be good vehicles for discussing the probabilities involved. The main point is to accustom students to the idea that relatively rare events, when attempted often enough, will eventually happen.

272

*Random guesses.* Many incidents of coincidence and remarkable predictions can be readily explained by randomness. For example, analyze the horoscope from the newspaper or the predictions of a seer. Point out that the probabilities of correctly making random predictions are increased with vagueness because more outcomes can be considered favorable.

*Random words.* If you have a computer connection in the classroom, program your computer to randomly generate letters. Display the letters and find words among them. One excellent example of how to do this phenomenon without a computer is the game of Boggle. Sixteen cubes each of which have a letter on each side, are shaken and arranged in a 4 × 4 square. The object of the game is to find words that can be constructed using contiguous letter cubes. This game shows the relatively common occurrence of finding words among random letters. Of course, Boggle allows you to seek words among letters in a square array rather than just a sequential list generated by a typewriter, but it's still compelling.

*Buffon's needle.* If possible, use a computer and run a Web program that shows the needle being dropped many times and that records the number of times the needle hits a line, the number of attempts, and the ratio. Display the number $2/\pi$ so students can watch the ratio approach it. The ratio should settle down and converge to $2/\pi$. Another way to get the data is to have each student do the experiment at home 100 times, counting how many times the needle hits the line. Collect their data in class or have a student compile the data and present the result. This method is compelling in that students are all contributing to the final result; they might be amused to see that their random work converged. It might also be worth pointing out that their individual results are probably not as good approximations as the whole group's data combined.

*Random walks.* Show, either using a computer or with a table, all the possible walks of length 1, 2, 3, . . . , 10. Plot the walks on a graph to show the nice curve representing the destinations at the ends of the $2^{10}$ different paths. Color those destination points that at some time during the walk returned to the origin, and point out the large percentage of the paths that did return.

### Sample Test Questions

(The Mindscapes are another good source of test questions. All test questions below are available electronically on the Test Bank CD-ROM.)

- Suppose you are in a receiving line meeting hundreds of people. As each one walks up you say, "Isn't your birthday ___?" (and you just pick a random date). Write an expression showing the probability that you will get at least one birthday correct if the receiving line is 300 people long.

- Suppose you have the 26 letters of the alphabet on separate cards in a hat. You pick out a card, write down the letter, put the card back in the hat, mix up the cards, pick out another card, and so on. Write an expression for the probability that on your first six draws your letters spell HAMLET.

- Suppose you flip a coin. If it lands heads up, you go 1 unit left on a line; if it lands tails up, you go 1 unit right. What is the probability that after three flips you will be 3 units from where you started?

- Suppose the World Series is about to begin. Some clairvoyants get together and decide to each predict *different* win-loss patterns for the seven possible games during the coming series. That is, one might predict that the first team will win the first three games, then lose two, then win the next. How many clairvoyants would be needed before being certain that one of them would correctly predict the future of the World Series?

## Sample Lecture Notes for Section 7.3, "Random Thoughts" (1 day)

Question of the Day (to be written on the board for students to think about and discuss before class begins): *What's the most amazing coincidence you or someone you know has ever experienced?*

10 minutes—Ask students to describe their coincidence experiences. Respond to each story with an amazed, "What are the chances?!?"

10 minutes—Suggest several probabilities: 1 in 10, 1 in a 100, 1 in a 1000, and so on. Ask students what probability value would qualify an event for "amazing coincidence" status. If they decide 1 in a 1000 is amazing, do the analysis in the text (page 543) to show that such an event would have a probability of occurring once in a year equal to nearly 1/3. (If they think amazing coincidences should have smaller probability, do the analysis for 1 in a 1000 anyway, just to confirm their intuition.)

15 minutes—Refer to the stock scam on text pages 534–544. Pick a student in the class, say Ed, and pose the following scenario: Ed receives an e-mail one day from the Can't Lose Investment Co. predicting that ACME stock will go up next week. The prediction turns out to be correct! The next week Ed gets another prediction, which also turns out to be correct. This continues for 10 weeks, with 10 correct predictions. The Can't Lose Investment Co. offers its continuing, never-fail investment advice newsletter for the low price of $500, with a money-back guarantee. Ask Ed if he would sign up. Ask the rest of the class how many would sign up. Reveal the scheme. Though the scam is a trick, reinforce the idea that the behavior of any particular stock will be reflected in some sequence of ups and downs over 10 weeks. With all the stock analysts making predictions, the probability that someone will get it right legally may be quite high.

5 minutes—Describe the typing monkeys scenario. Ask for the probability of typing the letter *a*. Now ask for the chances that one short line will be typed (for example, "To be or not to be" or "Et tu, Brute"). Point out that even though the set of all sequences of that length is very large, we would expect every possible sequence to occur eventually if there were enough monkeys typing for long enough. (Acknowledge that "eventually" might be longer than the predicted age of the universe, but that's mathematics!)

10 minutes—Present the Random Journeys scenario (text pages 547–548). Once students accept the idea that the tape recorder will eventually be found, ask, "What's the probability that you will return to your starting position?" Continue the discussion until students recognize that this question is the same as asking for the probability that a flipped sequence has the same number of heads as tails. They will then easily recognize that this probability must be 1.

**Section 7.4  Down for the Count** [Systematic counting]. This section introduces some counting techniques, such as those needed to count the probability of winning the lottery. We do not formally define and make distinctions between combinations and permutations. Instead, we present them as ideas and strategies. We believe that if explained with various basic examples without mentioning threatening notation and formalism, students will have a more solid grasp of the ideas.

The section also discusses the distinction between "and" and "or" and how counting is affected by that distinction. We do not abstract and axiomatize the issues via set theory or logic. Rather, we believe that an informal exposition leads to a clearer, firmer understanding of the issues. The counting techniques are used to demonstrate the probability of the questions posed by Gombauld, whose dice games played a role in the early development of probability theory. The discussion of DNA fingerprinting and the use of DNA evidence in criminal trials is a section that demonstrates counting techniques and the "and/or" issue. Finally, a different kind of counting question is posed and answered, namely, how to count tigers in the wild. The capture-recapture technique gives an example of probability in action.

---

*General Themes*

Looking at simple cases can reveal patterns as well as methods for describing general phenomena.

---

*Mathematical Underpinnings.*  (*Caveat:* These mathematical overviews are usually for the instructor only. The symbols and detailed proofs are not designed for class presentations because mathematical symbols and terminology often present a formidable barrier for students.)

This section gives accessible presentations on how to count basic permutations and combinations with and without repetition. Following is a summary of the results, using formal terminology that the text avoids.

Suppose a process can be completed in two steps, the results of which have no impact on each other. If the first step can be completed in $m$ ways and the second step can be completed in $n$ ways, then the entire process can be completed in $mn$ ways. This is sometimes called the *Product Rule*, which can also be stated in terms of the size of a Cartesian product of two finite sets: $|A \times B| = |A||B|$. This rule can be generalized to processes with any finite number of steps and is applied when counting the total number of outcomes of such experiments as tossing coins or rolling dice. This rule is also

277

used in the jai alai example, in which there are eight ways to complete (win) each of six steps (games), for a total of $8^6$ outcomes. The text also includes an example in which a die is rolled and a penny flipped. Because there are two results from tossing a penny and six results from rolling a die, the total number of outcomes is $2 \times 6 = 12$. Examples from earlier sections include the number of outcomes from flipping a coin six times ($2^6$ outcomes) or rolling a pair of dice ($6^2$ outcomes).

Another application of the Product Rule is as follows: Suppose a set $A$ has $n$ objects. Then the number of ways to arrange $r$ objects chosen from $A$, with repetition allowed, is $n^r$. For example, the number of ways to construct a five-letter "word" using the English alphabet is $26^5$, where "word" means an arbitrary string of five letters that need not have meaning. Another example occurs in the birthday problem. The number of possible birth date selections for a group of 10 people is $365^{10}$ (assuming no one was born on February 29).

If we want to count the number of arrangements of $r$ objects chosen from a set with $n$ elements without allowing repetition, then we are counting *permutations*. The number of permutations of $r$ elements chosen from a set of $n$ elements is the product $n(n - 1)(n - 2) \ldots (n - r + 1)$, or $n!/(n - r)!$. This quantity is sometimes denoted $P(n, r)$. Thus, the number of five-letter "words" composed of *distinct* letters from the English alphabet is $26 \times 25 \times 24 \times 23 \times 22 = 26!/21! = P(26, 5)$. Similarly, the number of possible birth date selections for a group of 10 people, no 2 of whom share a birthday, is $P(365, 10) = 365 \times 364 \times \cdots \times 356$ (again assuming no one was born on February 29).

In the case of lottery tickets, the numbers selected must be distinct but can be chosen in any order—we are counting subsets or selections from a set, not arrangements or permutations. Such a selection is called a *combination*. Because each combination of $r$ elements yields $r(r - 1)(r - 2) \ldots (3)(2)(1) = r!$ permutations, the number of permutations must equal $r!$ times the number of combinations. Thus, the number of combinations of $r$ elements from a set of $n$ elements is $n(n - 1)(n - 2) \ldots (n - r + 1)/r(r - 1)(r - 2) \ldots (3)(2)(1) = n!/(n - r)!r!$, often denoted $C(n, r)$. This value is the familiar binomial coefficient $\binom{n}{r}$, read "$n$ choose $r$," or the number of ways to *choose* $r$ elements from a set of $n$ elements. It is the coefficient of $x^r$ in the expansion of $(x + 1)^n$, because there are $C(n, r)$ ways to choose $x$ rather than 1 from exactly $r$ of the $n$ factors in this product. In the lottery example in the text, the number of possible lottery tickets equals the number of ways to choose six distinct numbers from 1 to 50, which is simply $C(50, 6)$.

Note that students can master these elementary counting techniques without using the formal notation $P(n, r)$ or $C(n, r)$. Of course, permutations and combinations are only the tip of the counting iceberg. Interested students could explore more complex counting techniques in an independent project.

The section continues with a discussion of finding the probability that two unrelated events occur. The text avoids the subtle matter of independent events, and instead allows students to rely on common sense to determine when they can multiply probabilities.

For many students, counting combinations is harder, or less intuitive, than counting permutations. In the Sample Lecture Notes, the topic of combinations is omitted to allow more time to discuss two applications: using DNA evidence and estimating population size. Counting arrangements with repetition is central in examining DNA evidence, as discussed in the text. The section ends with a counting technique used to estimate the size of a set that may otherwise be difficult or impossible to count, such as the size of a population of one species of fish in a lake. This technique offers a great opportunity to discuss proportions.

---

*Our Experience*

*Time spent:* One 50-minute session (or longer if you wish to do a thorough treatment)

*Emphasized items:* Counting tigers

*Items omitted or treated lightly:* Most of the hard counting

*Dangers:* Counting is a challenging issue on which students can easily get bogged down and frustrated.

*Remarks:*

1. We often omit this section in class, because it is an outstanding group research project with good poster session possibilities.
2. This section has several conceptual challenges for students. You will have to decide how deeply you want them to understand such issues as the counting techniques. You might choose to point out only a few of the examples instead of trying for real depth of understanding of the counting techniques themselves.

*Sample homework assignment:* Read the section and do Mindscapes I.3; II.6, 7, 8, 12, 15, 18; III.28, 31.

---

### Sample Class Activities

*Combinations.* In the section, a case is presented of betting on six games of jai alai, each with eight contestants. Because there are eight players per game, there are $8^6$ different possible combinations of winners. In class, you could ask three groups of four students to come to the front of the room. Suppose each group of four has a contest. How many combinations of winners are possible? Having the sets of three step forward and then developing a systematic method for counting could show why you must multiply $4 \times 4 \times 4$ to get all possible combinations.

*Play a lottery.* Take six balls numbered 1 through 6. Each player must choose three correct numbers to win. Ask students to figure out how many possible bets there are. Then ask a student to take three balls at random from the six. Write down the numbers and replace the balls. Draw three more. Write them down. Repeat this process a few more times. Then point out that some of the groups of three are not in ascending order. For each set of three numbers you have written down, write the other five permutations of those three numbers. Then point out that there are $6 \times 5 \times 4$ different ways to draw three numbers from the six, but that each set of three is counted $3 \times 2$ times. Hence, you must divide. Ask students what their probability is of winning. Finally, play the game. Ask each student to submit one or more plays and draw three numbers randomly. You will typically have a winner. Present the winner(s) with a silly prize, such as a bag of chips, or a doughnut (a low-fat doughnut, of course).

*And vs. or.* To explain the distinction between *and* and *or* when used in probabilistic statements, you may find it useful to do several examples involving characteristics of people in the class. For example, you could say, "Please stand if you have brown hair and a blue shirt." Then say, "Please stand if you have brown hair or a blue shirt." This type of example drives home the point that *and* cuts down on the number of people who fit, whereas *or* includes many more people. Thus, it would be easier to find someone who is tall, dark, or handsome rather than tall, dark, and handsome.

Give one die to each pair of students. Ask the class to compute various probabilities: the probability that student A rolls a 6 and student B rolls a 2; the probability that student A rolls a 6 or student B rolls a 2; the probability that student A or B rolls a 6 or 2. Use the table of dice outcomes from Section 7.2 to get the count.

*Counting the opposite.* Frequently, when counting permutations and combinations, it is valuable to use the technique of counting what is missing rather than what is there. Counting how many rolls of four dice contain at least one six is difficult. But counting how many rolls of four dice all avoid a six is easy. This example, which is done in the text, is a good discussion topic.

Other good dice examples come from the game of Yahtzee. You can also refer to the coincidence ideas from Section 7.3. Another good example is to use a crime detection story. You have several witnesses who see various characteristics of the assailant. Police catch a culprit who matches the witnesses' descriptions. What is the probability of him not being the actual assailant? Play this game in class: Have the class choose characteristics: hair color, approximate shoe size, tall or short, and so on. Then secretly choose a person at random to be the assailant. One by one, do the following: Have everyone with the hair color stand. From that group, have everyone sit who has a different shoe size, and so on. When you have been through a few characteristics, the lone assailant will be left standing.

*Counting tigers.* Try the jawbreakers experiment in class, giving a prize of jawbreakers to those who get close. Another way to do this experiment is to have a bag of unmarked pennies. Mark 50 other pennies, add them to the bag, and mix thoroughly. Extract a certain number of pennies, count the marked ones, and deduce an estimate for the total number of pennies.

### Sample Test Questions

(The Mindscapes are another good source of test questions. All test questions below are available electronically on the Test Bank CD-ROM.)

- How many different 3-letter "words" are there (they don't have to be real words) made from a 26-letter alphabet? (Letters can repeat.)

- What is the probability of rolling a pair of dice three times and never getting doubles?

- The lottery in an extremely small state consists of picking two different numbers from 1 to 10. Ten numbered Ping-Pong balls are dropped into a fishbowl, and two are selected. Suppose you bet on 2 and 9.

  **a.** What is the probability that at least one of your numbers matches the ones selected?

  **b.** What is the probability that both of your numbers match the ones selected?

- Suppose the course offerings at The University of Texas at Austin are simplified. Only 20 classes are offered, each meets at a different hour each week in the stadium, and there is no enrollment limit on any course. Each student is required to take exactly four courses. There are no prerequisites, and every student can take any class. UT at Austin has about 48,000 students.

**a.** If the students choose independently, how many different course schedules are possible?

**b.** Will some students have to have the exact same program or not? Justify your answer.

- You have a jar of copper pennies. You want to estimate how many there are, so you take 50 steel pennies, put them in the jar, and shake the jar vigorously until all the pennies are mixed together. Then you reach into the jar and pick out 100 pennies at random. You find that 10 of them are steel. Estimate the number of copper pennies in the jar. Explain your reasoning.

## Sample Lecture Notes for Section 7.4, "Down for the Count" (1 day)

Question of the Day (to be written on the board for students to think about and discuss before class begins): *Is DNA evidence reliable?*

10 minutes—Begin by acknowledging that answering questions about probability often boils down to counting, and *counting is hard!* Look at several examples of simple counting problems that illustrate arrangements with repetition allowed. For example, ask students:

- How many outcomes are possible if you toss a coin 10 times?

- How many outcomes are possible if you roll a die five times?

- How many four-digit PINs are possible if you can use any digit or letter?

Emphasize that these are counting problems in which order matters and repetition is allowed. A certain number of choices are available at each step and the total number of arrangements equals the product of the number of choices at each step.

5 minutes—Continue with examples of arrangements in which repetition is not allowed, such as possible outcomes of a race at a swim meet or arrangements of students in the first row of the classroom. Include related probability questions, such as: What's the probability that Ed sits in a particular seat?

10 minutes—Ask students how many outcomes are possible if they roll a die and toss a coin. List the 12 outcomes on the board. Ask for the probability of rolling a 3 and flipping tails. Then ask for the probability of rolling an even number and flipping heads. Be sure they compute each probability as the quotient of the number of desired outcomes over total outcomes. Relate the results to the product of probabilities for the die roll and coin toss as separate events. Give the formula for computing such probabilities in general, and then do more examples. Caution students that multiplying probabilities is fraught with subtlety and should be done only if they are confident that the events in question are unrelated.

15 minutes—Discuss the DNA example from the text (page 563). If time allows, do a few similar DNA calculations, varying the number of genes and alleles. Acknowledge that the reliability of DNA testing rests on

many nonmathematical factors. Focus the discussion on numerical matters. Use the topic to reinforce this section's counting techniques, as well as probability.

10 minutes—Ask students how they would estimate the number of fish in a lake. Discuss the Counting Tigers example from the text (page 564). Ask students to suggest other applications of the technique. Finish with more examples.

**Section 7.5  Collecting Data Rather than Dust** [Gathering data]. The last three sections in the chapter look at data: collecting data, describing data, and interpreting data. This section examines issues that arise in collecting data, such as getting false data from a survey, avoiding sample bias, or using too small a sample size. One of this section's goals is to encourage students to keep various cautions in mind when they hear statistics.

---

*General Themes*

Gaining insight into a particular phenomenon requires collecting accurate data.

---

*Mathematical Underpinnings.*  (*Caveat:* These mathematical overviews are usually for the instructor only. The symbols and detailed proofs are not designed for class presentations because mathematical symbols and terminology often present a formidable barrier for students.)

Students may not appreciate how difficult data collection can be. Getting reliable answers to potentially embarrassing questions is a compelling topic. The coin-flipping method of data collection is an example of *randomized response*. Both the one-coin and the two-coin approaches offer interesting opportunities for students to think through the logic of a technique as well as to apply some important skills in basic algebra. Explain that although these methods sound very appealing, they effectively reduce the actual data sample by a factor of 3/4 or even 1/2, thus requiring data collectors to interview more subjects. This added expense might make it more efficient to use written, anonymous surveys rather than face-to-face interviews. Polling by telephone, however, may require a randomized response method to improve the accuracy of the data. As pointed out in the text, using a random-response method also allows for the simultaneous collection of data from a large group of subjects in public.

The two-coin approach to randomized response allows, in theory, for complete privacy of the subject's response to a yes/no question. Here's a condensed, generalized version of the technique presented in the text: Before answering the question, the subject flips two coins and, without the interviewer seeing, notes whether the results show two heads. If two heads are showing, the subject answers the question falsely. With any other combination showing, the subject answers truthfully. Because the interviewer doesn't see the outcome of the coin toss, she or he doesn't know if the subject is lying or telling the truth and, thus, doesn't know the true answer to the question for any individual. The interviewer records the total number of yes and no answers. Using probability and algebra, a good estimate of the actual number of subjects whose true behavior corresponds to an answer of yes can be estimated as follows.

Suppose there are $n$ subjects. Using the same variables as those in the text, suppose that a total of $4d$ subjects would answer yes truthfully, and $4s$ would answer no truthfully. Using the two-coin random-response method, approximately 1/4 of all subjects will toss two heads, so 1/4 of the yes subjects and 1/4 of the no subjects will lie. Thus, $3d + s$ subjects will give a yes answer, and $3s + d$ subjects will give a no answer. Now, suppose the total number of yes answers recorded is $a$ and the number of no answers recorded is $b$. Solving the simultaneous equations $3d + s = a$ and $3s + d = b$ will give values for $d$ and $s$. The interviewer can then conclude that approximately $4d$ subjects would answer yes if telling the truth, and approximately $n - 4s$ would answer no.

Sample bias is another issue with many engaging examples. The text presents the classic failure of the *Literary Digest* prediction of the 1936 presidential election. Another example can be found in the data gathered for *The Hite Report on Female Sexuality,* published in 1976. Researcher Shere Hite's work on women's experiences in relationships, marriage, and sex, though eye-opening in many ways, was criticized in part for the manner in which she collected data. Her survey contained open-ended questions, requested essay-style answers, and encouraged responders to answer only those questions they wished to answer. Her mailing list was derived in large part from membership lists of feminist organizations, thus perhaps skewing her data away from women with more conservative lifestyles or opinions. Only 4.5% of her 100,000 surveys were returned. Those who responded did not reflect many typical qualities of the general population of women in the United States at the time. For instance, the percentage identifying themselves as writers was much larger than in the overall population. Similarly, the percentage of responders who identified themselves as holding blue-collar jobs was much lower than in the overall population. Hite's research cannot be entirely dismissed on these grounds, but any claim that her results offer a representative sample of U.S. women's experiences at the time must be questioned.

Emphasize to students that much of the challenge in research, especially in the social sciences, lies in gathering reliable data. Urge them to be skeptical when presented with statistical claims, especially with regard to how data were gathered.

### Sample Class Activities

*Sensitive data.* Try the embarrassing question method. Ask students whether they have used an illegal drug in the past 72 hours (or as a milder alternative, whether they like probability and statistics)—yes or no? It is critical to spend the necessary time to describe the entire data collection procedure of flipping a coin to decide on an answer. Everyone in the class must understand and believe that by answering either way, no one will know the actual reality for any individual student. Stress that *no one* else must see how each student's coin lands!

*Homework activity—sampling bias.* Ask students to go to the library on Saturday morning at 9:00 a.m. to conduct a survey asking: (1) How many hours per week do you study? (2) How many generations of your ancestors have lived in the United States? Then have students do the same survey at a football game or keg party. Ask them to describe and explain any disparity.

*Sample size.* Ask 10 students to state what brand of toothpaste they use. Discuss what they can conclude, based on the results of your survey, about the toothpaste preferences for the entire school.

## Sample Test Questions

(The Mindscapes are another good source of test questions. All test questions below are available electronically on the Test Bank CD-ROM.)

- The criminal justice system wants to collect data about repeat offenders. They gather 1000 ex-cons who were released two years ago and who have not been arrested since then. The court wants to know how many ex-cons have committed felonies since their release. Of course, none of those present want to answer such a question. The interviewer asks everyone in the room to secretly flip a dime. Then he asks them to raise their hands if they flipped a head or committed a felony; 620 people raise their hands. Estimate how many ex-cons committed felonies since their release. To receive credit, you must clearly show all work leading to your conclusion.

- To find out how many students are unhappy with the dining hall food, you distribute a survey at the dinner line. What possible bias could result in your data?

- Describe some of the reasons why the data collected in *The Hite Report on Female Sexuality* are considered suspect.

288

## Sample Lecture Notes for Section 7.5, "Collecting Data Rather than Dust" (1 day)

Question of the Day (to be written on the board for students to think about and discuss before class begins): *Would you answer the following question honestly in public: Have you been drunk in the past 48 hours?*

5 minutes—Begin by acknowledging that understanding human behavior and many other phenomena requires good data. Ask students to brainstorm in small groups about what makes it difficult to gather reliable data. Write their suggestions on the board. Some examples of student responses include: subjects may lie, too many subjects, sample may be biased, hard to design good questions, hard to get surveys returned.

5 minutes—Focus on the Question of the Day. (You can also ask students to suggest an embarrassing question; sometimes answering no is what's embarrassing.) Ask students if they would be more likely to answer an embarrassing question in public if people knew there was a one-in-four chance they were lying. Emphasize that only the students themselves would know if they were telling the truth.

20 minutes—Introduce the two-coin randomized response method. Decide on an embarrassing question on which to poll the class. Have each student flip two coins and note his or her results without telling anyone. Pose the question and have students raise their hands for yes and no according to the prescribed method. Analyze the data following the technique in the text, using tables to display the data.

10 minutes—As a follow-up, apply the two-coin method to collect responses to a benign question, such as, Do you have a younger brother? Analyze the data. Then ask students to answer the question honestly, without flipping coins. Compare values and discuss.

10 minutes—Finish with a discussion of examples of other pitfalls of data, such as biased samples.

**Section 7.6  What the Average American Has** [Describing data]. This section focuses on describing data. The notion of measures of central tendency is introduced; the mean and median are described, as well as variation from the mean. The section also discusses various methods of presenting data, including graphs and pie charts. Students are cautioned on how such presentations can be either helpful or misleading. The section ends with a discussion of various distributions.

This section begins with a provocative, humorous fact about what the average American has. One of the many important points this statistical fact raises is that there might be no element in the entire data set that actually has the properties of the average. It also illustrates how statistical statements can be used to lead people to certain erroneous conclusions. The issue makes for a lively and amusing discussion.

---

*General Themes*

Determining the key features of a large mass of data is powerful and perilous.

---

*Mathematical Underpinnings.*  (*Caveat:* These mathematical overviews are usually for the instructor only. The symbols and detailed proofs are not designed for class presentations because mathematical symbols and terminology often present a formidable barrier for students.)

This section describes both quantitative and qualitative means to present or summarize a set of data. The two basic parameters for measuring central tendency, mean (or average) and median, are easy for students to grasp. Students also enjoy discussing various visual displays of data. What can be trickier for them to grasp is measuring the extent to which values in a data set vary from the center. Although most students appreciate the importance of this quality, computing an appropriate parameter is more technical than for the mean or median. Such a parameter should reflect some sort of average amount by which data values differ from the mean, but we can't simply average these differences because positive and negative values would cancel out and give a misleading result. To avoid unwanted cancellation, the differences are squared before being averaged. Here's a formal definition for a data set with values $x_1, x_2, x_3, \ldots, x_n$ and a mean value of $m$: The *standard deviation* of this distribution is the value $s$ given by

$$s = \sqrt{\frac{1}{n-1}\sum_{i=1}^{n}(x_i - m)^2}.$$

As described in the text, for a normal distribution, $s$ has the property that about 68% of the data will have a value within $s$ of the mean, about 95% will be within $2s$ of the mean, and about 99.7% will be within $3s$ of the mean. (These results follow from calculating the area under the distribution curve that lies over the intervals $[m - s, m + s]$, $[m - 2s, m + 2s]$, and $[m - 3s, m + 3s]$, respectively.)

It may seem curious that the sum in the formula for $s$ is divided by $n - 1$ and not by $n$. To statisticians, the distribution in question is just a sample from a larger population, which has its own mean and standard deviation that are only approximated by the mean and standard deviation of the sample. Using techniques beyond the scope of these notes, one can show that dividing by $n - 1$ instead of $n$ makes $s$ a better approximation to the standard deviation of the entire population. (This is a classic topic in a course on mathematical statistics.)

Another topic in the section is the way data can be distributed. One of the distributions mentioned in the text is too technical for most students. The Poisson distribution models many real phenomena in which an event occurs randomly, such as the number of hits a Web site receives in a minute or the number of earthworms in a square foot of a garden. The classic historical example of a Poisson distribution measured the number of soldiers in the Prussian army killed in a year by a horse kick. The actual distribution for a particular phenomenon depends on the average number of events per time interval. If the average number of events per time interval is $\lambda$, then the probability that you will see $k$ events in a random time interval is given by the formula $P(k) = \lambda^k e^{-\lambda}/k!$. If $\lambda$ is large, the Poisson distribution will approximate the normal distribution, but for small values of $\lambda$, the Poisson distribution will appear similar to the graphs below.

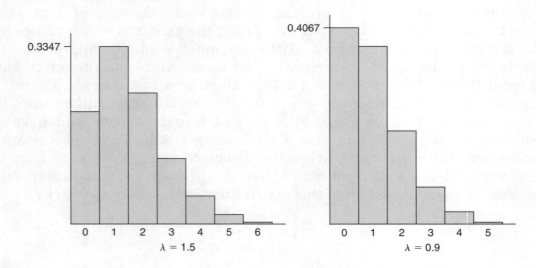

The $x$-axis gives the number of events in a particular time interval, and the $y$-axis gives the probability that exactly $x$ events will occur in a random interval. A motivated student could explore the Poisson distribution as part of an independent project.

---

*Our Experience*

*Time spent:* One 50-minute session

*Emphasized items:* Lakeside School

*Items omitted or treated lightly:* Distributions

*Personal favorite topic or interactions:* Lakeside School

*Remarks:* Pitfalls are great sources of insight, and statistical pitfalls are particularly entertaining and illuminating. Students are quite responsive to the issues in this section.

*Sample homework assignment:* Read the section and do Mindscapes I.4; II.6, 7, 10; III.21, 24.

---

### Sample Class Activities

*Student survey.* Have students compute the Class Survey (text pages 601–602) before beginning this section. Collect the information and distribute the total class data to students. Use this data set to illustrate various statistical issues. Comment on issues of missing data (when some students do not answer all the questions) and issues of correlation.

*Average pet.* Ask students how many pets they each have had. Compute the average. In general, this average will be badly skewed by someone who has had an enormous number. Ask what measure of central tendency would be more appropriate in this case. Perhaps they will conceive of the median.

*Measure up.* Select one student in the class. Have a number of other students measure his or her height. Display the data in various ways. Discuss the idea of a normal distribution.

*E-mail tally.* Ask students to estimate the number of e-mails they get in a day. Gather the data and make a rough histogram. Analyze using quartiles. Use the example to illustrate the five-number summary of a data set.

*Picturing data.* Have students bring in displays of data (graphs, pie charts, and so on) from newspapers or magazines. Select a few to critique with the class.

### Sample Test Questions

(The Mindscapes are another good source of test questions. All test questions below are available electronically on the Test Bank CD-ROM.)

- Here's a histogram showing the number of students who get different numbers of e-mail messages in a given day.

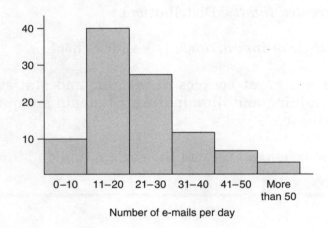

Number of e-mails per day

Estimate the mean and median number of messages received.

- Here's a chart of gas mileage (MPG) for various cars.

| Make and Model | MPG (highway driving) |
|---|---|
| Honda Insight | 66 |
| VW Passat | 24 |
| Chevy Malibu | 30 |
| Pontiac GTO | 21 |
| Toyota Celica | 36 |
| Land Rover Range Rover | 16 |

Compute the mean and the median MPG.

- Suppose you must choose a high school for your child. You are interested in your child getting rich so that he or she can support you in your old age. Explain what you can deduce from statistics about the mean and median of the incomes of graduates of various schools.

- **a.** The median is the best measure of central tendency when there is an outlier. True or False?

  **b.** The median is always less than the mean. True or False?

  **c.** Give an example of a data set where the mean is greater than the median. Compute the mean and median for your data set.

  **d.** Give an example of a data set in which the median is greater than the mean.

## Sample Lecture Notes for Section 7.6, "What the Average American Has" (1 day)

Question of the Day (to be written on the board for students to think about and discuss before class begins): *What do these numbers have in common: 3.23; 0.360; 82; 1.08; 2,500,000?* [Note to instructors: To make this truly realistic, the third number should be the average score on the last quiz or test you gave in class.]

10 minutes—Display some lists of numbers (for example, heights of students in the class, scores on a test, prices of homes sold recently in your neighborhood, number of runs a baseball pitcher has allowed in each of the previous 20 innings pitched). Ask students to describe a single number that might best summarize the data in each case. Define *mean*. Do more examples. Suggest that numbers given in the Question of the Day might be averages: a grade point average, a batting average, a class's quiz average, an ERA (earned run average), an "average" income in a community, respectively.

10 minutes—Present the Lakeside School story. Discuss the failings of the mean as a *measure of central tendency* in this case. Ask for a different value that could be used to measure the center. Define *median*. Do more examples.

15 minutes—Refer again to the numbers in the Question of the Day. Ask students for other suggestions about describing the data. Lead the discussion to pictures and graphs. Display several histograms for one data set, using different scales on the *x*- and *y*-axes. Discuss which image best conveys the variation in the data. Display several histograms showing a bell curve. Introduce *normal distribution*.

15 minutes—Now display several normal distributions that have the same mean but different amounts of variation from the mean. Ask students how they might describe the differences numerically. Suggest quartiles or quintiles, and display corresponding graphs. Announce that there is a single parameter, the *standard deviation,* used to measure the "spread" in a normal distribution. Give several examples, pointing out that, in each case, about 2/3 of the data lie within 1 standard deviation from the mean and that nearly all the data lie within 3 standard deviations from the mean. (Acknowledge that you used a spreadsheet or calculator to find the standard deviation in each case.)

**Section 7.7 Parenting Peas, Twins, and Hypotheses** [Surprising implications of data]. This section presents fundamental ideas used in interpreting data. The first illustration is the famous example of Mendel's peas and the conclusions he was able to draw from statistics and probability. An extended analysis of the spinning pennies example demonstrates the idea of hypothesis testing, while the results of a political poll demonstrate the idea of a confidence interval. The section also includes a discussion of twin studies, a revisitation to the circle of questions involved in coincidences, and a brief discussion of the logical fallacy *post hoc, ergo propter hoc*.

---

*General Themes*

Using statistical methods can help distinguish between events that occur randomly and events that have underlying causes.

---

*Mathematical Underpinnings.* (*Caveat:* These mathematical overviews are usually for the instructor only. The symbols and detailed proofs are not designed for class presentations because mathematical symbols and terminology often present a formidable barrier for students.)

By examining Gregor Mendel's pea experiments, data about a pair of twins separated at birth, and the Dizzy Lincoln experiment from Section 7.1, this section raises questions about how we might decide when something happens due to random chance and when it might have some underlying cause. The idea of hypothesis testing is introduced informally through a detailed presentation of the penny-spinning experiment, which is an excellent demonstration for class use. Point out to students that they can make inferences about data by using graphs rather than by calculating probabilities. A similar example illustrating hypothesis testing is given in the Sample Class Activities and the Sample Lecture Notes.

Emphasize to students that when analyzing a phenomenon or population, statisticians always formulate a hypothesis, called the *null hypothesis*. They test the null hypothesis by doing experiments—gathering data about the phenomenon or population they are trying to study. In the case of spinning pennies, the population can be viewed as the set of all outcomes (number of heads) from spinning $n$ pennies. Following the approach in the text, the null hypothesis for this formulation of the problem is that half the pennies should land heads up.

The section also introduces the idea of statistical significance. Measuring statistical significance depends on many factors: the distribution of the data, how many variables are involved, the size of the sample, and so on. If we restrict our discussion to normal distributions, however, the related ideas of *confidence interval* and *margin of error* are quite accessible. These concepts are mentioned in the text. What follows is a bit more detail.

Consider the experiment of tossing a fair coin 100 times and recording the number of heads. The possible outcomes form a normal distribution, with a graph shaped in the familiar bell curve, as shown.

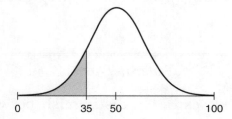

This distribution arises from $n = 100$ trials of the same experiment: tossing a fair coin. Each trial has probability $p$ of success (heads) and probability $1 - p$ of failure (tails). The results of any particular trial do not depend on the results of the previous trials. (Experiments satisfying all these conditions are called *Bernoulli trials*.) The resulting distribution will be a normal distribution with mean $np$ and standard deviation $\sqrt{np(1-p)}$. (These formulas are derived using *expected value,* which is introduced in Chapter 8.)

For our 100 coin tosses, the mean of the distribution is 50 and the standard deviation is $s = \sqrt{100(1/2)(1-1/2)} = 5$. As described in the Sample Lecture Notes for Section 7.6, we know that approximately 68% of all experiments will give outcomes that lie within 1 standard deviation of the mean and that approximately 95% of all experiments will give outcomes that lie within 2 standard deviations of the mean. Thus, if we repeated the 100 coin toss experiment many times, we would expect to get between $50 - s$ and $50 + s$ heads about 68% of the time and between $50 - 2s$ and $50 + 2s$ heads about 95% of the time. To put it another way, if we perform the experiment just once, we are 95% confident that the number of heads will be between 40 and 60. The interval [40, 60] is called a *confidence interval*. The value ±10 is the *margin of error.*

There are many implications of the topics in this section. Encourage interested students to explore these ideas in a project or independent reading.

Please note that the Sample Lecture Notes for this section emphasize quantitive topics: hypothesis testing and confidence intervals. You may wish instead to focus on the broader applications to reasoning implied in the section, such as the fallacy of *post hoc, ergo propter hoc.*

```
                          Our Experience

Time spent: One 50-minute session

Emphasized items: Hypothesis testing and confidence intervals

Items omitted or treated lightly: Mendel's peas

Personal favorite topic or interactions: Confidence intervals

Remarks: This section has many good topics for independent projects.

Sample homework assignment: Read the section and do Mindscapes I.4; II.10,
12, 15.
```

### Sample Class Activities

*Sweet sampling.* Bring in an opaque bowl or bag of small wrapped candies of two colors, say red and green (peppermints work well). Make sure there are significantly more of one color, say red, than the other. State the hypothesis that there are equal numbers of red and green candies in the bag. Have students test this hypothesis by drawing out, one by one, 50 candies. They should return each candy before drawing the next. Record the number of reds. Mark their results on a prepared graph of the distribution for the number of red candies in a sequence of 50 chosen from a bag with equal numbers of red and green.

*Lincoln is still dizzy.* Distribute 100 pennies among pairs of students. Have them spin the pennies and record the number of heads. Pool the data and mark the results on a graph of the distribution that presumes heads will show half the time. Discuss whether the assumption is reasonable.

*Polling politicos.* Ask students to bring to class or recall the results of political polls they've heard, including the margin of error. Discuss the significance of hypothetical changes in the polling results that are within the margin of error or are larger.

*Mendel's peas.* Ask students to create a diagram showing the expected generations of pod colors, starting with one green-green gene parent and one yellow-yellow gene parent. For the first few generations, ask students to compute the probabilities of randomly choosing a green-green plant, a green-yellow plant, and a yellow-yellow plant. Then ask them the probability of

301

choosing a green-podded plant, given that green is dominant. This exercise can help connect the relationship between probability and statistics. You may wish to ask students to work together in class on one or more of the variations of Mendel's experiment found in the Mindscapes. The three-parent alien question (Mindscape III.21) also provides many good variations of the inheritance question.

*Finding soul mates.* Have the entire class answer the 10-question survey from this section. Vote by a raise of hands to find "amazing" coincidences. Discuss how likely or not such coincidences actually are.

*Eerie.* As a class, make up a short survey of no more than five unusual or exotic questions such that a number (at least two) people in the class would have the exact same answers. Now ask how likely it would be for two random people to have all the answers match? Follow up by asking, Suppose that we create a survey of 10,000 equally unusual or exotic questions for these two random people. How likely is it that they will have at least five matched answers?

*Fallacies.* Have the class make up or find several interesting, humorous, or newsworthy illustrations of *post hoc, ergo propter hoc*. For homework, have students find real examples of *post hoc, ergo propter hoc* either in advertising or in news stories. Discuss some in class.

### Sample Test Questions

(The Mindscapes are another good source of test questions. All test questions below are available electronically on the Test Bank CD-ROM.)

- Suppose you have four flowers that have their color genes as follows: two red-white and two red-red. Suppose that red is dominant, so all your flowers are red. If you pick two flowers at random and breed them, what is the probability that the resulting flower will be white?

- With a margin of error of 5% and a 95% confidence interval, a recent poll claims that 47% of Americans support the president's policy toward broccoli growers. The next week, the poll shows support is now 51%. Should broccoli growers jump for joy? Explain.

- Suppose you flip a coin 50 times and get heads only 20 times. Do you think your coin is fair? Explain.

- Between 1997 and 1998, the GPA of the women students at UT at Austin went up and the GPA of the male students also went up; however, the overall GPA of all students went down. Explain what happened.

## Sample Lecture Notes for Section 7.7, "Parenting Peas, Twins, and Hypotheses" (1 day)

Question of the Day (to be written on the board for students to think about and discuss before class begins): *If you flip a coin 100 times and see heads only 41 times, how confident are you that your coin is fair?*

5 minutes—Pull a penny out of your pocket and announce to the class that you're going to toss it 100 times and record the number of heads. Then decide that this would take too long. Instead, you'll have 10 students each toss a penny 10 times and pool the results. Distribute pennies, conduct the experiment, and collect the data.

10 minutes—Ask the class to assume that the results are from 100 tosses of a single penny. Based on the results, ask students if they would consider the penny fair. (If you got very close to 50 heads, ask students if they feel 100% sure that the penny is fair.) Ask how many heads students would have to get to feel 95% sure that the penny is fair. Write down several suggestions. Motivate the idea of a confidence interval. Point out that to be 95% confident in a result, they should expect that when repeated many times, an experiment should yield that result about 95% of the time.

10 minutes—Remind students about the normal distribution. Display a bell curve for the 100 coin toss experiment. Point out that the mean is 50. Reintroduce the idea of standard deviation. Announce that for this distribution, the standard deviation is 5; so, as discussed in Section 7.6, about 2/3 (68%) of the outcomes lie within 5 units of 50 and nearly all (99.7%) lie within 15 units of 50. Announce that 95% of results lie within 10 units of 50. Thus, repeating the 100 coin toss experiment many times with a fair coin should give between 40 and 60 heads about 95% of the time. Conclude that experimental results inside the 95% confidence interval would seem to support the hypothesis that the coin is fair.

5 minutes—Announce that the class will now consider another scenario in which they will test a hypothesis. Hold up an opaque bag and say that it's filled with red and green peppermint candies. The goal is to determine whether there are equal numbers of each color. (Acknowledge that one approach would be to empty the bag and count, but suggest that you want to develop a technique that could be applied to a more general situation, such as checking the sizes of two fish populations in a lake.) Write the hypothesis on the board: *The bag contains equal numbers of red and green.* Then sketch or display the binomial distributions of the number of red candies drawn in a sequence of 10, 20, or 50.

5 minutes—Ask for suggestions about how to test the hypothesis, leading to experiments. Have a student draw 10 candies (one-by-one with replacement; no peeking!), and note the number of reds. Mark the results on a distribution graph. Ask students if the experiment supports the hypothesis.

10 minutes—Have another student draw 10 more candies. Combine the results with the first drawing and mark them on a second distribution. Discuss the implications for the stated hypothesis. Have three more students draw 10 candies each, combine with previous results, and mark on a third distribution. Discuss. Then pass the candies around!

5 minutes—Ask students which of the three experiments (10, 20, or 50 candies) was more convincing. Discuss the importance of sample size. Finish by acknowledging that statisticians have very sophisticated techniques for testing hypotheses.

# Chapter 8   Deciding Wisely

**8.1   Great Expectations** [Expected value]
**8.2   Risk** [Deciding personal and public safety]
**8.3   Money Matters** [Compound interest]
**8.4   Peril at the Polls** [Voting]
**8.5   Cutting Cake for Greedy People** [Dividing scarce resources]


## I.   Chapter Overview

This chapter looks at decision making. Although underlying techniques range from computation and probability (expected value), to logic (voting), to geometric reasoning (dividing scarce resources), each section emphasizes the importance of starting with simple examples to build understanding. All the topics raise questions relevant to most people at some point in their lives, underscoring the important role mathematics can play in making choices in everyday life.

---

**Dependencies**
Sections 8.1 and 8.2 use material on probability from Sections 7.1 and 7.2. One example at the end of Section 8.3 (Compound interest) uses expected value from Section 8.1. The bulk of Section 8.3 and all of Sections 8.4 and 8.5 are completely independent.

---

**Sections to Assign as a Reading Assignment or Term Project**
Sections 8.4 and 8.5 are particularly appropriate for term projects. Any sections that are not done in class are good paper or extra credit topics.

---

*Reminders*
1. Class activities are suggested for each section. These can be done with students working in groups of two or three and then gathering their responses or as one global class discussion.
2. Before introducing a new topic, engage students with enticing questions, which they quickly discuss with neighbors; solicit responses within a minute. Use the responses as the springboard for the introduction.
3. We have included more sample class activities than you will probably want to use. Choose only the ones that are right for you and your class.
4. Each section can be treated in more or less depth. It is frequently a good idea to omit difficult technicalities in order to treat the main ideas well and then move on.

## II. Section-by-Section Instructional Suggestions

**Section 8.1 Great Expectations** [Expected value]. The idea of expected value is presented in the context of gambling games, the lottery, and a philosophical paradox known as Newcomb's Paradox.

---

*General Themes*

Considering choices in quantitative ways can help us make important decisions.

---

*Mathematical Underpinnings.* (*Caveat:* These mathematical overviews are usually for the instructor only. The symbols and detailed proofs are not suitable for class presentations because mathematical symbols and terminology often present a formidable barrier for students.)

This section presents a simple formula for expected value using the outcomes and associated probabilities of a game or experiment in which outcomes have a numerical value. A more formal definition uses the idea of a random variable: A *random variable, X*, is a variable that takes on values equal to outcomes from a particular game or experiment. In the text, because all experiments for this section have only a finite number of outcomes, any associated random variable could take on only a finite number of values. Such a random variable is called a *discrete random variable*. Random variables can also take on values from a continuum, such as a real number chosen at random from between 0 and 1, in which case the random variable is a *continuous random variable*.

Expected value is traditionally defined in terms of a given random variable. For example, if $X$ is the random variable corresponding to the roulette experiment introduced at the beginning of the section, then $X$ can take on the value \$1 (if Ms. Scarlett wins) or –\$1 (if Ms. Scarlett loses). We calculate the expected value of $X$ as

$$E(X) = (\$1)\text{Prob}(X = \$1) + (-\$1)\text{Prob}(X = -\$1) = (\$1)\left(\frac{18}{38}\right) + (-\$1)\left(\frac{20}{38}\right) = -\$\frac{2}{38}.$$

In general, the expected value of a discrete random variable $X$ is

$$E(X) = \sum x \cdot \text{Prob}(X = x),$$

where the sum is taken over all possible values of $X$. Expected value is the mean of all possible outcomes, given their respective probabilities. Thus, it is sometimes denoted $\mu$.

306

Newcomb's Paradox is a great topic for many reasons. The idea of a logical paradox is itself a compelling one. The two choices offered in Newcomb's Paradox seem equally appealing and yet cannot both occur. Students often enjoy the struggle of trying to reconcile this seeming inconsistency. Some students may decide that there is no paradox—taking only the Zero-or-Million-Dollar Box is the clear choice, at least when the psychologist's prediction record is 90% accurate. Whatever your students think, this topic offers not only a great example of expected value but also an opportunity to broaden the discussion of a quantitative topic to include issues of human behavior, free will, and causality. There are many analyses and discussions of Newcomb's Paradox on the Web. Encourage your students to browse these analyses to expand the discussion further.

---

*Our Experience*

*Time spent:* One-and-a-half 50-minute sessions

*Emphasized items:* Newcomb's Paradox and insurance issues

*Personal favorite topic or interactions:* Having students discuss and vote on Newcomb's Paradox.

*Remarks:* Students are intrigued by the idea that insurance policies are games of chance that can be analyzed accordingly.

*Sample homework assignment:* Read the section and do Mindscapes I.3; II.7, 13, 15, 22.

---

### Sample Class Activities

*Dice game.* Take a die, and propose a game to a member of the class. The class member pays $1 to play. You then roll the die. If the number is a 6, you pay the player $5. If you do not roll a 6, you keep the money. To make the game go faster, bring 10 or more dice and roll them all at once. Repeat the experiment 10 or 20 times, each time simply recording the number of 6's. Compute how much your opponent owes you and vice versa. Compute the expected value of playing this game. The key to making this an effective demonstration is that the expected value per game becomes increasingly accurate as the number of trials increases. A computer model is also ideal. Or you can have students do the experiment at home and simply bring the data to class.

*Roulette.* A standard roulette wheel has the numbers 1–36 plus 0 and 00. The payoff for betting a dollar on a number and getting that number is $36 (your $1 bet plus $35 winnings). Point out that the 0 and 00 in roulette are what account for the casino always winning in the long run; otherwise, the payoffs would make it an even money (*fair*) game. In other words, if there are 38 equally likely outcomes, and they pay off only 36 times the bet on a number, then the expected value is negative for the player. For example, if $1 is bet on each number, including 0 and 00, $36 would be returned, and the bank would keep $2. Have students compute the expected value for each $1 bet to see how much the bank would win if $1,000,000 were bet during a day.

*Expected value of studying.* Pose the following situation to students: Suppose you are in a class where you are expected to learn 1000 facts. On the test, 100 of those facts will be asked. It takes you 6 minutes to learn a fact. What is the expected value of studying for an hour in terms of points gained?

*Newcomb's Paradox.* Newcomb's Paradox is a great topic for lively discussions. After a careful presentation of the situation, solicit student opinions. In general, some students will not see any reason to choose the one-box option. Make the one-box option more attractive: Describe a scenario of a line of 100 participants, with the one boxers emerging wealthy and the two boxers poor. Picturing expected value as the aggregate outcome is valuable.

Perhaps the most important goal of the discussion is for students to try thought experiments to find a way to understand an issue. In particular, encourage students to think of alternative variations that may help clarify the issues. For example, imagine there is much more or much less money in either the transparent box or the opaque box. Other variations include being poor or rich or waiting years between the time the money is placed in the box and the time the decision is made. Newcomb's Paradox is a great vehicle for encouraging students to explore a question by modifying the premises.

Ending the discussion with a vote is often illuminating. Usually it is possible to gauge the class and strengthen the supporting statements on the one-box or the two-box side as necessary to end the day with the class almost exactly evenly divided between one boxers and two boxers. This example is a good illustration that there may not be one correct answer, and the complete resolution of the paradox remains unknown.

*Homework.* Ask students to bring examples from their lives of times when expected value could have played a role.

## Sample Test Questions

(The Mindscapes are another good source of test questions. All test questions below are available electronically on the Test Bank CD-ROM.)

- Suppose you play a game consisting of someone handing you a dime and a quarter. After you flip the dime and quarter, you keep all the coins that come up heads. What is the expected value of this game?

- You play a simple game with a friend. You flip two coins. If they both come up heads, your friend gets a dollar; otherwise, you keep the money your friend bet. How much should you charge your friend to play so that the game is a fair game?

- You find a raffle ticket on the ground. The prize is $1000, and there are 2000 tickets. What is the expected value of your found ticket?

- You own a $5000 car. The probability that your car will be stolen next year is 0.02, but the probability that your car will be broken into and the radio stolen is 0.10. The damage of such a break-in and theft is $200. Cheatem's Insurance Company offers you a policy that would cover both of the above thefts for a cost of only $150.

  **a.** What is the expected value of this insurance policy?

  **b.** What is a fair price for such a policy?

- Paradoxes describe situations that can be viewed from two different angles to give contradictory impressions. In Newcomb's Paradox, the plausible idea of taking the contents of both boxes because the money is there or it is not is pitted against the also plausible argument based on the idea of expected value. In this discussion, suppose the experimenter is right 90% of the time. You are presented with two boxes: one transparent box containing $100 and one opaque box containing either $0 or $10,000,000. Suppose you varied the situation by increasing the amount of money in the transparent box. At what dollar figure would the paradoxical aspect of Newcomb's Paradox cease to exist? Please explain briefly.

## Sample Lecture Notes for Section 8.1, "Great Expectations" (2 days)

### Day One

Question of the Day (to be written on the board for students to think about and discuss before class begins): *If your bicycle is worth $1000, does it make sense to buy theft insurance that costs $50 per year?*

10 minutes—Describe the following game, to be played with a standard deck of cards including two jokers. (This game is very similar to the roulette game described in the text, with the two jokers representing the green slots 0 and 00.) You pay $5 to play, then draw a card. If it's a black card, you win $10; if it's a red card or a joker, you win nothing. Each time you play, you replace the card you drew, and the deck is reshuffled. Have students work in small groups to guess what the average winnings would be if they played the game *a lot*—say, 5400 times.

5 minutes—Write some guesses on the board, then analyze the game. Because 26 of the 54 cards are black, on average about 2600 of the 5400 games will result in winning $10, for a total of $26,000. However, 28 of the cards are red or a joker, so on average the other 2800 games will yield $0 in winnings. Because you pay $5 to play each time, the total cost is $27,000. Thus, your net winnings, on average, will be –$1000. Not a good game!

10 minutes—Point out that the *average loss per game* was $1000/5400, or $5/27. Introduce the idea of *expected value*. In this case,

$$EV(\text{game}) = (\text{Net gain if black}) \times (\text{Probability of drawing black})$$
$$+ (\text{Net gain if red or joker}) \times (\text{Probability of drawing red or joker})$$
$$= (\$10 - \$5) \times (26/54) + (\$0 - \$5) \times (28/54) = -\$5/27.$$

Do several more examples, including games with positive, negative, and zero expected value and games with more than two outcomes.

5–10 minutes—Introduce the following game: The player rolls a fair die. If it comes up 1, she wins $1; if it comes up 2, she wins $2; and so on. Have students compute the expected value of the payoff. Ask what the player should pay for the game to be fair. If desired, distribute dice to pairs of students and have them play the game several times. Pool the resulting data to reinforce the idea that the average payoff will be close to the expected value if the game is repeated many times.

10–15 minutes—Ask students to discuss in small groups how the Question of the Day relates to expected value. Offer suggestions if needed. For example, set some probability that the bicycle will be stolen; assume that if the bicycle is stolen, then the insurance policy will pay $1000; compute the expected

311

value of the insurance policy; and so on. Compute several expected values based on different theft probabilities, such as,

$$EV\text{(insurance policy)} = \text{(Payoff)(Bike stolen)} + \text{(Payoff)(Bike not stolen)}$$

$$= (\$1000 - \$50)(0.10) + (-\$50)(0.90) = \$50.$$

(Note that the policy is fair if the probability of theft is 5%.)

5 minutes—Conclude with the observation that an insurance policy, like a game of chance, could be seen as a good deal if it has a positive expected value. If time permits, ask students what factors other than expected value might be important in considering an insurance policy. (For example, if their bike were stolen, could they afford to replace it or get along without it?)

## Day Two

Question of the Day (to be written on the board for students to think about and discuss before class begins): *If someone offered you a box that contained $1000, would you ever turn it down?*

10 minutes—Introduce Newcomb's Paradox. Be very clear from the start that the person running the show is a wealthy psychologist with excellent predictive powers. Acknowledge that, of course, she keeps her predictions private, using it only to decide what to place in the Zero-or-Million-Dollar Box. When discussing expected value, be clear that because there are two possible choices, each has its own expected value:

$$EV\text{(Zero-or-Million-Dollar Box only)} = (\$1,000,000)(0.9) + (\$0)(0.1)$$

$$= \$900,000$$

$$EV\text{(choosing both boxes)} = (\$1000)(0.9) + (\$1,000,000)(0.1)$$

$$= \$101,000.$$

Be sure all students understand the scenario, and then let the discussion begin.

15 minutes—Have students discuss the choices in small groups and then share responses with the class. If most of the class seems to be leaning toward taking both boxes, describe the scene in the text: 100 people in line ahead of you do the experiment; for 90 of them, the psychologist predicts their choice correctly. If most people seem to be leaning toward taking only the Zero-or-Million-Dollar Box, raise the question of free will. It's particularly fun to see a class evenly divided on this paradox.

If time permits, ask students to consider how their choice might change if the dollar amounts or probabilities were different. Ask if there are other factors that could affect the decision, such as being well off or desperately poor.

312

**Section 8.2 Risk** [Deciding personal and public safety]. This section uses expected value and other techniques to analyze risk. Examples include HIV testing and airline safety. Other examples illustrate ways to measure loss of life expectancy or cost per life saved, two additional techniques that should provoke discussion. Throughout the section, examples illustrate that people often overestimate the riskiness of some choices and underestimate the costs of others.

---

*General Themes*

Careful analysis can put risks into perspective and help lead to wise choices in both behavior and public policy.

---

*Mathematical Underpinnings.* (*Caveat:* These mathematical overviews are usually for the instructor only. The symbols and detailed proofs are not suitable for class presentations because mathematical symbols and terminology often present a formidable barrier for students.)

Two of the major examples in this section are HIV testing and airline safety. The first offers an excellent example of conditional probability; the second, an example of unintended consequences.

The HIV testing example in the text uses the following data:

- Of the 280,000,000 Americans, 500,000 are HIV-positive;

- 95% of those with HIV will test positive (475,000 out of 500,000);

- 99% of those without HIV will test negative (276,705,000 out of 279,500,000).

As acknowledged in the text, the percentages suggest that the HIV test used is very reliable. Further analysis shows that because being HIV-positive is quite rare, a positive test result requires careful interpretation. In particular, the third data item above is equivalent to stating that 1% of those without HIV will test positive (2,795,000 out of 279,500,000), so the test will result in a large number of false positives. Under these assumptions, what are the chances that someone with a positive test result actually has HIV? The analysis is quite straightforward if conditional probability is used, though this approach may be too quantitative for many students.

Here's a definition of conditional probability: Given two events, $A$ and $B$, in a probability space, the *conditional probability of B given A*, denoted $P(B|A)$, is $P(B|A) = \dfrac{P(A \cap B)}{P(A)}$. In the HIV example, event $A$ is "positive test result," and event $B$ is "has HIV." In a condensed version of the computation in the text (page 648), we compute the relevant probabilities as follows:

$$P(A \cap B) = \frac{475,000}{280,000,000} \approx 0.00169643$$

$$P(A) = \frac{475,000 + 2,795,000}{280,000,000} \approx 0.0116786.$$

Thus, we have $P(B|A) = \dfrac{0.0016963}{0.0116786} \approx 0.14526$, or less than 15%.

This example clarifies the definition of conditional probability. If events $A$ and $B$ are being considered in a finite, equiprobable probability space, then $P(B|A) = \dfrac{|A \cap B|}{|A|}$. In the HIV example, this quotient is $\dfrac{475,000}{475,000 + 2,795,000} \approx 0.14546$, as computed above. This alternate formula for conditional probability follows simply from the formulas for probabilities in a finite, equiprobable space:

$$P(B|A) = \frac{P(A \cap B)}{P(A)} = \frac{\dfrac{|A \cap B|}{|S|}}{\dfrac{|A|}{|S|}} = \frac{|A \cap B|}{|A|}.$$

Airline safety is a compelling topic, though somewhat delicate since the events of September 11, 2001. Independent of terrorism, issues related to the safety of air travel are complex. The text example on unintended consequences of higher ticket prices resulting from safety improvements is quite provocative. It is also possible to look at air safety from entirely different perspectives. For instance, because most airplane accidents happen during takeoff or landing, measuring deaths per flight rather than deaths per mile might be a revealing statistic for comparing the safety of air travel versus other modes of travel. Because airplanes travel about 10 times faster than cars or trains (500 mph vs. 50 mph), you could also examine deaths per hour traveled in each mode of transportation. Some of these approaches have been undertaken by Arnold Barnett, a professor at MIT. Students interested in public policy will find many project or paper topics in this and related examples in the text.

### Sample Class Activities

*Phone numbers.* Suppose you took 10,000 telephone numbers, and you sought those that ended with 826. The probability of anyone having such a phone number is 1 in 1000. Suppose you told someone that they have the last two digits correct—26. What is the probability that that person has all three of the last digits correct? Ask students to discuss the analogy with the HIV-testing scenario.

*Conflicting tests.* Pose the following situation: You will give students two tests at the end of the semester. One test will be an all-day test with 1000 questions. The other will be a one-hour test with 100 comparable questions. Imagine that on the big test, a student scores 85%. On the shorter test, the same student scores 50%. What score should the student receive in the class? After discussion, you might encourage students to look at the totality of the 1100 questions.

*Accidents near home.* The comedian Steven Wright had a routine dealing with safety that went something like this: "I heard that most automobile accidents happen within 10 miles of your home. So I moved." Ask students why most automobile accidents happen within 10 miles of home and what information they would need before they could decide whether to be surprised at how high or low that number is.

### Sample Test Questions

(The Mindscapes are another good source of test questions. All test questions below are available electronically on the Test Bank CD-ROM.)

- In the universal HIV-testing scenario, we found that under the assumptions that 1 out of 1000 people is HIV-positive and that the test is

315

99% accurate, the chance of being HIV-positive after receiving a positive test result would be only about 9%. What would be the probability change if the incidence of HIV were 1 in 100 instead of 1 in 1000? Suppose the incidence of HIV were only 1 in 10,000. How would the probability change?

- Suppose you go to a school where the school colors are red and black. Everyone wears red or black every day, with 90% of the students wearing red. You see someone across the room. It's a little dark, so your ability to discern colors is only 80% accurate. You think the person you are looking at is wearing a black shirt. What is the probability that the person is really wearing a black shirt?

- Why might it be prudent to reduce the safety standards for airline travel?

- Suppose 1 in 10,000 schoolchildren dies each year from a certain disease. A vaccination program costing $4.5 million would prevent 90% of these deaths. Assuming there are 50 million schoolchildren in the United States, compute the cost per life saved if the vaccination program were implemented.

# Sample Lecture Notes for Section 8.2, "Risk" (1 day)

Question of the Day (to be written on the board for students to think about and discuss before class begins): *An HIV test is 95% accurate for infected people. Suppose your roommate's test result is positive. What are the chances your roommate has HIV?*

5 minutes—Announce that the topic of the day is risk. Ask students to suggest some choices involving risk to life and limb (for example, smoking, skydiving, unprotected sex, drinking and driving, and so on). Acknowledge that even everyday choices, such as driving to work or jogging along a road, include some risk. Using one of their suggestions, ask students what information they would need to assess the risk involved. (Choose an example that lends itself well to the data and probability approach of the section. For example, weighing the risk of jogging along a road might require data on the number of pedestrians hit by cars along that road.)

5 minutes—Move into a discussion about the need for data: The number of people who die or are injured each year related to a particular choice or circumstance, the effectiveness of a vaccine or a diagnostic test, and so on. Introduce the subject of HIV and AIDS. Pose the Question of the Day, and ask students to suggest probabilities. Write their guesses on the board.

15 minutes—Ask students what information they need to answer the Question of the Day carefully. Lead to relevant data from the text and to analysis of the test results:

- Of the 280,000,000 Americans, 500,000 are HIV-positive;

- 95% of those with HIV will test positive (475,000 out of 500,000);

- 99% of those without HIV will test negative (276,705,000 out of 279,500,000);

- 1% of those without HIV will test positive (2,795,000 out of 279,500,000).

Thus, of those with positive test results (475,000 + 2,795,000), only 475,000 (14.5%) have HIV.

(This is a good opportunity to demonstrate calculating with percentages. To reinforce these techniques, answer the question again with different percentages.)

5 minutes—Briefly discuss the implications of the HIV test example for making decisions about health care and the treatment of illness. Ask students what they would do if they tested positive for a rare illness.

15 minutes—Shift the discussion to airline safety. Ask students to guess how many people die each year in nonterrorist related airline accidents in the United States. Then ask them how many die in car accidents each year in the United States. Write their guesses on the board. Proceed with the example from the text on how increasing airline safety might have unintended consequences.

5 minutes—End with a more general discussion about how cost-benefit analysis might be used in public policy decisions. Display some of the data from the text on cost per life saved to provoke further discussion outside of class.

**Section 8.3 Money Matters** [Compound interest]. This section discusses compound interest, mortgage amortization, and annuities. The final presentation applies expected value and future value in an example illustrating decision making about investments.

---

*General Themes*

Looking at simple cases can reveal patterns that lead to greater understanding of repeating processes.

---

*Mathematical Underpinnings.* (*Caveat:* These mathematical overviews are usually for the instructor only. The symbols and detailed proofs are not suitable for class presentations because mathematical symbols and terminology often present a formidable barrier for students.)

This section's basic lesson is that the power of compound interest is really the power of exponential growth. Some students will have been told the formula for the *future value A* of a principle $P$ invested at an interest rate $r$ compounded $t$ times per year for $y$ years: $A = P(1 + r/100t)^{ty}$. However, we note that (1) they will not recall even seeing it, and (2) they will not know what it means (see text pages 702–703). In fact, exploring the effect of compound interest is a great way for students to compare different kinds of growth: additive, multiplicative, or exponential. Although the calculations reinforce algebra skills, it may be even more worthwhile to simply work through examples using a spreadsheet program or Web site calculator. Students can gain wisdom and confidence about financial matters by comparing examples, even if they don't understand every detail about where the formulas come from. Many valuable life lessons can be gained by exploring the terminology, questions, and issues involved in interest-related problems.

For students who are already comfortable with compound interest as applied to a single initial investment of principle, move on to annuities. The text gives a single example involving a college savings account (pages 668–669). More generally, suppose the final savings goal is $A$, the interest rate is $r$ compounded *monthly*, and payments will be made monthly for $y$ years. The monthly payment $P$ needed to reach the goal is $P = \dfrac{A(\frac{r}{100})}{(1+\frac{r}{100})^{12y}-1}$. Solving this equation for $A$ in terms of $P$ gives a formula for the future value of the annuity: $A = \dfrac{P[(1+\frac{r}{100})^{12y}-1]}{(\frac{r}{100})}$. Doing examples with annuities can help students appreciate the value of saving regularly for long-term goals.

Students may not yet appreciate the burden of a mortgage, but many should be familiar with student loans. The borrowing and repayment examples offer an accessible introduction to the rather complex matter of computing monthly loan payments. Deriving the general formula is complicated, though more algebraically confident students will be able to comprehend the amortization formula as follows. The monthly payment $P$ on a mortgage of $M$ dollars borrowed at $r$ percent over $n$ years is

$$P = \frac{M\left(1 + \frac{\frac{r}{100}}{12}\right)^{12n}\left(\frac{\frac{r}{100}}{12}\right)}{\left(1 + \frac{\frac{r}{100}}{12}\right)^{12n} - 1}.$$

Of course, students should realize that, in practice, these computations are usually done using spreadsheets. There are many mortgage calculation Web sites as well, which would make for useful class demonstrations. You may also consider demonstrating "Savings Calculator" and "Loan Calculator" from the Interactive Explorations CD-ROM. Emphasize that the important parameters are the amount borrowed, the interest rate, and the length of the loan.

Many students will have little or no experience with calculus; however, exponential growth is, of course, related to the most fundamental of all exponential functions: $e^x$. Returning to the future value formula of a one-time investment, $A = P(1 + r/100t)^{ty}$, we can take a limit of $A$ as $t$ goes to infinity to reflect interest *compounded continuously*. Computing this limit requires use of the natural logarithm and exponential functions, as well as l'Hôpital's Rule. Thus, it is typically reserved for a calculus class. As a reminder for instructors, we provide here the calculation with $P = 1$, $y = 1$, and $r$ given in decimal form. This derivation is definitely not appropriate for students.

To compute, $\lim_{t\to\infty}(1 + \frac{r}{t})^t$, we let $z = (1 + \frac{r}{t})^t$. Then we find $\lim_{t\to\infty}(\ln z)$ as follows:

$$\lim_{t\to\infty}(\ln z) = \lim_{t\to\infty}\ln(1 + \tfrac{r}{t})^t = \lim_{t\to\infty} t\ln(1 + \tfrac{r}{t}) = \lim_{t\to\infty}\frac{\ln(1 + \frac{r}{t})}{\frac{1}{t}}.$$

This last limit is an indeterminant form of the type 0/0, so we apply l'Hôpital's Rule:

$$\lim_{t\to\infty}\frac{\ln(1 + \frac{r}{t})}{\frac{1}{t}} = \lim_{t\to\infty}\frac{\frac{d}{dx}(\ln(1 + \frac{r}{t}))}{\frac{d}{dx}(\frac{1}{t})} = \lim_{t\to\infty}\frac{\frac{1}{(1 + \frac{r}{t})}(-\frac{r}{t^2})}{-\frac{1}{t^2}} = \lim_{t\to\infty}\frac{1}{(1 + \frac{r}{t})}(r) = r.$$

Thus, because $\lim_{t\to\infty}(\ln z) = r$, we have $\lim_{t\to\infty} z = e^r$. Therefore, if interest is compounded continuously, an initial deposit $P$ will grow to $Pe^{rt}$ in $t$ years.

320

Because banks tend to offer compound interest monthly or, perhaps, daily, this formula is more useful as an approximation or in theoretical applications.

Calculus also comes into play in more complex investment scenarios. The previous examples compute the future value of a fixed principle invested at the beginning of a time interval. A principle $P$ could instead be invested periodically throughout the time interval. To apply calculus, suppose that over a time interval from $a$ to $b$, you deposit $f(t)$ dollars *continuously* into an account paying an interest rate $r$ compounded continuously. The final account balance will be $\int_a^b f(t)e^{kt}dt$. If $f(t)$ gives the amount of a natural resource used and if $r$ is the rate at which use increases each year, then this formula can be used to approximate the accumulated use of the resource. Even students who have not studied calculus may appreciate knowing that the formulas in the text can be generalized to make complex computations in many fields other than accounting.

---

*Our Experience*

*Time spent:* One 50-minute session

*Emphasized items:* Compound interest, annuities, and loan payments

*Items omitted or treated lightly:* Opportunity costs and average annual rate of return

*Dangers:* Don't get too bogged down in calculations. Use a spreadsheet program to do many examples comparing different interest rates, different investment or repayment time periods, or variations in other parameters.

*Sample homework assignment:* Read the section and do Mindscapes I.2; II.6, 8, 12; III.16.

---

### Sample Class Activities

(*Note:* Many of these activities require a calculator or spreadsheet display.)

*Comparing investments.* (Calculators required.) Divide students into pairs, and give each pair $500 in play money (or tell them to imagine they have $500). Have them use a calculator to compute the amount they would have if they invested the money at different interest rates, compounded annually or quarterly, for one or two years. You could also assign different pairs of students different investment parameters, collect the data, and then compare the results as a class.

*Comparing strategies.* (Calculators required.) Have students work in pairs. Announce that each pair has to save $5000 (*A*) for the down payment on a car. Have students use the formula $P = \dfrac{A(\frac{r}{100})}{(1+\frac{r}{100})^{12y} - 1}$ to explore different values of *r* and *y* to discover differences in savings strategies (value of *P*) and constraints to reach their goal.

*Demonstration.* (Spreadsheet display required.) Demonstrate a spreadsheet or Web site calculator for computing future value, annuity payments, mortgage payments, and so on. Ask students to suggest the kinds of calculations they are curious about, such as saving for a car or repaying a student loan.

*The real world.* (Homework assignment.) Have students bring to class advertisements about mortgage rates, car loans, or savings opportunities such as CDs. Do comparisons in each category.

### Sample Test Questions

(The Mindscapes are another good source of test questions. All test questions below are available electronically on the Test Bank CD-ROM.)

(*Note:* For many of these questions, it makes sense to provide students with a sheet of formulas or to let them use programmable calculators. Even if formulas are provided, students still must demonstrate a reasonable understanding of the material in order to choose the correct formula and use it properly.)

- Which investment generates more interest: $1000 invested for 2 years at 3% compounded annually, or $1000 invested for 3 years at 2% compounded annually?

- If you invest $500 in an account that pays 2.5% interest compounded quarterly, how much money will be in your account at the end of 2 years?

- To buy a car, you borrow $8000 for 3 years at 6% annual interest. What is your monthly payment?

- You start a new job and decide to save the same amount of money each month for a down payment on a condominium. If you want to have $30,000 in 5 years, and you plan to invest at an annual interest rate of 4%, what should your monthly savings payment be?

## Sample Lecture Notes for Section 8.3, "Money Matters" (1 day)

(*Note:* This lecture will be enhanced by the use of a computer display with a spreadsheet program or prepared transparencies.)

Question of the Day (to be written on the board for students to think about and discuss before class begins): *Adam and Eve invest one penny in a bank account that pays 3% compounded annually. How much money will the account hold after 1000 years: $10,000? $100,000? $1 million? $1 billion?* (The correct answer is $68,742,400,000. See text pages 666–667.)

1 minute—Have students guess the answer to the Question of the Day. Write their guesses on the board for later discussion.

15 minutes—Introduce simple interest, and do a few examples. Move to compound interest, and work through several examples. (Include 3% compounded annually in at least one example.) Derive the basic formula $A = P(1 + r/100t)^{ty}$. If you can display a spreadsheet program, use it to demonstrate the effect of different interest rates, compounding intervals, and so on. Alternatively, you can have students use programmable calculators. Compute the true answer to the Question of the Day.

15 minutes—Pose another question: How much should be saved each month for 4 years, in an account paying 3% interest, to accumulate $5000 for a down payment on a car? Assume the bank compounds the interest monthly and the same amount is to be saved each month. Have students guess the answer: $100? $200? $300? Derive the formula for the monthly payment $P = \dfrac{A(\frac{r}{100})}{(1+\frac{r}{100})^{12y} - 1}$. (Or you can simply present the formula to the class, and use the extra time for more examples.)

10 minutes—Pose a related question: How much money students would have if they saved $100 each month for 4 years in an account paying an annual rate of 3% compounded monthly? Have students talk among themselves for a minute. Some should note that the previous formula can be used, solving for $A$ in terms of $P$. Do another example.

10 minutes—Finish with examples of loan payments. Encourage interested students to read examples in the text of how loan payments are calculated, but spend class time mainly on examples. Use a spreadsheet program or mortgage payment Web site to display examples comparing different parameter values. Include an example of credit card debt; for example, $10,000 to be repaid over 3 years at 18% interest.

**Section 8.4   Peril at the Polls**   [Voting]. This section analyzes a number of voting methods, including plurality voting, Borda count, and runoff elections, that can be used in elections with more than two candidates, thus demonstrating the inconsistencies that result. Arrow's Impossibility Theorem is presented, with a proof for a simple case.

---

*General Themes*

Using "majority rules" doesn't always determine a clear winner.

---

*Mathematical Underpinnings.*   (*Caveat:* These mathematical overviews are usually for the instructor only. The symbols and detailed proofs are not suitable for class presentations because mathematical symbols and terminology often present a formidable barrier for students.)

Politics can make strange bedfellows; elections can lead to logical conundrums. With only two candidates, choosing a winner is easy by following the old adage of "majority rules." With three or more candidates, however, confusion ensues. This section describes several voting methods that might be used for an election with three or more candidates and includes an example to illustrate how, with the same set of votes cast, each method produces a different "winner." These conflicting results should provoke lively discussion, even as they require students to clearly think through a set of rules.

Students must also consider the basic principles that an election method should satisfy. One set of principles leads to a fundamental result in voting theory: Arrow's Impossibility Theorem. In 1952, economist Kenneth Arrow proved that, given a reasonable set of assumptions, no voting method existed satisfying those assumptions. The assumptions are described informally in the text (page 694), along with a rigorous proof of Arrow's theorem in the case of three candidates and two voters. A general proof involves similar reasoning but requires more cases. Arrow was awarded the Nobel Prize in 1972 for his work.

Even with the gloomy implications of Arrow's Impossibility Theorem, the clear shortcomings of plurality voting should motivate students to look carefully at alternatives. One method used in some countries, such as Australia, is instant runoff voting (IRV), which can be implemented in several ways. In each case of IRV, every voter ranks all the candidates, then one or more candidates are eliminated, say those who receive the smallest number of (or no) first-place votes. The rankings are then reconfigured with the reduced list of candidates. As described in the text (pages 686–687), however, IRV has a peculiar property: It's possible that by improving their ranking of a candidate, individual voters could cause that candidate's rank to fall in the final outcome, which is why IRV fails the criterion known as *monotonicity*.

Condorcet voting seems to offer a lot of appeal. For one thing, it does satisfy monotonicity. As with IRV, every voter ranks all the candidates. If there is one candidate who is preferred over each of the other candidates when considered pairwise, then that candidate is the winner. This method would seem to allow voters to express their preferences most sincerely, without risk of wasting their vote. The difficulty arises when no candidate exists who beats every other candidate in a one-on-one race. The text presents an example of such an outcome (page 690), in which the collected voter preferences are not transitive. The four Cool Dice from the Manipulatives Kit offer another example. If you didn't discuss these dice when they were introduced in Chapter 1, now's a great time to do it.

If you do discuss Condorcet voting, you might mention that there are ways of resolving a nontransitive outcome, making this voting method one that some would prefer over plurality voting. However, the details are rather technical and make Condorcet voting a tough sell among voters.

---

*Our Experience*

*Time spent:* One 50-minute session

*Emphasized items:* Different voting methods

*Items omitted or treated lightly:* Proof of Arrow's Impossibility Theorem

*Sample homework assignment:* Read the section and do Mindscapes I.3; II.6, 10–15.

---

### Sample Class Activities

*Cool conundrum.* Have students bring in the Cool Dice from their kits and play the game in pairs. Tally the results on the board, and discuss.

*Who's the winner?* Write the following vote tally on the board. Have students discuss in small groups what they think the outcome of the election should be. (Each voter was asked to rank all three candidates.) Write each group's winner on the board, and discuss the results.

|                      | Candidate A | Candidate B | Candidate C |
|----------------------|-------------|-------------|-------------|
| First-choice votes   | 8           | 6           | 3           |
| Second-choice votes  | 0           | 5           | 12          |
| Third-choice votes   | 9           | 6           | 2           |

(*Note:* Plurality voting declares A the winner; instant runoff declares B the winner; throwing out the losers declares C the winner. Here are the vote tallies in a different format:

2 voters ranked A, B, C.

6 voters ranked A, C, B.

6 voters ranked B, C, A.

3 voters ranked C, B, A.)

*Class opinions.* Ask the class for suggestions of choices about which most people have an opinion (e.g., favorite soft drink, music group, sports team, or, to make it easier, color or flavor of ice cream). Write three or four choices on the board and tally votes using various methods. Analyze the results.

*Ranking wrangle.* In small groups, have students devise individual rankings of three teams that make it difficult to determine a consensus ranking.

*Real ranking.* Ask each student to bring in election or ranking results from an outside source. Have students work in small groups to share their examples. Then have each group present the example they found most interesting.

### Sample Test Questions

(The Mindscapes are another good source of test questions. All test questions below are available electronically on the Test Bank CD-ROM.)

- Here are the vote tallies from an election. Determine the winner using plurality, instant runoff, and Borda count voting methods. Explain your reasoning in each case.

|                      | Candidate A | Candidate B | Candidate C |
|----------------------|-------------|-------------|-------------|
| First-choice votes   | 5           | 6           | 4           |
| Second-choice votes  | 4           | 2           | 9           |
| Third-choice votes   | 6           | 7           | 2           |

- Suppose three voters rank four choices as indicated. Why will it be difficult for them to determine a winner? What is the name of this paradox?

Voter X: A, B, D, C

Voter Y: B, D, C, A

Voter Z: C, B, A, D

- Arrow's Impossibility Theorem assumes several principles that every voting method should satisfy. State two of these principles.

## Sample Lecture Notes for Section 8.4, "Peril at the Polls" (1 day)

Question of the Day (to be written on the board for students to think about and discuss before class begins): *How do you pick the winner of a democratic election?*

10 minutes—Display the vote tallies below. Ask students who they think should win the election.

| Candidate | A | B | C |
|-----------|---|---|---|
| Votes | 7 | 4 | 9 |

Encourage everyone to agree that C should win (plurality voting). Have students work in small groups to criticize this technique and discuss alternatives. Take student suggestions, and write the most promising ideas on the board. Include vote for two and Borda count methods in the list.

10 minutes—Follow up with scenarios that fit as many of the suggestions as possible, such as the following:

Vote for two:

| Candidate | A | B | C |
|-----------|---|---|---|
| First votes | 7 | 4 | 9 |
| Additional votes | 7 | 11 | 2 |

(Candidate B should win, having received the most votes altogether.)

Borda count:

| Candidate | A | B | C |
|-----------|---|---|---|
| First-choice votes | 7 | 4 | 9 |
| Second-choice votes | 7 | 11 | 2 |
| Third-choice votes | 6 | 5 | 9 |

(Candidate A should win, with a Borda count total of 39.)

10 minutes—If students haven't already suggested it, bring up instant runoff voting. Display the following data, and ask students to analyze the outcome of each election if in each case the two top vote getters had a runoff. (Have the class assume that liberals prefer liberal candidates and conservatives prefer conservative candidates when either is available.)

|              | Conservative |     | Liberal |     |     |
| ------------ | ------------ | --- | ------- | --- | --- |
| Candidates   | A            | B   | C       | D   | E   |
| Votes        | 8            | 7   | 6       | 5   | 5   |

|              | Conservative |     | Liberal |     |     |
| ------------ | ------------ | --- | ------- | --- | --- |
| Candidates   | A            | B   | C       | D   | E   |
| Votes        | 10           | 5   | 6       | 5   | 5   |

Discuss the possible downside, from the conservative point of view, to the increase in Candidate A's votes from the first scenario to the second.

15 minutes—Ask students to suggest the principles a good voting method should follow. Lead the discussion to those assumed by Arrow. State Arrow's Impossibility Theorem and discuss.

5 minutes—Finish with an acknowledgment that even though one ideal voting method doesn't exist, it is still worth comparing the pros and cons of methods that do exist. Discuss as time permits.

**Section 8.5 Cutting Cake for Greedy People** [Dividing scarce resources]. This section presents methods for fair division of desirable goods. The following questions are answered using arguments that seem primarily geometric: How should two, three, or even four people divide a cake so that each person prefers her or his piece to the piece anyone else received? How should two or three roommates decide who gets which room in an apartment and for how much rent?

---

*General Themes*

Applying mathematical techniques can allow us to settle disputes over how to share scarce resources.

---

*Mathematical Underpinnings.* (*Caveat:* These mathematical overviews are usually for the instructor only. The symbols and detailed proofs are not suitable for class presentations because mathematical symbols and terminology often present a formidable barrier for students.)

Students should find the ideas in this section quite compelling. Although cutting cake is really rather frivolous, the extension of the cake-cutting algorithm to dilemmas such as room and rent choices are highly relevant to many young people.

The mathematical result most fundamental to this section is the Intermediate Value Theorem.

**Intermediate Value Theorem:** *If $f$ is a continuous function on a closed interval $[a, b]$, then for any value $z$, $f(a) < z < f(b)$, there exists a value $c$, $a < c < b$, such that $f(c) = z$. (The result also holds if $f(b) < z < f(a)$.)*

That is, a continuous function must take on all values between its endpoint values.

In the case of cake-cutting, the values in the interval $[a, b]$ correspond to cutting positions on the cake, and the function $f$ gives the "value" of the cut. To be more precise, if Adam is cutting the cake with the knife at position $x$, let $f(x)$ be the value to Adam of the piece to the left of the knife. Say that $f(a) = 0$ (Adam gets no cake) and $f(b) = 1$ (Adam gets the entire cake). Set the remaining values so that for $a < x < b$, if $f(x)$ is the value of the left piece, then $1 - f(x)$ is the value of the right piece. Note that any knife position for which $f(x) = 1/2$ will yield two pieces that Adam considers equally desirable. (See figure.)

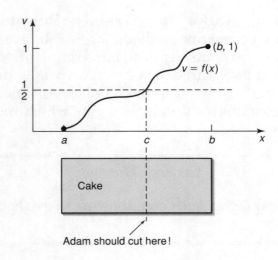

In more general cases, the Intermediate Value Theorem alone is not strong enough, so mathematicians bring out a big gun: the No Retraction Theorem.

**No Retraction Theorem:** *There is no continuous mapping of all points of the interior of a disk onto its boundary circle.*

This theorem can be used to prove the Brouwer Fixed Point Theorem (see Section 5.5). In the context of cake-cutting, this theorem guarantees that a preference diagram for a triangular cake will always have discrete regions bounded by lines (or curves) where the preference abruptly shifts from one piece to another.

Students interested in fair division will find many other related results by searching the Web. This section would make a great project topic.

---

*Our Experience*

*Time spent:* One 50-minute session

*Emphasized items:* Cake-cutting with two or three people

*Items omitted or treated lightly:* Cake-cutting with four people

*Sample homework assignment:* Read the section, and do Mindscapes I.2, 5; II.6, 13; V.24.

---

## Sample Class Activities

*Real cakes.* (Homework before class.) Ask students to bring in a photograph of a whole cake. Have them work in pairs to apply the moving-knife method to divide each cake into two pieces.

*Do you want ice cream with that?* Break students into groups of three. Distribute a copy of the triangle cake template, along with three copies of the cake template on transparencies, to each trio of students (Template 8.5.1). Have each student draw a preference diagram on a transparency. Ask each group to find a point from which they each prefer different pieces.

*Renter's roulette.* Display a floor plan for an apartment with three bedrooms of different sizes. Include other distinguishing features, such as small or large closets, creaky floors, sunny windows, and so on. Have students work in groups of three to decide how they would assign rooms and rent if the total monthly rent on the apartment were $1000. Have a few groups present their division results to the class.

*Pitching tents.* (Homework before class.) Have each student draw a simple map of a triangular campsite on which three backpackers are each to pitch a tent. Be sure students include varying features in the site, and ask them to bring in three copies. Distribute the copies to trios of students and have them apply the cake-cutting method to partition the campsite into three parts.

## Templates for Transparencies *(located at end of chapter):*

- Template 8.5.1—A triangular cake

## Sample Test Questions

(The Mindscapes are another good source of test questions. All test questions below are available electronically on the Test Bank CD-ROM.)

- Explain the moving knife method for dividing a cake between two people. Explain why there must always be a cut for which each person is at least as happy with their piece as they would be with the other piece.

- Below is a preference diagram for a cake. For each of the labeled points, indicate which piece you would prefer if the cake were cut from that point straight to each vertex.

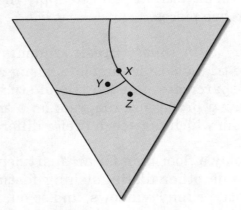

## Sample Lecture Notes for Section 8.5, "Cutting Cake for Greedy People" (1 day)

Question of the Day (to be written on the board for students to think about and discuss before class begins): *Can you always cut a cake so that everyone gets his or her favorite piece?*

10 minutes—Draw on the board or display a picture of an asymmetrical cake. Have students briefly discuss in pairs how they would divide the cake so that each person was happy with his or her piece. Find a pair that suggests the "I cut, you choose" method and have them come to the board to demonstrate their technique on the given cake.

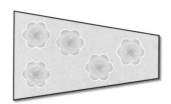

(*Note:* In this first example, be clear about the assumptions being made about cake in this section. See text pages 701–702.)

10 minutes—Briefly discuss why the "I cut, you choose" method will always yield a fair division for two people. Then have students discuss in small groups how they would divide the cake among three people. Take suggestions and lead the discussion to the "moving knife" approach. Have three students demonstrate. Discuss how this approach could result in someone being envious of another's piece of cake. State the Greedy Division Question.

15 minutes—Display a triangular cake, and describe the method underlying the Greedy Division Theorem (text pages 706–707). Do a detailed example, then have trios of students work on another example. (If you distribute transparencies, you could ask one or more groups to present their divisions to the class.) Use the example to suggest why the Greedy Division Theorem is true.

5 minutes—Ask students how they would use the triangle method to divide cakes of different shapes. Encourage students to think about whether the method could be extended to a group of four people.

10 minutes—Ask how many students have ever shared an apartment (or think they will someday.) Ask how many believe that all bedrooms in all apartments are equally desirable. Display a simple floor plan of a three-bedroom apartment corresponding to the example in the text (pages 709–711). Ask how many students think that the three roommates in this apartment should all pay equal rent. Work through the triangle method for the fair division of rent. If time permits, have students work in trios on an example.

# Template 8.5.1   A triangular cake

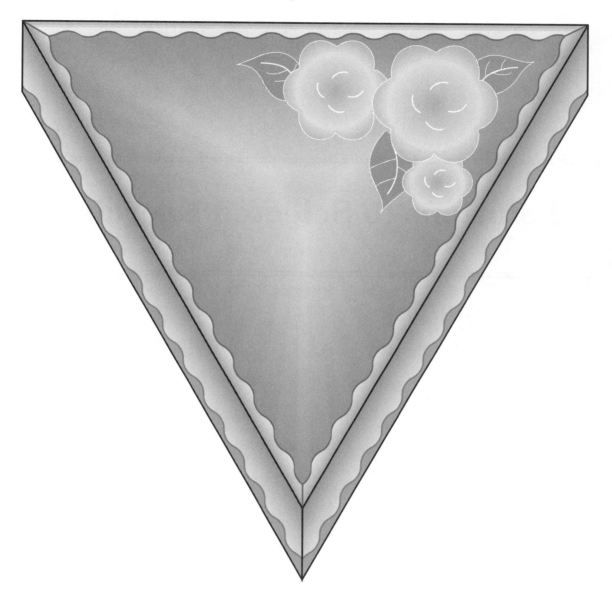

# Solutions to Selected Exercises

## 1.4 From Play to Power

**1. Late-night cash.** Simplify the problem by considering a specific case. Suppose Dave and Paul both have $100. If Dave gives Paul $5, then Dave and Paul have $95 and $105, respectively. Problem solved. Note that the amount Dave must give Paul is unrelated to the initial amount of money in each of their pockets. Though the $100 was extraneous information, it makes the problem easier to solve, and easier to explain!

**2. Politicians on parade.** If there were more than one honest politician, fact (b) would be false. The only possibility left is that there is exactly one honest politician; the remaining 99 are crooked.

**3. The profit.** As in "Late-night cash," we can clarify the problem by making it specific. Suppose that the dealer starts with $20 (the initial amount doesn't affect the final profit).

| After . . . | First purchase | First sell | Second purchase | Second sell |
|---|---|---|---|---|
| | $13 | $21 | $12 | $22 |

At the end of the transactions, she has made a profit of $2.

**4. The truth about . . .** What's the maximum possible number of cats? Because there are 56 biscuits, and each cat takes 5 of them, there are at most 11 cats. Now explore the possibilities. If there were exactly 11 cats, then there would be 1 biscuit left over. If there were only 10 cats, then there would be 6 biscuits left over, exactly enough to feed 1 dog! That's one solution. To find the only other solution, continue exploring alternatives.

| Number of Cats (5) | 11 | 10 | 9 | 8 | 7 | 6 | 5 | 4 | 3 | 2 | 1 | 0 |
|---|---|---|---|---|---|---|---|---|---|---|---|---|
| Number of Dogs (6) | 0 | 1 | 1 | 2 | 3 | 4 | 5 | 6 | 6 | 7 | 8 | 9 |
| Leftover Biscuits | 1 | 0 | 5 | 4 | 3 | 2 | 1 | 0 | 5 | 4 | 3 | 2 |

The two solutions are [10 cats, 1 dog] and [4 cats, 6 dogs].

**5. It's in the box.** There are two possible situations: (a) both signs are true, or (b) both are false. Let's assume first that both signs are true. B's sign says that box A contains a snake, and A's sign says that at least one of the two boxes contains money. So, in this situation, A contains a snake and B holds money. Now assume that both signs are false. If B's sign is false, then box A contains money. If A's sign is false, then both box A and box B must contain snakes. Because box A cannot contain both money and a snake, this whole situation is impossible—both signs can't be false. Because the second alternative is impossible, both signs are true, and box B holds a million dollars.

**6. Lights out.** At first glance this problem seems impossible. When you walk into the other room, you will see one of two possibilities. Either the light is on, or the light is off. Two possible outcomes can't resolve three possible situations! To solve the problem, we must find more information. The key to this problem is that real lights (most of them, anyway) become hot after an extended period of time. Furthermore, once hot, they take a long time to cool down. So you can obtain extra information by turning on one of the switches, waiting 10 minutes, and then turning off the switch.

Turn on switch #1. Wait 10 minutes. Turn off switch #1. Now turn on switch #2 and quickly go into the other room and put your hand on the light. There are now three possible outcomes.

| Outcome | The lucky switch |
|---|---|
| Light is on | #2 |
| Light is off and cold | #3 |
| Light is off and warm | #1 |

**7. Out of sight but not out of mind.** There is no mystery dollar. Slip paid $9, Spike paid $9, and Milly paid $7 for a total of $25, which was precisely the cost of the room. Chip's mistake lies in his addition of Milly's stolen money to the $27. The corrected logic: Each band member puts in $10, for a total of $30. Where does this $30 go? $1 worth of change goes to each of the band members, leaving $27 to be accounted for. Milly pocketed 2 of these dollars, and Chip took the remaining $25 for the cost of the room. All is accounted for.

**8. The cannibals and the missionaries.**

| | River | |
|---|---|---|
| CCCMMM <> | | |
| CMMM | | <>CC |
| CCMMM<> | | C |
| MMM | | <>CCC |
| CMMM<> | | CC |
| CM | | <>CCMM |
| CCMM<> | | CM |
| CC | | <>CMMM |
| CCC<> | | MMM |
| C | | <>CCMMM |
| CC<> | | CMMM |
| | | <>CCCMMM |

C = Cannibal, M = Missionary, <> = Boat

There are several solutions to this Mindscape. Notice the symmetry in the above solution; the first five trips are identical to the last five trips, but in reverse order.

**9. Whom do you trust?** Pocket states that Shlock and Wind are lying. Because there can be at most two liars, this would imply that the remaining three (Pocket, Greede, and Slie) are all telling the truth. Because Greede and Slie contradict each other, this is not possible. So we know for sure that Pocket is lying. There is room for at most one more liar, and once again, because Greede and Slie contradict each other, one of them is the other liar. It doesn't matter which one is the liar because by now we know that both Shlock and Wind are telling the truth. Shlock says that either Wind or Pocket leaked the story, and Wind denies leaking it himself. So it was Pocket who leaked the story.

**10. A commuter fly.** The trains are traveling toward each other at a rate of 70 mph. Because they were initially 210 miles apart, the trains will collide in 3 hr. Because the fly travels at 100 mph, the fly travels a total of 300 miles.

340

**11. A fair fare.** "How much should each person pay?" The term "should" isn't as precise as it sounds. We'll present two methods for divvying up the fare and let you be the judge: (1) Everyone should pay the same mileage rate. That is, because Mary travels twice as far as Dave does, she should expect to pay twice as much. There are 60 total passenger-miles $(10 + 20 + 30)$, and a total cost is $45. So each person pays $0.75 a mile. Bob pays $7.50, Mary pays $15, and Ivan pays $22.50. Everyone pays exactly half of what they would have paid had they traveled alone. (2) The fare for each mile should be split evenly among the passengers. All three passengers split the fare for the first 10 miles. Mary and Ivan split the fare for the second 10 miles, and Ivan pays for the last 10 miles himself. Bob pays $5, Mary pays $12.50, and Ivan pays $27.50. Ivan pays more per mile on average than Bob, but then again, Bob had to share his whole trip cramped with three other people, while Ivan spent the last 10 miles of his trip with plenty of room to stretch.

**12. Getting a pole on a bus.** Sarah gave Adam a box that was 4 ft long and 3 ft wide. The diagonal of the box was 5 ft, just long enough to store Adam's fishing pole. (Recall from the Pythagorean Theorem that the square of the length of the diagonal is the sum of the squares of the two sides: $5^2 = 3^2 + 4^2$.)

**13. Tea time.** Just as in "A commuter fly," there is a long, arithmetic-laden solution and a short, insightful solution. We started with 3 oz of cream in the creamer, and 3 oz of tea in the teacup. In the end, we still have 3 oz of liquid in both containers, but we don't know the relative amounts of tea and cream. The arithmetic-laden solution tries to find out exactly how much of each liquid is in each cup. We can avoid this by arguing only that there is as much cream in the teacup as there is tea in the creamer.

Suppose there are $C$ ounces of cream in the teacup. Because the teacup holds 3 oz, there are $(3 - C)$ oz of tea also in the teacup. Because we started with a total of 3 oz of tea, the remaining $C$ oz of tea must be in the creamer. So, the teacup is just as diluted as the creamer.

**14. A shaky story.** Let's refer to Sam's guests by the number of hands they shook. Specifically, G0 shook no hands, G1 shook one hand, up to G8, who shook eight other hands. Because there are 10 guests in all, G8 must have shaken hands with everyone but himself and G0. Because G8 did not shake hands with his spouse, the only person who could possibly be his spouse is G0. So, G0 and G8 are married. With this new information, we can simplify the problem.

The four remaining couples consist of the following eight people: Sam, G1, G2, G3, G4, G5, G6, and G7. Note that G8 shook hands with everyone in this new group. Because G7 shook hands with G8, he must have shaken hands with exactly six members of the new group. Similarly, G1 shook hands with no one else. We are in the same situation as before. G7 didn't shake hands with G1, and because he shook hands with everyone else, G1 is the only possible candidate for G7's spouse. So, G1 and G7 are married.

Continue simplifying the problem. Sam, G2, G3, G4, G5, and G6 all shook hands with both G7 and G8. A similar line of reasoning can show that G2 and G6 are married. A final simplification allows us to conclude that G3 and G5 are married. The remaining couple consists of Sam and G4. So Sam's wife shook exactly four hands. Is it possible to determine how many hands Sam shook?

**15. Murray's brother.** Begin by placing four stones on each side of the balance (see the first table). There are three possible outcomes, indicated by the three columns in the table. Either the left side is heavier, the right side is heavier, or the scale balances. The first two outcomes leave us in the same situation—we know the diamond is among eight of the stones, and if we knew which stone it was, we would also know whether the diamond was heavier or lighter. (The letters H and L below indicate this.) If the scale balances, then we know the diamond is one of the four remaining stones.

The second and third tables outline the argument needed to address these two situations. The last row in each table represents the third weighing needed for each situation. For each outcome of each weighing, make sure that you can identify the diamond and its relative weight.

? = Unknown stone
H = Potentially heavy stone
L = Potentially light stone
D = Dormant diamond
S = Stone

| First weighing | | ???? vs. ???? extra: ???? | |
|---|---|---|---|
| | | | |
| Outcomes | HHHH vs. LLLL extra: SSSS | LLLL vs. HHHH extra: SSSS | SSSS vs. SSSS extra: ???? |

| | | SSSS vs. SSSS extra: ???? | |
|---|---|---|---|
| Second weighing | | SSS vs. ??? extra: ? | |
| | | | |
| Outcomes | SSS vs. LLL extra: S | SSS vs. HHH extra: S | SSS vs. SSS extra: D |
| Third weighing | L vs. L extra: L | H vs. H extra: H | S vs. D |

| | | HHHH vs. LLLL extra: SSSS | |
|---|---|---|---|
| Second weighing | | HHL vs. HHL extra: LL | |
| | | | |
| Outcomes | HHS vs. SSL extra: SS | SSL vs. HHS extra: SS | SSS vs. SSS extra: LL |
| Third weighing | H vs. H extra: L | H vs. H extra: L | L vs. L |

## 2.1 Counting

### I. Developing Ideas

**1. Muchos mangos.** There are 5 layers with appoximately 18 mangos each, for a total of 90.

**2. Packing balls.** In the bottom of the box, line up as many balls as will fit from one corner to another corner along one edge. Because the box is a perfect cube, this gives the number of layers of balls that will fit when stacked vertically. Multiply this number by the number of balls that fit in a single layer to get your estimate.

**3. Alternative lock.** With five CDs, each can have its own shelf. With one more CD, some shelf must have two according to the Pigeonhole principle.

**4. The Byrds.** The answer to both questions is "no." If each shelf had 3 (or fewer) CDs, then the total number of CDs would be (at most) 15.

**5. For the birds.** The Pigeonhole principle states that if you have more pigeons than you have pigeonholes and every pigeon must be in some pigeonhole, then there must be at least one pigeonhole with more than one pigeon.

### II. Solidifying Ideas

**6. Treasure chest.** The weight of the bills alone is enough to consider rejecting the offer. Let's *underestimate* the weight in the following way: A typical ream of laser printer paper (500 sheets) definitely weighs more than 5 lb. So a single sheet of paper weighs at least $5/500 = 1/100$ lb. Now, we can almost fit 6 one dollar bills into the area contained by one $8.5" \times 11"$ piece of paper, so a single dollar bill weighs more than $1/600$ lb. Therefore, a million bills weigh $1,000,000/600$, or roughly 1666 lb! Wait for a better offer.

**7. Order please.** States in the United States, honest congressmen (debatable), cars, telephones on the planet, people, grains of sand

**8. Penny for your thoughts.** How many pieces would be placed on the last square of the checkerboard? The number of gold pieces doubles with each square, and there are 64 total squares. The first square has $1 = 2^0$ pieces, the second square has $2 = 2^1$ pieces, the third square has $4 = 2^2$ pieces, and so on. So, the 64th square has $2^{63}$, or more than $9 \times 10^{18}$, gold pieces. Notice, too, that the number of pieces on each square is one more than the sum of all the gold pieces on the previous squares. So, in total, there are $2^{64} - 1$ pieces.

**9. Twenty-nine is fine.** Two possible candidates: First, 29 is prime. Second, 29 happens to be the sum of three consecutive squares, $29 = 4 + 9 + 16$. Lest the number 27 feel left out, it should be noted that $27 = 3 \times 3 \times 3$ is a perfect cube. Is the set of prime numbers sparser than the set of perfect cubes? Does this make it less interesting in your eyes?

**10. Perfect numbers.** The next perfect number in line is $28 = 1 + 2 + 4 + 7 + 14$.

**11. Many fold.** To get started, let's estimate the width of an ordinary piece of paper by noting that packages of 200 sheets of paper are more than half an inch thick. So, a single piece of paper is at least $1/400$ in. thick. Now, after one folding, the paper is twice the original thickness. After 2 foldings, the paper is $4 = 2^2$ times as thick. After 50 foldings, the paper will be $2^{50}$ times as thick. The resulting paper is then $2^{50}/400$ inches thick. (That's more than $2.8 \times 10^{13}$ in. and more than 40 million mi!)

343

**12. Only one cake.** Because there are more people than possible birthdays, there must be at least two people who share the same birthday. To be more convincing, imagine that they all have different birthdays. Now select 366 people from the group. Because they all have different birthdays and because there are only 366 possible birthdays (including leap year), all the birthdays must be accounted for. The remaining four people must all share a birthday with someone else in the room.

**13. For the birds.** There must be some hole containing more than one pigeon. In the hairy-bodies question, the six billion people in the world play the role of the pigeons, and the 400 million hairs play the role of the holes. Just as there are at least two pigeons sleeping in the same hole, there are necessarily two people with the same total number of body hairs.

**14. Sock hop.** To guarantee one match, you need only pull out three socks. Either two will be black, or two will be blue. To get two matched pairs, you need at most five socks. However, to guarantee a black pair, you need to pull out 12 socks, because you might be unlucky and pull out all the blue socks first!

**15. The last one.** 19, 58, 29, 88, 44, 22, 11, 34, 17, 52, 26, 13, 40, 20, 10, 5, 16, 8, 4, 2, 1
The sequences for 11 and 22 are within the sequence above.
30, 15, 46, 23, 70, 35, 106, 53, 160, 80, 40, 20, 10, 5, 16, 8, 4, 2, 1

### III. Creating New Ideas

**16. See the three.** There are two ways to approach this question: (1) Count the numbers with 3's, or (2) Count the numbers without 3's. Method (1): There is only one number with three 3's in it, namely 333. How many numbers have exactly two 3's? There are 9 such numbers of the form $33x$, 9 of the form $3x3$, and 9 of the form $x33$, for a total of 27 doubles. How many numbers have exactly one 3 in them? Let's overcount by saying that there are 100 numbers of the form $xx3$, $x3x$, and $3xx$. Of the 300 numbers we counted, the 27 doubles twice and 333 three times. So, we have $300 - 27 - 2 = 271$. The corresponding proportion is 0.271. Method (2): A number with no 3 could have any of nine digits in each position, for a total of $9 \times 9 \times 9$, or 729 numbers. The remaining 271 numbers have a 3.

**17. See the three II.** There are two ways to approach this question: (1) Count the numbers with 3's, or (2) Count the numbers without 3's. Method (1): There is a nonobvious way to keep track of all the overcounting. There are 1000 numbers of the forms $xxx3$, $xx3x$, $x3xx$, and $3xxx$, for a total of 4000. There are 100 numbers of the form $xx33$, $x3x3$, $3xx3$, $x33x$, $3x3x$, and $33xx$, for a total of 600. We have 10 each of the form $x333$, $3x33$, $33x3$, and $333x$, for a total of 40 such numbers. And lastly, there is only 1 number with four 3's. Here's the trick: $4000 - 600 + 40 - 1 = 3439$ represents the number of numbers with 3's in them. The alternating signs account for all the overcounting! The corresponding proportion is 0.3439. Method (2): A number with no 3 could have any of 9 digits in each position, for a total of $9 \times 9 \times 9 \times 9$, or 6561 numbers. The remaining 3439 numbers have a 3.

**18. See the three III.** The proportion of million-digit numbers without a 3 is 9/10 raised to the millionth power.

**19. Commuting.** There are 100 people arriving at work between 8:00 and 8:30. Imagine slicing this time frame into 30 distinct intervals. Because we have got more people than intervals, at least two people will arrive within the same interval. This also means that their arrival times differ by less than a minute.

**20. RIP.** Within the next 100 years, the 6.2 billion people currently populating the Earth will die. If less than 50 million people died each year, then at the end of 100 years, only 5 billion people would have died. This contradiction shows that at some point more than 50 million people will die. Alternatively, you could say that the average number of people that will die each year is 6,200,000,000 / 100, or 62 million. And because this is the average, there must be at least some year in which more than 62 million people will die.

## IV. Further Challenges

**21. Say the sequence.** Reading the last number, "One 3, One 1, Two 2's, Two 1's" generates the next number in the sequence, namely, 13112221. "1" is read, "One 1," which becomes "11." This in turn is read, "Two 1's" or "21."

**22. Lemonade.** You have two choices for the first option (yes or no), two for the second option, two for the third option, and four choices for the color. Therefore, there are $2 \times 2 \times 2 \times 4 = 32$ different types of this particular model. Each of the 100,000 cars fits into one of these 32 categories. There is an average of $100,000/32 = 312.5$ cars per category, so some category must have at least this many cars. You are guaranteed to find at least 312 identical cars.

## 2.2 Numerical Patterns in Nature

### I. Developing Ideas

**1. Fifteen Fibonaccis.** 1, 1, 2, 3, 5, 8, 13, 21, 34, 55, 89, 144, 233, 377, 610

**2. Born $\varphi$.** The $\varphi$ symbol represents the infinitely long fraction expression $1+\cfrac{1}{1+\cfrac{1}{1+\cfrac{1}{1+\cfrac{1}{1+\cdots}}}}$.

It is also a solution to the equation $\varphi = 1+\dfrac{1}{\varphi}$.

One sequence of numbers that approaches $\varphi$ is the list of ratios of consecutive Fibonacci numbers.

**3. Tons of ones.** Simplifying, we see that $1+\cfrac{1}{1+\cfrac{1}{1}} = 1+\cfrac{1}{1+1} = \dfrac{3}{2}$.

**4. Twos and threes.** $2+\cfrac{2}{2+\cfrac{2}{2}} = 2+\cfrac{2}{2+1} = \dfrac{8}{3}$; $3+\cfrac{3}{3+\cfrac{3}{3}} = 3+\dfrac{3}{4} = \dfrac{15}{4}$

**5. The family of $\varphi$.** If $x = 2+\dfrac{1}{x}$, multiply through by $x$ to get $x^2 = 2x + 1$. So $x^2 - 2x - 1 = 0$. This does not factor, so we use the quadratic formula to get $x = \dfrac{-(-2)\pm\sqrt{(-2)^2 - 4(1)(-1)}}{2(1)} = \dfrac{2\pm\sqrt{8}}{2} = 1\pm\sqrt{2}$. Similarly, multiply $x = 3+\dfrac{1}{x}$ through by $x$ to get $x^2 = 3x + 1$. So $x^2 - 3x - 1 = 0$. Again, this does not factor, so we use the quadratic formula to get $x = \dfrac{-(-3)\pm\sqrt{(-3)^2 - 4(1)(-1)}}{2(1)} = \dfrac{3\pm\sqrt{13}}{2}$.

### II. Solidifying Ideas

**6. Baby bunnies.**

| Month | 1 | 2 | 3 | 4 | 5 | 6 | 7 | 8 |
|---|---|---|---|---|---|---|---|---|
| Adults | 1 | 1 | 2 | 3 | 5 | 8 | 13 | 21 |
| Babies | 0 | 1 | 1 | 2 | 3 | 5 | 8 | 13 |
| Total | 1 | 2 | 3 | 5 | 8 | 13 | 21 | 34 |

After each month, the total number of bunnies becomes the number of mature bunnies for the next month. Because all the mature bunnies produce offspring, the number of mature bunnies during one month becomes the number of new offspring in the next month. Each row contains the same sequence of numbers, but the sequences are offset from one another. Note the connection to Fibonacci.

## 7. Discovering Fibonacci relationships.

| $n$ | 1 | 2 | 3 | 4 | 5 | 6 | |
|---|---|---|---|---|---|---|---|
| $(F_n)^2$ | 1 | 1 | 4 | 9 | 25 | 64 | . . . |
| $(F_{n+1})^2$ | 1 | 4 | 9 | 25 | 64 | 169 | . . . |
| **Sum** | 2 | 5 | 13 | 34 | 89 | 233 | |
| | $F_3$ | $F_5$ | $F_7$ | $F_9$ | $F_{11}$ | $F_{13}$ | |

Note that we are getting all the odd Fibonacci numbers. This leads to the formula $(F_n)^2 + (F_{n+1})^2 = F_{2n+1}$.

## 8. Discovering more Fibonacci relationships.

| $n$ | 1 | 2 | 3 | 4 | 5 | 6 | |
|---|---|---|---|---|---|---|---|
| $(F_{n+1})^2$ | 1 | 4 | 9 | 25 | 64 | 169 | . . . |
| $(F_{n-1})^2$ | . | 1 | 1 | 4 | 9 | 25 | . . . |
| **Difference** | . | 3 | 8 | 21 | 55 | 144 | |
| | | $F_4$ | $F_6$ | $F_8$ | $F_{10}$ | $F_{12}$ | |

Now we're getting all the even Fibonacci numbers (see Mindscape II.7.). More compactly, $(F_{n+1})^2 - (F_{n-1})^2 = F_{2n}$

## 9. Late bloomers.

| Month | 1 | 2 | 3 | 4 | 5 | 6 | 7 | 8 | 9 | 10 |
|---|---|---|---|---|---|---|---|---|---|---|
| **Mature** | 1 | 1 | 1 | 2 | 3 | 4 | 6 | 9 | 13 | 19 |
| **New babies** | 0 | 1 | 1 | 1 | 2 | 3 | 4 | 6 | 9 | 13 |
| **Old babies** | 0 | 0 | 1 | 1 | 1 | 2 | 3 | 4 | 6 | 9 |
| **Total** | 1 | 2 | 3 | 4 | 6 | 9 | 13 | 19 | 28 | 41 |

As in Mindscape II.6, each row contains the same sequence of numbers (though shifted). If $T_n$ represents the total number of bunnies at the end of the $n$th month, then $T_n = T_{n-1} + T_{n-3}$.

**10. A new start.** 2, 1, 3, 4, 7, 11, 18, 29, 47, 76, 123, 199, 322, 521, 843. Because 843/521 = 1.61803 . . . , it looks like the ratio of consecutive numbers still approaches the golden mean.

If we start with −7 and 3, we get a negative sequence of numbers: −7, 3, −4, −1, −5, −6, −11, −17, −28, −45, −73, −118, −191, 309, −500. Yet the ratios still converge to the Golden Mean, (−500)/(−309) = 1.61812 . . . . We can view the Golden Mean as defined by this infinite process independent of the starting numbers. In Chapter 6, we will see that images and pictures can be defined in a similar manner.

## 11. Discovering Lucas relationships.

| $n$ | 1 | 2 | 3 | 4 | 5 | 6 | 7 | 8 | |
|---|---|---|---|---|---|---|---|---|---|
| $L_{n-1}$ | . | 2 | 1 | 3 | 4 | 7 | 11 | 18 | |
| $L_{n+1}$ | | 1 | 3 | 4 | 7 | 11 | 18 | 29 | 47 |
| **Sum** | . | 5 | 5 | 10 | 15 | 25 | 40 | 65 | |
| | | $g_1$ | $g_2$ | $g_3$ | $g_4$ | $g_5$ | $g_6$ | $g_7$ | |

$L_{n-1} + L_{n+1} = g_{n-1}$, where $g_n$'s are constructed as the Lucas numbers are, but with the first two numbers being 5 and 5 instead of 2 and 1. You can write the answer in terms of Fibonacci numbers by noting that $L_n = F_n + F_{(n-2)}$.

**12. Still more Fibonacci relationships.**
See the solution to Mindscape II.11. By the same reasoning, we find that the sum is the Lucas sequence starting with 3 and 4.

| $F_{n-1}$ | . | 1 | 1 | 2 | 3 | 5 | 8 | 13 | 21 |
|-----------|---|---|---|---|---|---|---|----|----|
| $F_{n+1}$ | 1 | 2 | 3 | 5 | 8 | 13 | 21 | 34 | 55 |
| **Sum** | . | 3 | 4 | 7 | 11 | 18 | 29 | 47 | 76 |

**13. Even more Fibonacci relationships.**

| $F_{n+2}$ | 2 | 3 | 5 | 8 | 13 | 21 | 34 |
|-----------|---|---|---|---|----|----|----|
| $F_{n-2}$ | . | . | 1 | 1 | 2 | 3 | 5 |
| **Difference** | . | . | 4 | 7 | 11 | 18 | 29 |

Note that we get the same sequence as in Mindscape II.12. This is because $F_{n+2} - F_{n-2} = F_{n-1} + F_{n+1}$, which is straightforward to prove using the definition of Fibonacci numbers.

**14. Discovering Fibonacci and Lucas relationships.**

| $F_n$ | 1 | 1 | 2 | 3 | 5 | 8 | 13 | 21 | . . . |
|-------|---|---|---|---|---|---|----|----|-------|
| $L_n$ | 2 | 1 | 3 | 4 | 7 | 11 | 18 | | . . . |
| **Sum** | 3 | 2 | 5 | 7 | 12 | 19 | 31 | | . . . |

See Mindscape II.11 for more insights into these types of sequences.

**15. The enlarging area paradox.** If you look closely, you'll notice that the pieces don't line up exactly. Note that the little triangle with sides 3 and 8 appears to be similar to the big "triangle" with sides 5 and 13. If this were true, then the corresponding ratios would be equal; but 8/3 isn't 13/5. Because these are ratios of consecutive Fibonacci numbers, the ratios are close, which is why this is a convincing trick.

**16. Sum of Fibonacci.** Start with the largest Fibonacci number smaller than the given number and work your way backward.
$52 = 34 + 13 + 5$
$143 = 89 + 34 + 13 + 5 + 2$
$13 = 13$
$88 = 55 + 21 + 8 + 3 + 1$

**17. Some more sums.** $43 = 34 + 8 + 1$; $90 = 89 + 1$; $2000 = 1597 + 377 + 21 + 5$; $609 = 377 + 144 + 55 + 21 + 8 + 3 + 1$

**18. Fibonacci nim: The first move.** After mentally expressing 52 as a sum of nonconsecutive Fibonacci numbers ($52 = 34 + 13 + 5$), you remove five sticks from the pile.

**19. Fibonacci nim: The first move II.** Because $90 = 89 + 1$, you need only remove one stick.

**20. Fibonacci nim: The first move III.** Noting that $609 = 377 + 144 + 55 + 21 + 8 + 3 + 1$, we remove only one stick.

**21. Fibonacci nim: The next move.** After the friend removes four, there are nine sticks left. Because $9 = 8 + 1$, we remove one stick to keep ourselves in a winning position.

**22. Fibonacci nim: The next move II.** A total of 26 sticks have been removed, leaving 24. Express 24 as a sum of nonconsecutive Fibonacci numbers ($24 = 21 + 3$) and remove 3 sticks.

**23. Fibonacci nim: The next move III.** A total of 24 sticks has been removed, leaving 66. Because $66 = 55 + 8 + 3$, you can keep your winning position by removing only 3 sticks.

## III. Creating New Ideas

**26. Discovering still more Fibonacci relationships.**

| $F_n$ | | 1 | 1 | 2 | 3 | 5 | 8 | 13 |
|---|---|---|---|---|---|---|---|---|
| $F_{n-1}$ | . | | 1 | 1 | 2 | 3 | 5 | 8 |
| $F_{n+1}$ | | 1 | 2 | 3 | 5 | 8 | 13 | 21 |
| $F_{n+1} \times F_{n-1} - (F_n)^2$ | | | 1 | −1 | 1 | −1 | 1 | −1 |

Surprisingly, the expression is $(-1)^n$ (1 if $n$ is even, and −1 if $n$ is odd!).

**27. Finding factors.** Mindscape II.13 showed us that $F_{2n} = (F_{n+1})^2 - (F_{n-1})^2$, which can be factored as the product $(F_{n+1} - F_{n-1}) \times (F_{n+1} + F_{n+1})$. This means that except for 3, which is $F_2$, none of the Fibonacci numbers with even index ($F_{2n}$) are prime.

**28. The rabbits rest.**

| Month | 1 | 2 | 3 | 4 | 5 | 6 | 7 | 8 | 9 |
|---|---|---|---|---|---|---|---|---|---|
| Kids | | 1 | 0 | 1 | 1 | 1 | 2 | 2 | 3 | 4 |
| Parents | | 0 | 1 | 0 | 1 | 1 | 1 | 2 | 2 | 3 |
| Old parents | | 0 | 0 | 1 | 0 | 1 | 1 | 1 | 2 | 2 |
| Total | | 1 | 1 | 2 | 2 | 3 | 4 | 5 | 7 | 9 |

Thus, $T_n = T_{n-3} + T_{n-2}$, with $T_1 = T_2 = 1$, and $T_3 = 2$

**29. Digging up Fibonacci roots.** Once again, the limiting ratio is the Golden Mean. See the solution to Mindscape IV.38 for an argument to why this is so.

**30. Tribonacci.** 0, 0, 1, 1, 2, 4, 7, 13, 24, 44, 81, 149, 274, 504, 927; $504/274 = 1.8394 \ldots$; $927/504 = 1.8392 \ldots$. In fact, the limiting ratio converges to a solution of the equation $x^4 - 2x^3 + 1 = 0$.

**31. Fibonacci follies.** There are nine sticks left and we would like to take one stick ($9 = 8 + 1$) to get us back in a winning situation. Because we are allowed to take one stick, we're in good shape. But a mistake was made during the second move. We should have taken one stick instead of two. Had our friend taken two instead of one, we would still be losing.

**32. Fibonacci follies II.** There are 18 sticks left ($18 = 13 + 5$). Because our friend took only two sticks, we can't take the five sticks we need to get back to a winning position. We take one stick and hope our friend makes a mistake. On the previous move, we should have taken two sticks instead of three.

**33. Fibonacci follies III.** There are 18 sticks left, and we can remove at most 4 from the pile. Because $18 = 13 + 5$, we can't follow our winning strategy. The strategy doesn't apply here because we started with a number of sticks that equaled a Fibonacci number. The moral? When you're starting with a Fibonacci number, graciously let your friend play first.

**34. A big fib.** Let's suppose F is the $k$th Fibonacci number: $F = F_k$. Then $N$ lies between $F_k$ and $F_{k+1}$. Because $N < F_{k+1}$, then $N - F_k < F_{k+1} - F_k$; but this last difference is $F_{k-1}$, because $F_{k-1} + F_k = F_{k+1}$. Putting this together, we have $N - F_k < F_{k-1}$.

**35. Decomposing naturals.** Let's prove this inductively: We know that we can trivially express 1 as a sum of distinct, nonconsecutive Fibonacci numbers. We know that we can do this for the first few natural numbers. Now let's assume that we can do this for all numbers less than $k$, and try to show that we can also write $k$ as such a sum. If $k$ is a Fibonacci number, then we're done; if it isn't, then we will grab the largest Fibonacci number smaller than $k$, call it F, and use it in our sum. Because $(k - F)$ is less than $k$, we can invoke the induction hypothesis to express $(k - F)$ as a sum of distinct, nonconsecutive Fibonacci numbers. We add F to the list to complete the problem. Mindscape III.34 shows $(k - F)$ is smaller than the next smaller Fibonacci number, so by adding F to the previous list, we still have a set of nonconsecutive Fibonacci numbers.

## IV. Further Challenges

**36. How big is it?** In other words, can the ratio $F_{k+1}/F_k$ equal 2? Rewrite the ratio as was done in the text (page 53), $(F_k + F_{k-1})/F_k = 1 + F_{k-1}/F_k$. Because the Fibonacci numbers keep getting bigger, the last fraction is always less than or equal to 1, which implies that the original expression is less than or equal to 2. It is only equal to 2 when $F_{k-1} = F_k$, that is, when $F_{k+1} = 2$.

**37. Too small.** Let $F_k$ and $F_{k-1}$ represent F and the Fibonacci number immediately preceding F. The largest Fibonacci numbers less than F are $F_{k-2}$ and $F_{k-1}$, and their sum is precisely F, which is less than $N$. The sum of any other distinct Fibonacci numbers would be even smaller.

**38. Beyond Fibonacci.** 0, 1, 2, 5, 12, 29, 70, 169, 408, 985, 2378, 5741, 13,860, 33,461, 80,782. 80,782/33,461 = 2.414213562 . . . . Assume that consecutive ratios tend to some mystery number $x$. Use the recurrence relation to rewrite the quotient: $F_{n+1}/F_n = (2F_n + F_{n-1})/F_n = 2 + 1/(F_n/F_{n-1})$. The left side tends to $x$, while the right side tends to $2 + 1/x$. So $x$ satisfies the equation $x = 2 + 1/x$. Equivalently, $x^2 = 2x + 1$, whose positive solution is $1 + \sqrt{2}$.

**39. Generalized sums.** No such luck. 3 = 1 + 2, which involves consecutive generalized Fibonacci numbers. If we write 3 = 1 + 1 + 1, then we haven't used distinct generalized Fibonacci numbers. Even if we didn't mind using consecutive numbers, we would still run into problems. For example, 4 can be written as 1 + 1 + 1 + 1, 2 + 1 + 1, or 2 + 2, all of which duplicate numbers.

**40. It's hip to be square.** Rewrite the fraction under the square root in the following way: $F_9/F_7 = F_9/F_8 \times F_8/F_7$. Note that the last two fractions are ratios of consecutive Fibonacci numbers, and so should closely approximate $\varphi$, the Golden Mean. Therefore, the original fraction tends to $\varphi^2$; when we take the square root, we just get $\varphi$ back.

## 2.3 Prime Cuts of Numbers

### I. Developing Ideas

**1. Primal instincts.** 2, 3, 5, 7, 11, 13, 17, 19, 23, 29, 31, 37, 41, 43, 47

**2. Fear factor.** $6 = 2 \cdot 3$, $24 = 2 \cdot 2 \cdot 2 \cdot 3$, $27 = 3 \cdot 3 \cdot 3$, $35 = 5 \cdot 7$, $120 = 2 \cdot 2 \cdot 2 \cdot 3 \cdot 5$

**3. Odd couple.** No, $n + 1$ will be an even number greater than 2; thus, it will have 2 as a factor. If $n = 1$, the $n + 1 = 2$, which is prime.

**4. Tower of power.** The first 10 powers of 2 are 2, 4, 8, 16, 32, 64, 128, 256, 512, 1024. The first 5 powers of 5 are: 5; 25; 625; 3125; 15,625.

**5. Compose a list.** The list of even numbers starting with 4 contains no primes. The list of powers of 5 starting with 25 contains no primes.

### II. Solidifying Ideas

**6. A silly start.** It's a personal choice, but 51 has my vote. Nothing about the number screams that its factors are 3 and 17. Another favorite is $91 = 7 \times 13$.

**7. Waiting for a nonprime.** When $n = 4$, the resulting number is 25, which isn't prime. In fact, most of the time, the constructed number won't be prime. The next prime number doesn't occur until $n = 11$.

**8. Always, sometimes, never.** By definition, a product of two numbers is not prime, so "Never" is the answer to both questions.

**9. The dividing line.** Sometimes; for example, $8/4 = 2$ is prime, but $16/4 = 4$ is not.

**10. Prime power.** No; raising to a power stands for repeated multiplication; thus, the resulting number would be represented as a product of numbers, which is a definite giveaway of its nonprime status.

**11. Nonprimes.** Other than 2, all the even numbers are nonprimes. So, there are infinitely many numbers that are not prime.

**12. Prime test.** No; the crux of the definition of *prime* is that no other numbers other than 1 and $n$ divide into $n$. For example, 1 and 4 both divide into 4 evenly, but 4 is not prime. The numbers 1 and $n$ will always divide into $n$ evenly, for any number $n$.

**13. Twin primes.** (3, 5), (5, 7), (11, 13), (17, 19), (29, 31), (41, 43), (59, 61), (71, 73), (101, 103), (107, 109), (137, 139), (149, 151), (179, 181), (191, 193), (197, 199)
Do you think it becomes harder and harder to find twin primes as we look at larger and larger prime numbers? Or does their distribution appear random?

**14. Goldbach.** $4 = 2 + 2$, $6 = 3 + 3$, $8 = 3 + 5$, $10 = 3 + 7$, $12 = 5 + 7$, $14 = 3 + 11$, $16 = 3 + 13$, $18 = 5 + 13$, $20 = 3 + 17$, $22 = 3 + 19$, $24 = 5 + 19$, $26 = 3 + 23$, $28 = 5 + 23$, $30 = 7 + 23$
Note that as the numbers get larger, there are more ways to express the number as a sum of two primes. $32 = 3 + 29$, $11 + 19$, $13 + 17$, and so on.

**15. Odd Goldbach.** The smallest counterexample is 11. The sum of two odd primes is an even number; so if we are to represent 11 as such a sum, 2 will be one of the primes. The other number is then $11 - 2 = 9$, but 9 isn't prime.

**16. Still the 1.** The harder question is, "Are any of these prime?" We can describe each element in the list by its number of digits. If it has an even number of digits, then the number is divisible by 11. If it has 3, 6, 9, . . . digits, then the number is at least divisible by 111. So, the only candidates for prime numbers are those with lengths that are prime! By computer search, the first three primes in the sequence have 19, 23, and 317 digits.

**17. Zeros and ones.** $1001 = 13 \times 11 \times 7$ is the first nonprime in the sequence.

**18. Zeros, ones, and threes.** This sequence includes several primes, but they are still few and far between. The first nonprime is $1003 = 17 \times 59$. The next three primes on the list are $10^5 + 3$, $10^6 + 3$, and $10^{11} + 3$.

**19. A rough count.** The Prime Number Theorem states that the number of primes less than $10^{10}$ is about $10^{10}/\ln(10^{10})$, or just over 400 million.

**20. Generating primes.** The first nonprime given by the sequence $n^2 + n + 17$ occurs for $n = 16$. The resulting number is $289 = 17 \times 17$. The first nonprime for $n^2 - n + 41$ occurs for $n = 41$. (What are the factors of the corresponding number?)

**21. Generating primes II.** These are the Mersenne primes; the first nonprime of this form is $2^4 - 1 = 15$.

**22. Floating in factors.** The answer is the product of the three smallest prime numbers, $2 \times 3 \times 5 = 30$.

**23. Lucky 13 factor.** Call the mystery number $X$. The first statement allows us to express $X$ as $13A + 7$ for some unknown $A$. The number less one, $X - 1 = 13A + 6$, is still not divisible by 13. If we subtract 7, then we get $X - 7 = 13A$, which is divisible by 13. So the answer is 7.

**24. Remainder reminder.** As in Mindscape II.23, we write the original number as $X = 13A + 7$. Adding 22 yields $X + 22 = 13A + 7 + 22 = 13A + 29 = 13A + (13 \times 2) + 3 = 13(A + 2) + 3$. So 13 goes into our new number $A + 2$ times with a remainder of 3.

**25. Remainder roundup.** As in Mindscapes II.23, 24, write $X = 91A + 52$. $X + 103 = 91A + 155$. Recognize that $91 = 7 \times 13$ and $155 = 22 \times 7 + 1$, so that we can write $X + 103 = 7(13A) + (7 \times 22) + 1 = 7(13A + 22) + 1$. Final answer is 1.

### III. Creating New Ideas

**26. Related remainders.** The first line allows us to write our two numbers, $X$ and $Y$, in the following way: $X = 57A + r$, and $Y = 57B + r$. So, $(X - Y) = 57A - 57B = 57(A - B)$, and 57 definitely divides this number. Because $57 = 3 \times 19$, 3 and 19 will also divide $(X - Y)$. Suppose we divide two numbers by some integer $m$. The two numbers will have the same remainder upon division if and only if $m$ is a factor of the difference.

**27. Prime differences.**

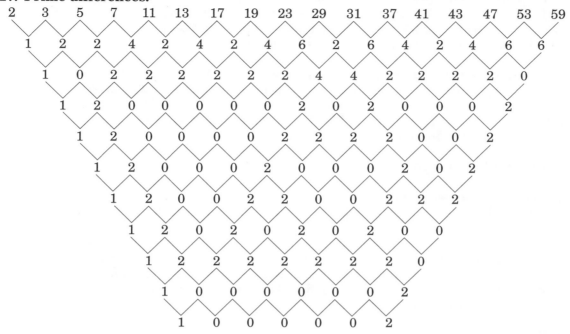

It appears that the first number of each row will be a one.

**28. Minus two.** If a prime number less two is also prime, then we call those numbers "twin primes." For example, 5 and 7 are twin primes, but 9 and 11 aren't. See Mindscape II.8.

**29. Prime neighbors.** Because 2 is the only even prime, 2 and 3 are the only primes that differ by one.

**30. Perfect squares.** There are 6 perfect squares less than 36, 12 less than 144, and in general, $n$ perfect squares less than $n^2$. Turning this around, we have an estimate of $\sqrt{N}$ perfect squares less than $N$.

**31. Perfect squares versus primes.** Using the results from Mindscape III.30, there are roughly $\sqrt{1,000,000,000} = 31,622$ perfect squares less than a billion. An estimate of the number of primes in this range is 1 billion/ln(1 billion), or just over 48 million. Because 48 million/32 thousand = 1500, there are roughly 1500 more prime numbers than perfect squares. So, yes, perfect squares are less common.

**32. Prime pairs.** This is the same question as in Mindscape III.29. If $p$ is a prime greater than 2, then $p$ is odd and $p + 1$ is even. Because 2 is the only even prime, $p + 1$ isn't prime.

**33. Remainder addition.** The two remainders are one and the same. We don't know how many times $n$ goes into $A$, but let's call it something, say $p$, so that we can express $A$ in equation form: $A = np + a$. Similarly, $B = nq + b$, where $p$ and $q$ are unknown. Finally, let's let $a + b = ns + c$, where both $s$ and $c$ are unknown. This means that the remainder after dividing $a + b$ by $n$ is $c$. Now, $A + B = np + a + nq + b = n(p + q) + (a + b) = n(p + q) + (ns + c) = n(p + q + s) + c$, which means that $A + B$ will also have a remainder of $c$ when divided by $n$.

**34. Remainder multiplication.** As we did in Mindscape III.33, write $A = np + a$, $B = nq + b$, and finally $ab = ns + c$. So, $AB = (np + a)(nq + b) = npnq + npb + naq + ab = nnpq + npb + naq + (ns + c) = n(npq + pb + aq + s) + c$. This means that dividing $AB$ by $n$ leaves a remainder of $c$, which is precisely the remainder when $ab$ is divided by $n$.

**35. A prime-free gap.** By exhaustively looking at the difference between successive primes, you will find that the first string of six nonprimes appears between 89 and 97. But it's more interesting to examine the hint. Let $M = 2 \times 3 \times 4 \times 5 \times 6 \times 7$. Both $M$ and $M + 2$ are divisible by 2, so $M + 2$ isn't prime. Similarly, $M + 3$ is divisible by 3, $M + 4$ is divisible by 4, and so on, up to $M + 7$, which is divisible by 7: $(7! + 1) = 5041$.

## IV. Further Challenges

**36. Prime-free gaps.** Mindscape III.35 shows that you can find six composite numbers in a row after the number $7! + 1$. Similarly, you can find $n$ composite numbers in a row after $n! + 1$.

**37. Three primes.** Other than 2 and 3, you cannot have two consecutive integers where both of which are prime, because one of them would be even, and the only even prime is 2. So, to have three in a row, 2 must be included. This leaves us with $\{1, 2, 3\}$ or $\{2, 3, 4\}$, neither of which contains purely prime numbers.

**38. Prime plus three.** In this problem, 11 is included as a distraction. Except for 2, all primes are odd; if you add 3 to an odd prime, you get an even number bigger than 2, so the sum won't be prime. If the 11 were replaced with a 2, the answer wouldn't change.

**39. A small factor.** Take 60 for example: $\sqrt{60} = 7.745 \ldots$. Let's suppose that 60 has only two prime factors, $p$ and $q$, so that $60 = pq$. If both $p$ and $q$ were greater than 7.745, then $pq$ would be greater than $7.745 \times 7.745 = 60$. This is impossible because $pq$ is equal to 60, not greater and not less. So, one of the factors has to be less than or equal to $7.745 \ldots$. By the same reasoning, one of the factors has to be greater than or equal to $7.745 \ldots$. Similarly, if a number $N$ has three factors, then its smallest prime factor is less than the cube root of $N$.

**40. Prime products.** If $N = 2 \times 5 \times 17 + 1$, then 2 doesn't divide $N$, because 2 goes into $N$ $5 \times 17$ times with a remainder of 1. Similarly, 5 and 17 don't evenly divide into $N$. All the prime factors of $N$ are necessarily different from 2, 5, and 17. (However, $N$ doesn't need to be prime.) Suppose there were only finitely many primes, $p_1, p_2, p_3, \ldots, p_L$. We can arrive at a contradiction to this statement by forming the product $N = p_1 p_2 p_3 \ldots p_L + 1$. The number $N$ is larger than any of the primes $p_1, p_2, p_3, \ldots, p_L$, so it can't be prime. Therefore, $N$ can be expressed as a product of primes. By the argument above, all the factors of $N$ must be different from $p_1, p_2, p_3, \ldots, p_L$, which contradicts the idea that the primes $p_1, p_2, p_3, \ldots, p_L$ were a complete list.

## 2.4 Crazy Clocks and Checking Out Bars

### I. Developing Ideas

**1. A flashy timepiece.** Twelve hours after 3:00, your watch will again show 3:00. Because $14 = 12 + 2$, in 14 hours your watch will show 5:00, 2 hours after 3:00. Because $25 = 2 \times 12 + 1$, in 25 hours your watch will show 4:00, just 1 hour after 3:00. Because $240 = 20 \times 12$, in 240 hours your watch will show 3:00 again.

**2. Living in the past.** Twenty-four hours before 8:00, your watch showed 8:00. Because $10 + 2 = 12$, 10 hours earlier, it showed 10:00. 25 hours earlier, it showed 7:00. 2400 hours earlier, it showed 8:00.

**3. Mod prods.** $16 \equiv 2 \pmod 7$; $24 \equiv 3 \pmod 7$; $16 \times 24 = 384 \equiv 6 \pmod 7$; $[16 \pmod 7 \times 24 \pmod 7] = 2 \times 3 = 6$. The last two quantities are equal.

**4. Mod power.** $7 \equiv 1 \pmod 3$; $7^2 \equiv 1 \pmod 3$; $[7 \pmod 3]^2 = (1)^2$, which equals $7^2 \pmod 3$. $7^{1000} \pmod 3 \equiv [7 \pmod 3]^{1000} \equiv 1^{1000} \equiv 1 \pmod 3$.

**5. A tower of mod power.** $13 \equiv 2 \pmod{11}$; $13^2 \pmod{11} \equiv 169 \pmod{11} \equiv 4 \pmod{11}$. Note that $[13 \pmod{11}]^2 = 13^2 \pmod{11}$. Also, $13^3 \pmod{11} \equiv [13 \pmod{11}]^3 \equiv 2^3 \pmod{11} \equiv 8 \pmod{11}$. Finally, $13^4 \pmod{11} \equiv [13 \pmod{11}]^4 \equiv 2^4 \pmod{11} \equiv 16 \pmod{11} \equiv 5 \pmod{11}$.

### II. Solidifying Ideas

**6. Hours and hours.** Because $96 = 8 \times 12$, the clock will complete 8 full revolutions after 96 hours, leaving the hand positions unchanged. Because $1063 = 88 \times 12 + 7$, after 1063 hours the clock will spin completely around 12 times, and then spin 7 more hours' worth, leaving the hands at 5:45. Because $-23 = -2 \times 12 + 1$, 23 hours before 7:10 the clock read 8:10. Similarly, 108 hours earlier, the clock read 7:10, because $-108 = -9 \times 12$.

**7. Days and days.** $3724 = 532 \times 7$ and $365 = 52 \times 7 + 1$. So, in 3724 days, it will still be Saturday, whereas the 365th day from now will fall on a Sunday.

**8. Months and months.** Express each number as a simpler number mod 12. ($219 = 18 \times 12 + 3$; $120{,}963 = 10{,}080 \times 12 + 3$; $-89 = -7 \times 12 - 5$; or $-8 \times 12 + 7$) 219 months from now will be October (July $+ 3$), and so will 120,963 months from now. Because $-89$ divided by 12 has a remainder of $-5$, we need only go back 5 months to February.

**9. Celestial seasonings.** Compute $3 \times 0 + 1 \times 7 + 3 \times 1 + 1 \times 7 + 3 \times 3 + 1 \times 4 + 3 \times 0 + 1 \times 0 + 3 \times 0 + 1 \times 2 + 3 \times 1 + 1 \times 8 = 43$. Because the sum is not evenly divisible by 10, it is not a correct UPC. The corresponding sum for the next two codes is 40 and 42, respectively. So the second code is the correct one.

**10. SpaghettiOs.** (See Mindscape II.9.) Because the sums are 41, 49, and 50, respectively, the third number is correct.

**11. Progresso.** (See Mindscape II.9.) Because the sums are 50, 24, and 68, respectively, the first number is correct.

**12. Tonic water.** (See Mindscape II.9.) Because the sums are 50, 51, and 19, respectively, the first number is correct.

**13. Real mayo.** If the covered digit were $D$, then the sum would be $55 + 3D$. The goal is to find a value for $D$ that will make the sum divisible by 10, which is easily found by trial and error. $D = 5$ is the only digit between 0 and 9 that works. (Note that we are solving the equation $55 + 3D \equiv 0$ (mod 10), or equivalently, $3D \equiv 5$ (mod 10).)

**14. Applesauce.** (See Mindscapes II.13–18.) If $D$ represents the missing digit, then the computed sum is $54 + D$. The only way to make the sum a multiple of 10 is to choose $D = 6$.

**15. Grand Cru.** (See Mindscapes II.13–18.) If the missing digit were 0, the sum would be 89. If the missing digit were $D$, the sum would be $89 + 3D$. We want to make this sum evenly divisible by 10. Because $89 \equiv 9$ (mod 10), we need $3D \equiv 1$ (mod 10), and the only solution is $D = 7$.

**16. Mixed nuts.** (See Mindscapes II.13–18.) Suppose that the missing digit were $K$. The sum would be $80 + K$. To make this sum a multiple of 10, we only need to choose $K = 0$; indeed, this is the missing digit.

**17. Blue chips.** (See Mindscapes II.13–18.) Just as in Mindscape II.16, the sum is $50 + 3M$, where $M$ is the covered digit. The only way to make this sum divisible by 10 is to choose $M = 0$.

**18. Lemon.** (See Mindscapes II.13–18.) If $M$ is the missing digit, then the sum is $49 + M$; so, $M = 1$.

**19. Decoding.** There are three unknown digits, the 9, the 1, and the 7. Because each digit could be one of two different numbers, we have eight possible combinations to try: 903068823517, 903068823511, 903068823577, 903068823571, 403068823517, 403068823511, 403068823577, 403068823571. Of all these numbers, only 903068823577 is a valid code. This is your best guess.

**20. Check your check.** Look up your bank code on your check. Use the technique on page 88 to verify that it is a valid bank code.

**21. Bank checks.** As with Mindscapes II.13–18, let the missing digit be $D$, and compute the sum. The resulting sum for the first bank code is $170 + 9D$, so $D$ must be 0 to keep the sum divisible by 10. The second sum is $136 + 9D$. We need $9D \equiv 4$ (mod 10), so $D = 6$.

**22. More bank checks.** (See Mindscape II.21.) With the missing digit represented by $D$, the sums are $171 + 9D$ and $84 + 9D$, respectively. So, the correct codes are 6 2 9 1 0 0 2 7 1 and 5 5 0 3 1 0 1 1 4. In the second example, $84 \equiv 4$ (mod 10), so we need $9D \equiv 6$ (mod 10). The only value of $D$ that satisfies this equation is $D = 4$.

**24. Whoops.** In each example, two changes were made that canceled each other out. In the first code, the 9th and 11th digits were switched. Because the sum is computed by multiplying the 9th digit by 3 and the 11th digit by 3, the sum doesn't change. Similarly, for the second example, the 3rd and 8th digits are changed. Instead of the sum being $\ldots + 3 \times 1 + \ldots + 1 \times 2 + \ldots$, we have $\ldots + 3 \times 0 + \ldots + 1 \times 5 + \ldots$, where the $\ldots$ represents parts of the sum that are unchanged. Because $3 \times 1 + 1 \times 2 = 3 \times 0 + 1 \times 5$, the sum remains unchanged.

**25. Whoops again.** (See Mindscape II.24.) In the first code, the 1st and 4th digits are changed; instead of $7 \times 0 + \ldots + 7 \times 7 + \ldots$, we have $7 \times 7 + \ldots + 7 \times 0 + \ldots$. Thus, the sum remains unchanged. The same type of mistake occurs in the second example where the 6th and 9th terms are interchanged. Because the 6th and 9th terms are both multiplied by the same weight, 9, the total sum will remain unchanged.

### III. Creating New Ideas

**26. Mod remainders.** $129 = 9 \times 13 + 12$, so 12 is the remainder when 129 is divided by 13. We can also say $129 \equiv 12 \pmod{13}$. A quick way to see this is $129 = 130 - 1 = 10 \times 13 - 1 = 9 \times 13 + 13 - 1 = 9 \times 13 + 12$. You would spin around 13 times and then move the clock ahead 12 hours more.

**27. More mod remainders.** $2015 = 287 \times 7 + 6$. So, $2015 \equiv 6 \pmod{7}$. If $m$ divided by $n$ gives a remainder $r$, then we can say $m \equiv r \pmod{n}$. If we had a clock with $n$ hour positions (0 through $n - 1$), then after moving the hour hand of the clock $m$ places, the hand will be sitting in the $r$th position.

**28. Money orders.** Because 6830910275 is divisible by 7, the check digit is 0.

**29. Airline tickets.** We have $10061559129884 = 1437365589983 \times 7 + 3$, so the check digit is 3.

**30. UPS.** (See Mindscapes III.28–29.) $84200912 = 12028701 \times 7 + 5$, so the check digit is 5.

**31. Check a code.** Check the identification number on your example using the technique in Mindscape III.28 or III.29.

**32. ISBN.** Verify this check method for the ISBN of the textbook.

**33. ISBN check.** The first code has a sum of $152 + D$, where $D$ is the check digit. Because $152 \equiv 8 \pmod{11}$, we want $8 + D \equiv 0 \pmod{11}$; that is, $D = 3$. Similarly, the second code has sum $107 + D$. To solve the equation $107 + D \equiv 0 \pmod{11}$, note that $107 = 9 \times 11 + 8$, simplifying our equation to $8 + D \equiv 0 \pmod{11}$. Once again the check digit is 3.

**34. ISBN error.** The current number corresponds to a sum of $192 \equiv 5 \pmod{11}$. If the first and second digits were transposed, what would the new sum be? The old sum can be expressed as $10 \times 3 + 9 \times 5 + \ldots$, and the new sum would be $10 \times 5 + 9 \times 3 + \ldots$. The difference between the two sums is $10 \times (5 - 3) + 9 \times (3 - 5)$, or $(5 - 3) \times (10 - 9) = (5 - 3) = 2$. Without recomputing the sum, we can deduce that the new sum would total 194. Notice the pattern—if we interchange the $i$th and $(i + 1)$th digits, the difference will be $(d_{i+1} - d_i)$. We want to find a difference that is equal to 6 so that the new sum will be $192 + 6 \equiv 198 \pmod{11}$. The only pair of adjacent digits that differs by 6 is in the fifth and sixth digits. The correct number is 3-540-60395-6.

**35. Brush up your Shakespeare.** Find a book with a play by Shakespeare and check its ISBN using the technique described in Mindscape III.32.

### IV. Further Challenges

**36. Mods and remainders.** Example: 23 divided by 7 is 3 with a remainder of 2. When performing long division, 7 is outside the division sign, 23 is inside, 3 is on top, and 2 is at the very bottom, the last number computed. This means that $23 = 3 \times 7 + 2$, or equivalently $23 \equiv 2 \pmod{7}$. In terms of clocks, if we had a mod 7 clock with hand initially at 0, then moving 23 units is equivalent to spinning around completely 3 times and then moving 2 more units. This puts the clock hand in position 2. Generalize this example, with $n = qm + r$, where $q$ is the quotient and $r$ is the remainder.

**37. Catching errors.** Extreme examples are usually simpler to understand. Because the weights of any adjacent pair of digits are either 3 or 1, let's focus on transposing the first two digits. 1600000001 and 6100000001 are both valid. Similarly, transposing the first digits of the following will still result in a valid number: 2700000007, 3800000003, 4900000009, 5000000005, and so on. In summary, if the difference between adjacent digits is divisible by 5, transposing the digits still represents a valid code.

**38. Why three?** The key insight is that you get 10 different remainders (mod 10) when multiplying by 3 and only 5 when multiplying by 6. Turns out that any number relatively prime to 10, such as 1, 3, 7, 9, 11, 13 . . . , will provide 10 distinct remainders. With 10 distinct remainders, each digit contributes a different amount to the total sum. For example, when multiplying by 6, both 4 and 9 contribute the same because $4 \times 6 = 24 \equiv 4 \pmod{10}$ and $9 \times 6 = 54 \equiv 4 \pmod{10}$. If the 4 were scratched out, you could only tell that the number was either a 4 or a 9. With 10 different remainders, you can always recover the covered digit.

**39. A mod surprise.** It's surprising that $n^4 \pmod 5$ is 1 for every number. Section 2.5 discusses this in detail.

**40. A prime magic trick.** The mystery number that you write down will always be 1, so don't play the trick too many times on the same person.

358

## 2.5 Public Secret Codes and How to Become a Spy

### I. Developing Ideas

**1. What did you say?** THIS IS THE CORRECT MESSAGE.

**2. Secret admirer.** The message encodes to B WXAU GXL.

**3. Setting up secrets.** The numbers $p = 7$ and $q = 17$ are both prime because each has no factors other than itself and 1. The number $m = (p - 1)(q - 1) = 6 \times 16 = 96$. The number $e = 5$ has no factors in common with $m = 96$. Finally, $5 \times 77 - 96 \times 4 = 385 - 384 = 1$.

**4. Second secret setup.** The numbers $p = 5$ and $q = 19$ are both prime because each has no factor other than itself and 1. The number $m = (p - 1)(q - 1) = 4 \times 18 = 72$. The number $e = 11$ has no factors in common with $m = 72$. Finally, $11 \times 131 - 72 \times 20 = 649 - 648 = 1$.

**5. Secret squares.** We find $2^2 = 4 \equiv 1 \pmod 3$; $3^2 = 9 \equiv 0 \pmod 3$; $4^2 = 8 \equiv 2 \pmod 3$; $5^2 = 25 \equiv 1 \pmod 3$. As you use successive integers, the result (mod 3) cycles through the pattern 1, 0, 2, 1, 0, 2, . . . .

### II. Solidifying Ideas

**6. Petit Fermat 5.** The expressions are all of the form $k^{(p-1)} \pmod p$; thus, by Fermat's Little Theorem, they are equal to 1 (mod $p$). For example, $4^4 = (4^2)^2 = (16)^2 = 1^2 = 1 \pmod 5$, where the second-to-last equality results because $16 \equiv 1 \pmod 5$.

**7. Petit Fermat 7.** As in Mindscape II.6, the numbers are all of the form $k^{(p-1)} \pmod p$, and so are equal to 1 (mod 7).

**8. Top secret.** The encoded word is $2^7 \pmod{143} \equiv 128$. To decode the number, raise the encrypted information to the 103rd power and compute the remainder (mod 143).

**9. Middle secret.** You don't need to compute $3^7$ explicitly. Because $3^5 \equiv 243 \equiv 100 \pmod{143}$, $3^6 \equiv 3 \times 100 \equiv 14 \pmod{143}$ and $3^7 \equiv 3 \times 14 \equiv 42 \pmod{143}$. As in Mindscape II.8, the information is decoded by the computation $42^{103} \equiv 3 \pmod{143}$.

**10. Bottom secret.** We need to compute $10^7 \pmod{143}$ $(10^7 = 10{,}000{,}000)$. Because $10{,}000{,}000/143 = 69930$ with a remainder of 10, we have $10^7 \pmod{143} \equiv 10$. The original 10 can be recovered by computing $10^{103} \equiv 10 \pmod{143}$. Even though the encoded number is identical to the original number, it's still a secret because you are the only person who knows that the two numbers are one and the same.

**11. Creating your code.** Note first that $m = (3 - 1)(5 - 1) = 8$. Because $e$ must be relatively prime to $m$, we need only consider the values $e = 1, 3, 5,$ and 7. For each possible value of $e$, find $d$ and $y$ that satisfy $de - 8y = 1$. For example, for $e = 1$, fill in the following blanks: _ $\times 1 - 8 \times$ _ $= 1$. Because $1 \times 1 - 8 \times 0 = 1$, $(e = 1, d = 1)$ is a pair. Similarly, because $3 \times 3 - 8 \times 1 = 1$, $5 \times 5 - 8 \times 2 = 1$, and $7 \times 7 - 8 \times 6 = 1$, $(e = 3, d = 3)$, $(e = 5, d = 5)$, and $(e = 7, d = 7)$ are all pairs.

**12. Using your code.** HI becomes (08)(09). To use the coding scheme ($p = 3$, $q = 5$, $e = 3$, $d = 3$), we need to compute $8^3 \equiv 2 \pmod{15}$ and $9^3 \equiv 9 \pmod{15}$. So the code is (02)(09), or BI. Because $2^3 \equiv 8 \pmod{15}$ and $9^3 \equiv 9 \pmod{15}$, we get the original message back upon decoding. Note that you can only use the first 14 letters of the alphabet.

**13. Public secrecy.** Using $83^7 \equiv 8 \pmod{143}$, the encoded version is 8. You can decipher this message with the formula $8^{103} \equiv 83 \pmod{143}$.

**14. Going public.** You encode 61 by computing $61^7 \equiv 74 \bmod 143$, and you decode 74 by computing $74^{103} \equiv 61 \pmod{143}$.

**15. Secret says.** Use $38^{103} \equiv 103 \pmod{143}$ to obtain the original message, 103.

### III. Creating New Ideas

**16. Big Fermat.** The hint asks you to recall that $5^6 \equiv 1 \pmod 7$. This means that $(5^6)^k \equiv 1^k \equiv 1 \pmod 7$ for any integer $k$. In particular, because $600 = 6 \times 100$, it is convenient to choose $k = 100$, giving us $5^{600} \equiv (5^{6 \times 120}) \equiv (5^5)^{120} \equiv 1^{120} \equiv 1 \pmod 7$. Similarly, because $1,000,000 = 10 \times 100,000$, $8^{1,000,000} \equiv 1 \pmod{11}$.

**17. Big and powerful Fermat.** (See also solution to Mindscape III.16.) Our building block is the formula $5^6 \equiv 1 \pmod 7$. After dividing 668 by 6, we can represent $668 = 6 \times 111 + 2$. Therefore,
$$5^{668} \equiv 5^{6 \times 111 + 2} \equiv 5^{6 \times 111} \times 5^2 \equiv (5^6)^{111} \times 25 \equiv 1^{111} \times 4 \equiv 4 \pmod 7.$$

**18. The value of information.** You would have to answer the following questions: Who am I keeping this from? How much time would they be willing to spend trying to break the code? With their resources, what size numbers can they factor in that time? As a reference point, you might note that Maple, a standard mathematical computer package, can factor the product of two 29-digit primes in 30 seconds on a Linux workstation! For every 3 digits you tack on, Maple takes *twice* as long to complete the factorization. With two 32-digit primes, it takes 1 minute; 35-digit primes, 2 minutes; 50-digit primes, 1 hour. (How large would the primes need to be for Maple to require 100 years' worth of computation time?)

**19. Something in common.** Because $p$ divides $n$, $p$ will also divide $n^{p-1}$, so that when we divide $n^{p-1}$ by $p$, the remainder will be 0. This means, exactly, $n^{p-1} \equiv 0 \pmod p$.

**20. Faux pas Fermat.** Fermat's Theorem doesn't hold here because 6 isn't prime. We get $1^5 \equiv 1$, $2^5 \equiv 2$, $3^5 \equiv 3$, $4^5 \equiv 4$, $5^5 \equiv 5 \pmod 6$. However, $1^2 \equiv 1$, $2^2 \equiv 4$, $3^5 \equiv 3$, $4^5 \equiv 4$, $5^5 \equiv 1 \pmod 6$. We also have $1^6 \equiv 1$, $2^6 \equiv 1$, $3^6 \equiv 0$, $4^6 \equiv 1$, $5^6 \equiv 1$, $6^6 \equiv 0$, $7^6 \equiv 1$, $8^6 \equiv 1 \pmod 9$. Note that the pattern of 1's is broken only when the base shared a factor with 9. Note also that there are six numbers sharing no factors with 9 and the exponent is also 6. Similarly, there are two numbers relatively prime to 6 (1 and 5), and when these are raised to the second power, they equal 1 (mod 6). Yes, if $k$ and $m$ are relatively prime and $n$ is the number of relatively prime numbers less than $m$, then $k^n \equiv 1 \pmod m$. For prime numbers $p$, there are $p - 1$ numbers less than $p$ that are also relatively prime to $p$ (why?). Thus, this is a generalization of Fermat's Little Theorem.

### IV. Further Challenges

**21. Breaking the code.** We know only the public numbers $e$ and the product $pq$, but we want to find $d$, the decoding number. After getting $p$ and $q$, construct $m = (p - 1)(q - 1)$. The numbers $e$ and $m$ satisfy the following equation: $\_e - \_m = 1$, where the blanks are integers and the first blank represents the decoding number $d$. Here is an outline of the systematic process that is the Euclidean Algorithm. Divide $e$ into $m$, getting remainder $r_1$. Divide $r_1$ into $e$, giving $r_2$, and so on, until you get down to a remainder of $r_n = 1$. Now go backward, and express 1 as a linear combination of $r_{n-1}$ and $r_n$. Then, because $r_n$ can be expressed as a linear combination of $r_{n-1}$ and $r_{n-2}$, we can write 1 as a linear combination of $r_{n-1}$ and $r_{n-2}$. Repeat this

360

process until you have written 1 as a linear combination of $e$ and $m$. At this point, you will have filled in the blanks and found $d$.

**22. Signing your name.** Joseph writes a separate message, "Really, this is me Joseph. Pork kidneys are the wave of the future." He then scrambles it by decoding it as he would for any incoming encoded messages and inserts the new text into his letter. You decode this secret message by encoding it with his public keys as if you were going to send him a secret note. When you raise the text to the $e$ power, his original message will appear. Irving Satan would have to know Joseph's private key to forge this extra personal message.

## 2.6 The Irrational Side of Numbers

### I. Developing Ideas

**1. A rational being.** A rational number is a number that can be expressed as a fraction—the ratio (or quotient) of two whole numbers.

**2. Fattened fractions.** $6/24 = 1/4$, $15/9 = 5/3$, $-14/42 = -1/3$, $125/10 = 25/2$, $-121/11 = -11$

**3. Rational arithmetic.** $\frac{1}{2} + \frac{5}{2} = \frac{6}{2} = 3$; $\quad \frac{1}{2} - \frac{2}{3} = \frac{3}{6} - \frac{4}{6} = -\frac{1}{6}$; $\quad \frac{1}{2} \times \frac{6}{5} = \frac{6}{10} = \frac{3}{5}$; $\quad \frac{\frac{1}{2}}{\frac{2}{3}} = \frac{1}{2} \times \frac{3}{2} = \frac{3}{4}$;

$\frac{\frac{5}{2} \times \frac{6}{5}}{\frac{2}{3}} = \frac{\frac{30}{10}}{\frac{2}{3}} = \frac{3}{1} \times \frac{3}{2} = \frac{9}{2}$

**4. Decoding decimals.** $0.02 = 2/100$, $6.23 = 623/100$, $2.71828 = 271{,}828/100{,}000$, $-168.5 = -1685/10$, $-0.00005 = -5/100{,}000$.

**5. Odds and ends.** The squares are 1, 4, 9, 16, 25, 36, 49, 64, 81, 100. The even numbers have even squares, and the odd numbers have odd squares.

### II. Solidifying Ideas

**6. Irrational rationalization.** No; $3\sqrt{2}/5\sqrt{2} = 3/5$, which is rational. The product or quotient of an irrational and a rational is always irrational, so both $3\sqrt{2}$ and $5\sqrt{2}$ are irrational. However, the quotient (or product) of two irrationals is not always irrational.

**7. Rational rationalization.** Yes; the quotient of two rationals is rational. If $a$, $b$, $c$, and $d$ are integers, then $(a/b)/(c/d) = ad/bc$, which is rational by definition; it is the quotient of two integers.

**8. Rational or not.** $\sqrt{2}/14$ is the only irrational number in the list. As the ratio of two integers, 4/9 is rational by definition. $1.75 = 1 + 3/4 = 7/4$, $\sqrt{20}/(3\sqrt{5}) = (\sqrt{20}/\sqrt{5})/3 = \sqrt{4}/3 = 2/3$, $3.14159 = 314{,}159/100{,}000$. We could reason that 3.14159 and 1.75 are rational because they each have a repeating decimal expansion ($1.7500000\ldots$, $3.13159000\ldots$).

**9. Irrational or not.** All but $\sqrt{3}/3$ are rational. $\sqrt{16}/20 = 4/5$, $12/7.5 = 120/75$, $-147 = -147/1$, $0 = 0/1$. $\sqrt{3}/3$ is the quotient of an irrational and a rational number and, therefore, is irrational.

**10. $\sqrt{5}$.** The proof is identical to the proof of the irrationality of $\sqrt{2}$, except the notions *even* and *odd* are replaced with *divisible by 5* and *not divisible by 5,* respectively. Assume $\sqrt{5} = b/c$, with $b$ and $c$ having no common factors. We have $b^2 = 5c^2$, implying $b$ is divisible by 5 (because 5 is prime and must therefore appear as one of the prime factors of $b$; we're using the uniqueness of the prime factorization here). Expressing $b = 5d$ gives $25d^2 = 5c^2$, or $5d^2 = c^2$, implying $c$ is *also* divisible by 5, a contradiction. This same idea can be applied to the square root of any prime number.

**11. $\sqrt{6}$.** Suppose $\sqrt{6} = b/c$ where $b$ and $c$ are natural numbers having no common factors. Then $b^2 = 6c^2 = 2 \times 3c^2$. So $b^2$ is even, and $b$ is even. So $b = 2n$. So $b^2 = (2n)^2 = 4n^2 = 2 \times 3c^2$. So $2n^2 = 3c^2$ and $c$ is even. Both $b$ and $c$ have 2 as a factor, contradicting our assumption that they have no common factor.

362

**12.** $\sqrt{7}$. Identical in spirit to Mindscape II.10.

**13.** $\sqrt{3}+\sqrt{5}$. An alternate style of proof: $(\sqrt{3}+\sqrt{5})^2 = (\sqrt{3}+\sqrt{5})(\sqrt{3}+\sqrt{5}) = 3+2\sqrt{3}\sqrt{5}+5 = 8+2\sqrt{15}$. First argue that $\sqrt{15}$ is irrational (see Mindscape II.15). Use the fact that a rational times an irrational is irrational to show that $2\sqrt{15}$ is irrational. Similarly, the sum of a rational and an irrational is also irrational, which implies that $8+2\sqrt{15} = (\sqrt{3}+\sqrt{5})^2$ is irrational. If $(\sqrt{3}+\sqrt{5})$ were rational, then $(\sqrt{3}+\sqrt{5})^2$ would be rational, too. Because $(\sqrt{3}+\sqrt{5})^2$ is *not* rational, $(\sqrt{3}+\sqrt{5})$ is not rational either; thus, completing the proof.

**14.** $\sqrt{2}+\sqrt{7}$. Model the text's proof that $\sqrt{2}+\sqrt{3}$ is irrational. Assume $\sqrt{2}+\sqrt{7} = a/b$ (in lowest terms). $(\sqrt{2}+\sqrt{7})^2 = 9+2\sqrt{14} = a^2/b^2$, so that $\sqrt{14} = (a^2/b^2 - 9)/2 = (a^2 - 9b^2)/2b^2$, contradicting the fact that $\sqrt{14}$ is irrational (see Mindscape II.15).

**15.** $\sqrt{10}$. We need a slight modification of the proof in Mindscape II.10. Begin the same way: Assume $\sqrt{10} = c/d$, with $c$ and $d$ having no common factors. Squaring gives $c^2 = 10d^2$. Because the right side is divisible by 5, the left side is divisible by 5 as well. This means that $c$ is divisible by 5. (If $c$ weren't divisible by 5, then $c^2$ wouldn't be divisible by 5 either.) Write $c = 5n$. Substituting gives $25n^2 = 10d^2$, or $5n^2 = 2d^2$. Now we must work harder to show that 5 divides $d$. Imagine writing out the prime factorization for the left and right sides of the equation. On the left, we have all the prime factors of $n$ (listed twice) and 5. On the right, we have 2 and all the prime factors of $d$ (listed twice). Because both sides represent the same number, we call upon the uniqueness of prime factorizations to argue that the list of primes on both sides are the same. Because the prime 5 appears on the left side, it must also appear on the right side. And because it can only come from the prime factorization of $d$, we must have that 5 is a prime factor of $d$. So, $d$ is divisible by 5, and we have our contradiction.

**16.** $1+\sqrt{10}$. If $1+\sqrt{10} = a/b$, then $\sqrt{10} = a/b - 1 = (a-b)/b$ is rational. Mindscape II.15 shows that this is not the case. This contradiction shows that our assumption was wrong, proving that $1+\sqrt{10}$ is irrational.

**17. An irrational exponent.** Assume that $E$ is rational, that is, $E = n/m$. Substituting gives $12^{n/m} = 7$, or $12^n = 7^m$. Look at the prime factorization of both sides of this equation. Because both sides represent the same number, the prime factorizations must agree (uniqueness of prime factorization). On the left, we have $n$ 3's and twice as many 2's. On the right we have $m$ 7's. These two lists can't be the same, regardless of the values of $n$ and $m$. This represents a contradiction, and because our only assumption was the rationality of $E$, we conclude $E$ is irrational.

**18. Another irrational exponent.** If $E = n/m$, then $13^n = 8^m$. The prime factorization of the left side is just a bunch of 13's, whereas the prime factorization of the right side is composed solely of 2's. Because this represents two different prime factorizations of the same number, we have derived a contradiction to the uniqueness of prime factorizations. Therefore, our assumption was wrong; $E$ is not rational.

**19. Still another exponent.** Identical in spirit to Mindscapes II.17 and II.18.

**20. Rational exponent.** $E = 2/3$. $8^{2/3}$ is equivalent to taking the cube root of 8 and then squaring the result. Apply the reasoning behind solutions to Mindscapes II.17 and II.18. Assume $E = n/m$, giving $8^n = 4^m$. The prime factorization of the left side has $3n$ 2's, and the right side has $2m$ 2's. But these two factorizations could be exactly the same if the number of 2's on both sides were the same. We need only $3n = 2m$, or $n/m = 2/3$.

**21. Rational exponent.** Because the notion of a prime factorization is relevant only to integers, let's rewrite the equation as $2^{E/2} = 2^{3/2}$. From this equation, it is immediately apparent that $E = 3$ works. So $E$ is rational after all.

**22. Rational sums.** The two rationals are $a/b$ and $c/d$, where $a$, $b$, $c$ and $d$ are all integers. $a/b + c/d = (ad + bc)/(bd)$. Because the product and sum of two integers is just another integer, we have expressed the sum as a ratio of two integers. So, the sum is rational.

**23. Rational products.** Let $a/b$ and $c/d$ represent the two rational numbers, where $a$, $b$, $c$, and $d$ are all integers. Because the product $(a/b)(c/d)$ can be expressed as the quotient of two integers, $ac/bd$, the product is rational.

**24. Root of a rational.** Rewrite $\sqrt{1/2}$ as $1/\sqrt{2}$, and use the fact that the quotient of a rational and an irrational is irrational. See Mindscape II.25 for an alternate approach to this problem.

**25. Root of a rational.** Adapt the "$\sqrt{2}$ is irrational" proof. Assume $\sqrt{2/3} = a/b$ (with no factors in common). Squaring gives $2b^2 = 3a^2$. At this point, it doesn't matter whether you choose 2 or 3, but you must stick with whatever you choose! (I'll choose 3.) The right side is divisible by 3; thus, $2b^2$ is divisible by 3. Because 3 is prime and because 2 isn't divisible by 3, $b^2$ must be divisible by 3. Again, because 3 is prime, we conclude that $b$ is divisible by 3. Writing $b = 3n$ and substituting it into the equation gives $18n^2 = 3a^2$, or $6n^2 = a^2$. Using the same reasoning, we conclude that 3 divides $a$. So 3 divides both $a$ and $b$, contradicting the fact that $a$ and $b$ had no common factors. We conclude that our original assumption was wrong, and therefore $\sqrt{2/3}$ is irrational.

### III. Creating New Ideas

**26. $\pi$.** Use an indirect proof. Assume that the sum is rational—that is, suppose that $\pi + 3 = a/b$, with $a$ and $b$ representing integers. Rewriting yields $\pi = a/b - 3 = (a - 3b)/b$, which contradicts the fact that $\pi$ cannot be written as the quotient of two integers. Because our only assumption was that the sum was rational, the contradiction allows us to conclude that the assumption was wrong. The only alternative is that the sum is irrational.

**27. $2\pi$.** As in Mindscape III.26, we prove this indirectly. If the product were rational, then we could write $2\pi = m/n$, where $m$ and $n$ were integers. Solving for $\pi$ then yields $\pi = m/2n$, which is the form of a *rational* number. In summary, we have shown that if $2\pi$ is rational, then $\pi$ is rational. Because we know that $\pi$ is *not* rational, we conclude $2\pi$ is also *not* rational.

**28. $\pi^2$.** This is identical in spirit to Mindscapes III.26 and III.27. If we assume that $\pi$ is rational, we can write $\pi = n/m$, where $n$ and $m$ are integers. So, $\pi^2 = n^2/m^2$, which means that we have expressed $\pi^2$ as a rational number, which is a contradiction. We must conclude that our original assumption is wrong. The only alternative is that $\pi$ is *not* rational!

**29. A rational in disguise.** One useful property of exponentials is $(x^a)^b = x^{ab}$. So, $(\sqrt{2}^{\sqrt{2}})^{\sqrt{2}} = \sqrt{2}^{(\sqrt{2}\sqrt{2})} = \sqrt{2}^2 = \sqrt{2}\sqrt{2} = 2$. After this simplification, we can easily classify it as a rational number.

**30. Cube roots.** Use the "$\sqrt{2}$ is irrational" proof as a template. Assume $\sqrt[3]{2}$ is rational—that is, $\sqrt[3]{2} = a/b$, where $a$ and $b$ have no common factors. *Cube* both sides, and multiply through by $b^3$ to get $2b^3 = a^3$. This implies that $a^3$ is even, which in turn implies that $a$ is even. This allows us to write $a = 2n$. Substituting gives $2b^3 = 8n^3$, or $b^3 = 4n^3$. By the same reasoning, we can argue that $b$ is even contradicting the fact that $a$ and $b$ share no common factors. Thus, our original assumption is wrong, which implies $\sqrt[3]{2}$ is irrational.

**31. More cube roots.** See Mindscapes II.11 and III.30. Assume $\sqrt[3]{3} = a/b$, where $a$ and $b$ are reduced to lowest terms. Cubing and rearranging gives $3b^3 = a^3$. This means that 3 divides $a^3$. Because 3 is prime, it must divide $a$ as well. (At the core of this reasoning is the Prime Factorization Theorem.) Write $a = 3n$, and substitute this into the last equation to get $3b^3 = 27n^3$, or $b^3 = 9n^3$. By the same reasoning, we can assert that because $b^3$ is divisible by 3, $b$ is also divisible by 3. But now both $a$ and $b$ share the common factor 3, which is a contradiction. Thus, our original assumption was wrong. Therefore, $\sqrt[3]{3}$ is irrational.

**32. One-fourth root.** This is identical in nature to Mindscape III.31.

**33. Irrational sums.** Not always; the not-so-satisfying counterexample is that $\pi$ and $-\pi$ are both irrationals, yet their sum is 0, which is rational. The numbers 1.01001000100001 . . . and 0.10110111011110 . . . are both irrational because the tail end of their decimal expansion can't be represented as a repeating segment. Their sum is 1.11111111111111 . . . = 10/9, a rational number. Keep in mind that sometimes the sum is irrational, for example, $\pi + \pi = 2\pi$.

**34. Irrational products.** Sometimes the product is irrational, as in $\sqrt{2}\sqrt{5} = \sqrt{10}$; but sometimes the product is rational, as in $\sqrt{2}\sqrt{2} = 2$. Another simple example: Both $\pi$ and $1/\pi$ are irrational, but their product is 1, which is rational.

**35. Irrational plus rational.** This is a generalization of Mindscape III.26. An indirect approach is best. Assume that the sum of an irrational and a rational is *rational,* and try to derive a contradiction. Let $q$ represent the irrational number, let $a/b$ represent the rational, and because we are assuming that the sum is rational, let $c/d$ represent the sum. We then have $q + a/b = c/d$. Solving for $q$ gives $q = c/d - a/b = (cb - ad)/db$. We've just expressed $q$ as the quotient of two integers, which contradicts the fact that $q$ represents an irrational number. This contradiction proves our assumption wrong. The only alternative is that the original sum is irrational.

## IV. Further Challenges

**36. $\sqrt{p}$.** (This generalizes Mindscape II.10.) Assume $\sqrt{p} = a/b$, where $a$ and $b$ share no common factors. Squaring and rearranging gives $pb^2 = a^2$, implying that $a^2$ is divisible by $p$. Because $p$ is prime, it appears in the prime factorization of $a^2$. Because the prime factorization of $a^2$ contains two copies of all the primes appearing in the prime factorization of $a$ (by the uniqueness of prime factorization), $p$ must also be in the prime factorization of $a$; thus, $a$ is also divisible by $p$. So we can write $a = pn$ and put this back into the equation, giving $pb^2 = p^2n^2$, or $b^2 = pn^2$. The exact same reasoning shows that $b$ is divisible by $p$, contradicting the fact that $a$ and $b$ share no factors. This contradiction implies that our assumption was wrong, which in turn means that $\sqrt{p}$ is irrational.

**37. $\sqrt{pq}$.** (This generalizes Mindscapes II.15 and IV.36.) Assume $\sqrt{pq} = a/b$, where $a$ and $b$ are in lowest terms. Squaring and rearranging gives $pqb^2 = a^2$, implying that $p$ divides $a^2$. Because $p$ is prime, we can use the reasoning given in Mindscape IV.36 to show that $p$ also divides $a$. Now replace $a$ with $np$ in the equation above: $pqb^2 = p^2a^2$ and so $qb^2 = pa^2$. Because $p$ divides the right side of the equation, it must also divide the left side of the equation. Equivalently, $p$ must appear as one of the primes in the prime factorization of $qb^2$. The prime factors of $qb^2$ are $q$ and all the prime factors of $b$ listed twice. Because $p$ is in the collection of primes, and because $p$ doesn't equal $q$, $p$ must be in the prime factorization of $b$. Thus, $p$

divides $b$, which is contrary to our assumption that both $a$ and $b$ share the common factor $p$. This contradiction proves that $\sqrt{pq}$ is irrational.

**38.** $\sqrt{p} + \sqrt{q}$. Break this problem into two cases. Case I: $p = q$. Case II: $p$ and $q$ are different. (Always do the easy one first!) If $p = q$, then we need only show that $2\sqrt{p}$ is irrational. By Mindscape IV.36, we know $\sqrt{p}$ is irrational, and because the product of a rational and an irrational is always irrational, we know that $2\sqrt{p}$ is also irrational (done with Case I). Now demand that $p$ and $q$ are different. Let's assume $\sqrt{p} + \sqrt{q}$ is rational. Try to derive a contradiction, that is, assume $\sqrt{p} + \sqrt{q} = a/b$. Squaring both sides yields $p + 2\sqrt{pq} + q = a^2/b^2$. Solving for $\sqrt{pq}$ gives $\sqrt{pq} = ((a^2/b^2) - p - q)/2$, which is the contradiction. The right side represents a rational number, but in Mindscape IV.37, we proved that the left side was irrational. This contradiction shows that our assumption about the rationality of $\sqrt{p} + \sqrt{q}$ was wrong. $\sqrt{p} + \sqrt{q}$ is irrational.

**39.** $\sqrt{4}$. Assume $\sqrt{4} = a/b$ and square both sides to get $4b^2 = a^2$. At this point we typically say, "4 divides $a^2$, so 4 must then also divide $a$." That's the mistake! For example, 4 divides 62, but 4 does not divide 6. All we can say is that because 2 divides $a^2$, then 2 also divides $a$. We can say this because 2 is prime. Because 2 appears in the prime factorization of $a^2$, it must appear in the prime factorization of $a$. This isn't enough to derive a contradiction. Write $a = 2n$; substituting gives $4b^2 = 4n^2$, or $b^2 = n^2$. We can't conclude anything about the factors of $b$.

**40. Sum or difference.** We want to show that either $a + b$ or $a - b$ is irrational. What's the alternative? What if this were *not* true? The only way the conclusion could be false is if both $a + b$ and $a - b$ were *both* rational. In other words, the world is divided into two situations. Situation I: At least one of $a + b$ or $a - b$ is irrational. Situation II: Both $a + b$ and $a - b$ are rational. Let's explore the consequences of the second situation. If $a + b = m/n$ and $a - b = r/s$, then (solving for $a$) $a = (m/n + r/s)/2$. This contradicts the fact that $a$ is irrational. So Situation II does not happen, and we are left with the fact that either the sum or the difference is irrational.

### 2.7 Get Real

### I. Developing Ideas

**1. X marks the "X-act" spot.** The X's on the number line below mark the approximate locations, from left to right, of the numbers −1.1, −0.55, −1/3, 0.9, 1.05, 3/2, and 2.3.

**2. Moving the point.** Simplifying, we get $10(3.14) = 31.4$,

$1000(0.123123\ldots) = 123.123123\ldots$, $10(0.4999\ldots) = 4.999\ldots$, $\dfrac{98.6}{100} = 0.986$,

$\dfrac{0.333\ldots}{10} = 0.0333\ldots$ .

**3. Watch out for ones!** Using long division, we find $1/9 = 0.111\ldots$ .

**4. Real redundancy.** If $M = 0.4999\ldots$, then $10M = 4.999\ldots$ . We find $10M - M = 9M$ and also $4.999\ldots - 0.4999\ldots = 4.5$. Then, $9M = 4.5$, so $M = 4.5/9 = 0.5$. Thus, $0.4999\ldots = 0.5$.

**5. Being irrational.** A number is irrational if it is not rational, that is, if it *cannot* be written as a ratio of two integers.

### II. Solidifying Ideas

**6. Always, sometimes, never.** Sometimes; by "an unending decimal expansion" we mean a number whose decimal doesn't end in a trail of zeros. All numbers ending in a trail of zeros are rational, but the converse is not true. for example, $9/7 = 1.285714285714285714285714285714285714285714285714285714285714285714\ldots$ is rational, but $1.0100100010000100000100000001\ldots$ is irrational.

**7. Square root of 5.** False; if the decimal expansion for $\sqrt{5}$ eventually repeated, we could use the ideas in the text to express $\sqrt{5}$ as the ratio of two integers and thus prove that $\sqrt{5}$ is rational. Because we proved $\sqrt{5}$ irrational in Section 2.6, this can't happen; so, the only alternative is that the decimal expansion for $\sqrt{5}$ does not repeat.

**8. A rational search.** The first nine digits of our mystery number are predetermined, $12.0345691\ldots$ . As long as the remaining decimals are not all zeros, the mystery number will sit between the given numbers. To keep the number rational, it's necessary to put a sequence of decimals that eventually repeats. $12.034569110000000\ldots$ and $12.034569133333333\ldots$ both work. We can also take the number halfway between the two, $12.034569150000\ldots$ .

**9. Another rational search.** Let's create, digit by digit, a number $X$ that sits in between the two numbers. The first six digits are 3.14159. Any smaller six-digit number would be less than 3.14159, and any larger six-digit number would be greater than 3.14159001, regardless of the remaining digits of $X$. The next three digits must be 0 to keep the number less than 3.14159001. After that, we are free to choose any repeating sequence or ending sequence, such as, $X = 3.141590005$.

**10. An irrational search.** Take an irrational number that you know and stick the nonrepeating decimals after 5.70. A simple and useful irrational number is $0.01001000100001000001\ldots$ . The irrationality follows because there is no fixed string of numbers that repeats forever. (There is an obvious pattern that repeats, but no fixed string repeats! See Mindscape II.25.) By the same reasoning 5.701010010001000010000001 . . . is irrational, and it lies between the two numbers above.

**11. Another irrational search.** Let's use an irrational number that we know and stick it on the tail of the smaller number. We proved the irrationality of $\sqrt{2}$ = 1.41421356237309504880168872421 . . . in Section 2.6. 0.000100001414213562373095048801688872421 . . . is therefore an irrational number sitting between the two given numbers.

**12. Your neighborhood.** The smallest number comes by inserting all zeros, 10.039800000, and the largest number is formed by inserting all nines, 10.039899999.

**13. Another neighborhood.** We don't just have five X's to replace, rather we have infinitely many X's to replace. Regardless, the smallest number is formed by replacing all the X's with 0's, 5.550100000 . . . = 5.5501, and the largest number is formed by replacing all the X's with 9's, 5.550199999 . . . = 5.5502.

**14. 6/7.** Use long division (or a calculator). 7 goes into 60 eight times, with a remainder of 4; bring down the 0. 7 goes into 40 five times, with a remainder of 5; and so on. 6/7 = 0.857142857142857142857142857143 . . . .

**15. 17/20.** Use long division as in Mindscape II.14, or rewrite the fraction in a simpler form by multiplying numerator and denominator by 5. 17/20 = 85/100 = 0.85.

**16. 21.5/15.** Either perform long division directly, 15 divided into 21.5, or first multiply the numerator and denominator by 10 to get rid of the pesky decimal in the numerator and reduce to lowest terms: 21.5/15 = 215/150 = 43/30. Now divide 30 into 43: 21.5/15 = 1.433333 . . . .

**17. 1.28901.** First write this as 1.28901/1.00000, and multiply both top and bottom by 100000 to get rid of the decimal. So, 1.28901 = 128,901/100,000. This method works for any decimal that stops.

**18. 20.4545.** Note that this decimal stops or ends in a trail of zeros (20.454500000 . . .). Thus, the method of Mindscape II.17 will work here, too. 20.4545 = 204,545/10,000. It isn't necessary to reduce the fraction to lowest terms, but if you were curious, $X$ = 40,909/2000.

**19. 12.999.** Because the decimal ends, we can eliminate the decimal by multiplying it by 1000. So, write 12.999 = 12.999/1 = (12.999/1)(1000/1000) = 12,999/1000.

**20. 2.22 . . .** $X$ = 2.22222 . . . . Because it has a segment of length 1 that repeats, multiply the number by 10 to shift the decimal by exactly one digit: $10X$ = 22.22222 . . . . Now subtract, $10X - X$ = 22.22222 . . . − 2.22222 . . . = 20. (Note that all digits to the right cancel exactly.) So, $9X$ = 20 and $X$ = 20/9.

**21. 43.12 . . .** The elusive number $X$ = 43.121212 . . . . Because there are two digits in our repeating segment, multiply $X$ by 100 to shift the decimals by two digits. $100X$ = 4312.121212 . . . . Subtracting gives $100X - X$ = 4312.121212 . . . − 43.121212 = 4269. Together we get $99X$ = 4269, or $X$ = 4269/99 = 1423/33. (Again, simplifying fractions is not necessary!)

**22. 5.6312 . . .** Follow the reasoning in Mindscape II.21. $X$ = 5.63121212 . . . , $100X$ = 563.12121212 . . . , $100X - X$ = 563.121212 . . . − 5.6312121212 . . . = 557.49. We still have a decimal number, but at least it stops. Solving for $X$ in $99X$ = 557.49 gives $X$ = 557.49/99. Now multiply both numerator and denominator of the fraction by 100 to eliminate the decimal: $X = 55{,}749/9900 = 18{,}583/3300$.

**23. 0.01 . . .** $X = 0.010101 \ldots$. Because the repeating segment has length 2, multiply by 100 to shift the decimal two digits to the left: $100X = 1.010101 \ldots$. Subtracting gives $100X - X = 1.010101 - 0.010101 = 1$, so that $99X = 1$, or $X = 1/99$.

**24. 71.2399 . . .** Note that this number can also be represented as 71.24, which is equal to $7124/1000$. However, we could still use the ideas from Mindscape II.22 to get this fraction. $X = 71.239999 \ldots$, $10X = 712.39999 \ldots$, $10X - X = 712.39999 \ldots - 71.239999 \ldots = 641.16$, so that $9X = 641.16$, or $X = 641.16/9 = 64{,}116/900 = 7124/1000 = 1781/25$.

**25. Just not rational.** This number has a pattern (one 0, one 1, two 0s, one 1, three 0s, one 1, etc.), but that does not mean it's rational. A decimal is rational if and only if there exists a fixed string of digits that repeats forever. This number has no repeating sequence. Suppose there was a repeating sequence of length $N$. If the repeating sequence were all 0's, then we'd end up with a rational number. If the repeating sequence were not all 0's, then eventually we would see a nonzero digit after every $N$ digits. But this isn't the case; we see arbitrarily large sequences of 0's, which implies that there is no repeating sequence. Thus, the number is irrational.

### III. Creating New Ideas

**26. Farey fractions.** Let $F_n$ be the collection of all rational numbers between 0 and 1 (we write 0 as 0/1 and 1 as 1/1) whose numerators and denominators do not exceed $n$. So, for example,

$$F_1 = \{0/1,\ 1/1\},\ F_2 = \{0/1,\ 1/2,\ 1/1\},\ F_3 = \{0/1,\ 1/3,\ 1/2,\ 2/3,\ 1/1\}.$$

$F_n$ is the set of $n$th *Farey fractions*. List $F_4$, $F_5$, $F_6$, $F_7$, and $F_8$. Make a large number line segment between 0 and 1 and write in the Farey fractions. (How can you generate $F_8$ using $F_7$?) Generalize your observations and describe how to generate $F_n$. (Hint: Try adding fractions a wrong way.)

$F_4 = \{0/1,\ \mathbf{1/4},\ 1/3,\ 1/2,\ 2/3,\ \mathbf{3/4},\ 1/1\}$
$F_5 = \{0/1,\ \mathbf{1/5},\ 1/4,\ 1/3,\ \mathbf{2/5},\ 1/2,\ \mathbf{3/5},\ 2/3,\ 3/4,\ \mathbf{4/5},\ 1/1\}$
$F_6 = \{0/1,\ \mathbf{1/6},\ 1/5,\ 1/4,\ 1/3,\ 2/5,\ 1/2,\ 3/5,\ 2/3,\ 3/4,\ 4/5,\ \mathbf{5/6},\ 1/1\}$
$F_7 = \{0/1,\ \mathbf{1/7},\ 1/6,\ 1/5,\ 1/4,\ \mathbf{2/7},\ 1/3,\ 2/5,\ \mathbf{3/7},\ 1/2,\ \mathbf{4/7},\ 3/5,\ 2/3,\ \mathbf{5/7},\ 3/4,\ 4/5,\ 5/6,\ \mathbf{6/7},\ 1/1\}$
$F_8 = \{0/1,\ \mathbf{1/8},\ 1/7,\ 1/6,\ 1/5,\ 1/4,\ 2/7,\ 1/3,\ \mathbf{3/8},\ 2/5,\ 3/7,\ 1/2,\ 4/7,\ 3/5,\ \mathbf{5/8},\ 2/3,\ 5/7,\ 3/4,\ 4/5,\ 5/6,\ 6/7,\ \mathbf{7/8},\ 1/1\}$

Create $F_8$ from $F_7$ by adding fractions that (a) have an 8 in the denominator and (b) are already in lowest terms. We don't add 6/8 because 3/4 is already in the list.

**27. Even irrational.** Stringing together all the even positive integers creates this number. Its irrationality follows from the same logic used to solve Mindscape II.25; that is, there exist arbitrarily large sequences of 0's. If the number were rational, then there would be a sequence of length $N$ that repeated (and this repeating sequence is obviously not all 0's). So, after the decimal starts repeating, every $N$ digits will contain a nonzero digit. Regardless of where we look in the decimal expansion, there will always be arbitrarily large sequences of 0's to the right. ($10^{10}$ has 10 zeros, $10^{100}$ has 100 zeros, etc.) This contradiction proves that our number is irrational.

**28. Odd irrational.** This problem is essentially identical to Mindscape III.27. Note that the number $10^{10} + 1$ is an odd number with nine 0's. $10^{100} + 1$ has 99 adjacent 0's. Because we have arbitrarily large sequences of 0's to the right (interspersed with nonzero digits), there can be no sequence of digits that repeats forever.

**29. A proof for $\pi$.** It may be that decimal expansion of a number repeats after the trillionth place. To prove the rationality or irrationality of $\pi$, we need to show that it repeats forever after some point or that it *never* repeats—neither of which can be done by looking at a finite number of digits. For example, $1/(10^{10^{12}} - 1)$ is a number that repeats every trillion digits. The 999,999,999,999 digits are 0's, but the trillionth digit is 1, and then it repeats!

**30. Irrationals and zero.** Build irrational numbers from something you *know* is irrational, such as $\sqrt{2}$. We showed that dividing irrational numbers by rational numbers leaves an irrational number. Therefore, $\sqrt{2}/2$, $\sqrt{2}/3$, $\sqrt{2}/4$, $\sqrt{2}/5, \ldots$ are all irrational numbers that get closer and closer to 0. So no, there is no smallest irrational number, just like there is no smallest rational number. Alternatively, because $\sqrt{2} = 1.1414213\ldots$ is irrational, the decimal expansion never repeats. Therefore, $0.01414213\ldots$, $0.001414213\ldots$, $0.0001414213\ldots$, are irrational numbers that get closer and closer to 0.

**31. Irrational with 1's and 2's.** In Mindscape II.25, we showed that $x = 0.01001000100001\ldots$ was irrational. By the same reasoning, $y = 0.21221222122221\ldots$ is also irrational. You could also argue that because $y = 2/9 - x$, the irrationality of $x$ implies the irrationality of $y$ because the sum of an irrational and a rational is always irrational. Finally, for fun, there is a more interesting, more random-looking irrational number with only 1's and 2's: List all the rational numbers, and apply Cantor's diagonalization argument with a rule like, "If the $n$th digit is a 1, put a 2, otherwise put a 1."

**32. Irrational with 1's and some 2's.** No; if only a finite number of 2's appeared in the decimal expansion, then after the last 2, the decimal tail would be all 1's and, therefore, a rational number.

**33. Half steps.** This is Zeno's paradox. You will land on the numbers 1/2, 1/4, 1/8, $1/2^4$, $1/2^5$, $1/2^6, \ldots, 1/2^n, \ldots$. The $n$th step takes you to $1/2^n$, so you will never get to 0 in a finite amount of time. You can get arbitrarily close, but you will never actually get there because $1/2^n$ doesn't equal 0 for any finite number $n$. The limit of this sequence of numbers is 0, but none of the numbers themselves are 0.

**34. Half steps again.** Suppose the left half of your segment has length $L$. $L$ may be small, but it is a positive number, and because the sequence $1/2^n$ tends to 0, there exists some $N$ such that $1/2^N < L$. This means that after $N$ steps, your segment will contain the origin. Note that your center will never hit the origin, but at least some part of you will get to where you want to go.

**35. Cutting $\pi$.** This is an alternate way of asking whether $\pi$ might be a rational number. Suppose we divided the interval into $N$ pieces. The endpoints land on $3 + 1/N$, $3 + 2/N$, $3 + 3/N, \ldots$, all of which are rational numbers. Because $\pi$ is irrational, there is no way that we can represent $\pi$ as $3 + I/N$ for any integers $I$ and $N$.

## IV. Further Challenges

**36. From infinite to finite.** How about using our favorite irrational number, $\sqrt{2}$? Because we proved it is irrational, we know that the decimal is unending and nonrepeating. By definition, its square is 2, which has a terminating decimal representation.

**37. Rationals.** Assume $x$ and $y$ are two different positive numbers and $y$ is bigger than $x$. The sequence 1/2, 1/3, 1/4, 1/5, . . . gets arbitrarily small; thus, for some number $N$, the value of $1/N$ is smaller than the difference $y - x$. Now imagine cutting up the real number line with hash marks every $1/N$ units apart. You mark 0, $1/N$, $2/N$, $3/N$, . . . . All these hash marks are on rational numbers, but at least one of the hash marks lies between the numbers $x$ and $y$ because $y - x$ is greater than $1/N$.

**38. Irrationals.** The argument used in Mindscape IV.37 could be used here as well. Instead of using hash marks at $1/N$, $2/N$, $3/N$, . . . , use hash marks at $1/N - \sqrt{2}$, $2/N - \sqrt{2}$, $3/N - \sqrt{2}$, and so on. But there is a simpler argument: Assume that $x$ and $y$ are positive real numbers with $x$ smaller than $y$, and let $N$ be such that $1/N$ is smaller than $y - x$. If $x$ is irrational, then $x + 1/N$ is also irrational and lies between $x$ and $y$; done. If $x$ is rational, then $x + (1/N)/\sqrt{2}$ is an irrational number between $x$ and $y$. (Because $\sqrt{2}$ is bigger than 1, $(1/N)/\sqrt{2}$ is less than $1/N$ and, thus, less than $y - x$.)

**39. Terminator.** Mindscapes II.17 and II.18 show how to express terminating decimals as fractions, namely, remove the decimal and divide by a power of 10 equal to the number of nonzero decimals. For example, $12.3456 = 123,456/10^4 = 123,456/10,000$. Because the only factors of 10 are 2 and 5, the only factors of the denominator $10^n$ are 2's and 5's, which completes the argument.

**40. Terminator II.** Let's turn the solution to Mindscape IV.39 around. We have a number of the form $x/(2^n 5^m)$ (such as, 101/400, where $400 = 2^4 5^2$). If the denominator is a power of 10, then we are done, because we can immediately write the terminating decimal. If not, we try to make the denominator a power of 10 by multiplying both the numerator and denominator by the right number of 2's or 5's. If $m > n$, that is, if there are more 5's than 2's, multiply by $2^{m-n}$, otherwise multiply by $5^{n-m}$. For example, $101/400 = 101/2^4 5^2$ (need 2 extra 5's) = $(25/25)(101/400) = 2525/10,000 = 0.2525$.

### 3.1 Beyond Numbers

#### I. Developing Ideas

**1. Still the one.** A one-to-one correspondence is a pairing of objects from two sets in such a way that each object from one set is paired with exactly one object from the other set.

**2. I get around.** Ed corresponds to the Saab; the Trail-a-Bike corresponds to Julia.

**3. Numerical nephew.** Ask your nephew what he gets if he adds one to his biggest number.

**4. Pile of packs.** Ask each student to pick up his or her backpack. If every student has a backpack, and every backpack is in the possession of one student, then there is a one-to-one correspondence between students and backpacks. If there are leftover backpacks or packless students, then you know the number of students is different from the number of backpacks.

**5. Bunch of balls.** Most cans of tennis balls contain three balls, so there will not be a one-to-one correspondence between cans and balls (unless you have very exclusive, individually packaged tennis balls).

#### II. Solidifying Ideas

**6. The same, but unsure how much.** Weighing two jars of pennies against each other indicates which jar has more pennies without giving any clue about the number of pennies in each jar. Similarly, the process of passing out tests to students tells you whether you have more tests, more students, or exactly the same number of tests as students. Other possibilities include seating people in an auditorium or counting poker chips.

**7. Taking stock.** If the correspondence were not one-to-one, we would have one of the following situations:
a. A symbol representing no company: This would be a misleading state of affairs. Because no one would buy stock in a company that didn't exist, you won't find extraneous symbols in the New York Stock Exchange.
b. A symbol representing two different companies: This poses a problem if you want to invest in one of the companies but not the other. Furthermore, how would the companies decide how to split the money?
c. A company with no symbol: It's in the company's best interest to have an exchange symbol because it makes the buying and selling of stocks easier.

**8. Don't count on it.** We easily see that there are the same number from the natural correspondence implied by the positioning of the symbols; each @ corresponds to the unique © immediately below it; alternatively, each © corresponds to the single @ immediately above it. To see the importance of this geometric correspondence, try determining whether there are more @'s or i's.
@@@@@@@@@@@@@@@@@@@@@@
iiiiiiiiiiiiiiiiiiiiii
Similarly, try determining (without counting) whether there are more C's than D's.
C    C    CCC    C    C     C     CC
DD   DD      DD    D   D    DD   D

**9. Here's looking @ ®.** As in Mindscape II.8, the geometric correspondence makes it easy to see that there are more ®'s than @'s.

**10. Enough underwear.** Deb can confidently say that she has enough underwear without knowing exactly how many. Deb has constructed a one-to-one correspondence between the days of the week and her underwear. Each day is matched up with exactly one pair of underwear and every pair that she packs is accounted for. Because of this correspondence, she can know the number of underwear in her suitcase by counting either the underwear or the days of the week. It is in this sense that these two sets are equivalent.

**11. 791ZWV.** We can phrase the question in the following way, "Is there a one-to-one correspondence between the cars in the United States and their license plate numbers?" Because a car in Mississippi might coincidentally have the same license plate number as a car in Texas, the answer is "no." This counterexample provides some insight into the problem, because there is a one-to-one correspondence between all the vehicles registered in a given state and their license plate numbers. We need only learn the vehicle's state of registration.

**12. 245-2345.** There is not a one-to-one correspondence between telephones and telephone exchanges (last seven digits), so you can't expect to easily track down the stranger. However, there is a one-to-one correspondence between working phones and their ten-digit phone number. You'll have to find the area code to contact the stranger.

**13. Social security.** Note that the question is not, "Is there a one-to-one correspondence between U.S. residents and all social security numbers?" The restrictive word "their" allows us to rephrase the question as, "Do all U.S. residents have a unique social security number?" Because no two people have the same social security number, the real question boils down to, "Are there U.S. residents who do not have social security numbers?" Because U.S. residents must apply for social security numbers on their own, there is not a one-to-one correspondence between residents and numbers.

**14. Testing one two three.** Of course the answer depends on the size and rowdiness of the class. If the class is moving around, it's probably easier to give each student an exam than to be assured of a proper count.

**15. Laundry day.** Forget about predicting the answer, just load up the washing machines one at a time. Stuff as many clothes as you can into a single washing machine and then place one quarter for each slot in the machine. Repeat. You'll have your answer when you run out of either clothes or quarters, and you didn't need to count at all.

### III. Creating New Ideas

**16. Hair counts.** We can overestimate the numbers of hairs on any single person's head in the following way: A human head will easily fit into a 1 foot by 1 foot by 1 foot box. The total surface area is thus less than 6 square feet, or (times $12 \times 12$) 870 square inches, or (times $2.54 \times 2.54$) 5600 square centimeters. The width of an average human hair is 20 microns, or 0.002 centimeters. If the hairs were perfectly lined up, a single square centimeter could hold 250,000 hairs ($1/0.002 = 500$, $500 \times 500 = 250,000$). We can then say that a human head has fewer than ($250,000 \times 5600$) 1.4 billion. Because there are more than 5 billion people on Earth, two people necessarily have the same number of hairs. The individual hairs on these two unknown people are then in one-to-one correspondence with each other. Until now, we have used the notion of one-to-one correspondence to prove that two sets are the same size. Now, we reverse the process; if two sets have the same size, we can place their elements into a one-to-one correspondence.

**17. Social number.** As in Mindscape III.16, we need information about the sizes of the sets involved. The number of allowable social security numbers is $10^9$. Because 10 billion is far greater than 250 million, these two sets cannot be in one-to-one correspondence. If a one-to-one correspondence existed between the two sets, then there would be the same number of elements in each set. Compare with Mindscape II.13.

**18. Musical chairs.** A one-to-one correspondence exists between people and chairs in that short time after the chairless person leaves and before the next chair is removed. The correspondence is best described by pairing each child with the last chair in which he or she was sitting.

**19. Dining hall blues.** We have no information about the precise number of students in the dining hall. Similarly, we don't know how many forks there are. We can say, however, that the number of students is greater than the number of forks. An underlying assumption is that no student took more than one fork. With this assumption, we have set up a correspondence between each fork and the student who is using that fork. Therefore, a one-to-one correspondence exists between the set of all forks and a subset of the students.

**20. Dorm life.** There *may* be a one-to-one correspondence between dorm rooms and students, but we are not guaranteed one. If the rooms were all doubly occupied, we would have more students than rooms. If enough rooms were unoccupied, then we could have more rooms than students. Finally, even if we had the same number of rooms as students, corresponding a student with his or her room wouldn't necessarily provide a one-to-one correspondence. Half the rooms could be empty, with the other half doubly occupied. Because we have no information about the relative sizes of the two sets, we can't conclude anything.

## IV. Further Challenges

**21. Pigeonhole principle.** Suppose you have 10 pigeons sleeping in 7 holes. Because you have more pigeons than holes, you can conclude that at least one of the holes contains more than one pigeon. Anyone claiming to have found a way to give all the pigeons private sleeping quarters is delusional. Such a claim implies a one-to-one correspondence between pigeons and some or all of the holes. Because we know the number of pigeons is greater than the number of holes, we will never be able to find such a one-to-one correspondence.

For infinite sets, we can't discuss "the number of elements" in the traditional way. We can only say, "the number of elements is infinite," which doesn't allow us to reason as we did above. With an infinite number of pigeons and an infinite number of holes, we will find that a second attempt may actually work.

**22. Mother and child.** First, let's assume that we are dealing with all the mothers and children that have lived up to this point, that is, we are dealing with a finite number of mothers and a finite number of children. Because every mother has at least one child, there are at least as many children as mothers. Because some mothers have more than one child, we know that the number of children is strictly larger than the number of mothers. This precludes any one-to-one correspondence between the two sets.

## 3.2 Comparing the Infinite

### I. Developing Ideas

**1. *Au natural.*** The natural numbers are the whole numbers 1, 2, 3, . . . , so called because they arise naturally when we count.

**2. *Au not-so-natural.*** The set {3, 6, 12, 15, 18, . . .} is the set of all natural numbers that are multiples of three. The set {1, 2, 3, 4, 5} is the set of natural numbers from 1 to 5. The set {1/2, 1/4, 1/8, 1/16, . . .} is the set of reciprocals of the powers of 2. (Alternatively, it's the set of powers of 1/2). The set {−1, −2, −3, −4, . . .} is the set of all negative integers. The set {1, 4, 9, 16, 25, 36, 49, 64, 81, 100} is the set of squares of the numbers from 1 to 10.

**3. Set setup.** The set of natural numbers less than 10 is the set {1, 2, 3, . . . , 10}. The set of all even natural numbers is the set {2, 4, 6, 8, . . .}. The set of solutions to the equation $x^2 - 4 = 0$ is the set {2, −2}. The set of all reciprocals of the natural numbers is the set {1, 1/2, 1/3, 1/4, 1/5, . . .}.

**4. Little or large.** The first and third sets are finite. The other two are infinite.

**5. A word you can count on.** The *cardinality* of a set is the "number" of elements in the set. If the set is finite, then the *cardinality* of the set really is the number of elements in the set. For infinite sets, we are more likely to speak of two sets having the *same cardinality,* meaning that their elements can be put into one-to-one correspondence.

### II. Solidifying Ideas

**6. Even odds.**

| E | 2 | 4 | 6 | 8 | 10 | 12 | ... | 2n | ... |
|---|---|---|---|---|----|----|-----|------|-----|
| O | 1 | 3 | 5 | 7 | 9 | 11 | ... | 2n − 1 | ... |

If we imagine the natural numbers in place on the real number line, then each even number is paired with the odd number to its left, and each odd number is paired with the even number to its right. (Why couldn't we pair each even number with the odd number to its right?)

**7. Naturally even.**

| N | 1 | 2 | 3 | 4 | 5 | 6 | ... | n | ... |
|---|---|---|---|---|---|---|-----|-----|-----|
| E | 2 | 4 | 6 | 8 | 10 | 12 | ... | 2n | ... |

Given an arbitrary natural number *n*, we find its mate by multiplying by 2. This shows that all the naturals are accounted for. To show that all the even numbers are accounted for, we reverse the process. Given an arbitrary even number *e*, we find its mate by dividing by 2.

**8. Fives take over.**

| N | 1 | 2 | 3 | 4 | 5 | 6 | ... | n | ... |
|---|---|---|---|---|---|---|-----|-----|-----|
| EIF | 5 | 15 | 25 | 35 | 45 | 55 | ... | 5(2n − 1) | ... |

Though a formula can be found, $\Delta (2n - 1)5$, it isn't really necessary (we'll see why in later problems). We can state the pairing in another way: Given an arbitrary natural number $n$, subtract 1 from the number, and tack on a 5 to the right. The reverse pairing is then straightforward. Given an arbitrary number in **EIF**, remove the 5, and add 1 to the remaining number.

### 9. Six times as much.

| ℕ | 1 | 2 | 3 | 4 | 5 | 6 | ... | n | ... |
|---|---|---|---|---|---|---|-----|---|-----|
| | | | | | | | | | |
| 6ℕ | 6 | 12 | 18 | 24 | 30 | 36 | ... | 6n | ... |

This correspondence is completely analogous to the pairing of the even natural numbers and the naturals in Mindscape II.7.

### 10. Any times as much.

| ℕ | 1 | 2 | 3 | 4 | 5 | 6 | ... | n | ... |
|---|---|---|---|---|---|---|-----|---|-----|
| | | | | | | | | | |
| **a**ℕ | a | 2a | 3a | 4a | 5a | 6a | ... | na | ... |

Note that Mindscapes II.7 and II.9 are special cases.

### 11. Missing 3.

| ℕ | 1 | 2 | 3 | 4 | 5 | 6 | ... | n   (n > 2) | ... |
|---|---|---|---|---|---|---|-----|-------------|-----|
| | | | | | | | | | |
| **TIM** | 1 | 2 | 4 | 5 | 6 | 7 | ... | n + 1 | ... |

This is the first problem whose solution can't be given by a "nice" formula. The correspondence ($n \Delta n + 1$) doesn't apply to every natural number $n$, only those greater than 2. Even though we treat 1 and 2 separately, we still have an explicit one-to-one correspondence. Every natural number is paired with a unique element of **TIM**, and every element of **TIM** is paired with a unique natural number.

### 12. One weird set.

| ℕ | 1 | 2 | 3 | 4 | 5 | 6 | 7 | ... | n (n > 5) | ... |
|---|---|---|---|---|---|---|---|-----|-----------|-----|
| | | | | | | | | | | |
| **OWS** | 1 | 3 | 5 | 7 | 8 | 10 | 11 | ... | n + 4 | ... |

Just as in Mindscape II.11, our correspondence treats the first five numbers separately. How can we describe where an arbitrary number in **OWS** goes? If the arbitrary number is {1, 3, 5, 7, or 8}, then we use the table above. Otherwise, we subtract 4 to find its match.

### 13. Squaring off.

| ℕ | 1 | 2 | 3 | 4 | 5 | 6 | 7 | ... | n | ... |
|---|---|---|---|---|---|---|---|-----|---|-----|
| | | | | | | | | | | |
| **S** | 1 | 4 | 9 | 16 | 25 | 36 | 49 | ... | $n \times n$ | ... |

## 14. Counting cubes (formerly crows).

| $\mathbb{N}$ | 1 | 2 | 3 | 4 | 5 | 6 | 7 | ... | $n$ | ... |
|---|---|---|---|---|---|---|---|---|---|---|
| $C$ | 1 | 8 | 27 | 64 | 125 | 216 | 343 | ... | $n \times n \times n$ | ... |

## 15. Reciprocals.

| $\mathbb{N}$ | 1 | 2 | 3 | 4 | 5 | 6 | 7 | ... | $n$ | ... |
|---|---|---|---|---|---|---|---|---|---|---|
| $R$ | 1 | 1/2 | 1/3 | 1/4 | 1/5 | 1/6 | 1/7 | ... | $1/n$ | ... |

The set $\mathbb{R}$ consists of all the rational numbers that can be written as 1 divided by a natural number. This correspondence is special in that one rule describes the pairing in both directions. Given a natural number $n$, we can find its mate by forming the reciprocal, $1/n$. Similarly, consider an arbitrary element in $\mathbb{R}$, say, 1/89. We find its mate by forming the reciprocal, $1/(1/89) = 89$.

**16. Hotel Cardinality (formerly California).** The hotel manager turns on the intercom and instructs all of his guests to move one room down the line. The person in room 1 moves to room 2, the person in room 2 moves to room 3, and so on. Now room 1 is vacant, and the weary traveler has a place to sleep. This Hotel Cardinality, so known as The Hilbert Hotel, introduces a new way to "see" one-to-one correspondences. Specifically, the problem shows that the set of hotel rooms, {1, 2, 3, 4, . . .} has the same cardinality as the set of people, {X, 1, 2, 3, 4, . . .}. The manager's instructions serve to illustrate the one-to-one correspondence.

**17. Hotel Cardinality continued.** No problem! The manager tells everyone to move two rooms down and lets the two new guests take the first two rooms.

**18. More Hotel C.** It is still possible to fit everyone. In addition to the manager's ability to talk to everyone at exactly the same time (there are some speed-of-light issues here), we also have to accept that people can switch between any two rooms in just a few minutes, regardless of the distance involved. That said, the manager saves the day by telling his guests that they will all have to relocate. To find their new room number, he tells them to write down their current room number (remember, some of these numbers are *big*!), and then multiply it by 2. After they relocate, there are now an infinite number of vacant rooms (the odd numbered rooms). He then lines up the travelers and gives the first in line a key to the first vacant room, the second in line gets a key to the second vacant room, and so on.

Before (travelers: T#; guests: G#)    T1, T2, T3, T4, T5, . . .

| Room # | 1 | 2 | 3 | 4 | 5 | 6 | ... |
|---|---|---|---|---|---|---|---|
| Occupant | G1 | G2 | G3 | G4 | G5 | G6 | ... |

After:

| Room # | 1 | 2 | 3 | 4 | 5 | 6 | ... |
|---|---|---|---|---|---|---|---|
| Occupant | T1 | G1 | T2 | G2 | T3 | G3 | ... |

**19. So much sand.** To really prove this, we must acknowledge that all grains of sand occupy at least some minimum volume $V$. It doesn't matter how small $V$ is, as long as it's not 0. An indirect argument is appropriate at this point: What if there were an infinite number of grains of sand on Earth? Just as 100 grains of sand requires a container with volume $100V$, an infinite number of grains would require a container with an infinite volume. Earth is no such container!

**20. Half way.** What is a line? How can we uniquely label or discuss all the line segments in the problem? We can quantify the problem by viewing the line as part of the real number line, say the interval from 0 to 1. But this introduces some ambiguity. Certainly, the points in the middle of the line correspond to points between 0 and 1, but are the points 0 and 1 also included? To make the problem specific, let's assume they are included.

After 1st cut: $[0, 1/2)$ $(1/2, 1]$   (consider the point 1/2 as part of the sawdust)
After 2nd cut: $[0, 1/4)$ $(1/4, 1/2)$ $(1/2, 1]$
After 3rd cut: $[0, 1/2^3)$ $(1/2^3, 1/2^2)$ $(1/2^2, 1/2)$ $(1/2, 1]$
After $n$th cut: $[0, 1/2^n)$ $(1/2^n, 1/2^{n-1})$ $\ldots (1/2, 1]$
After all cuts: $[0]$ $\ldots$ $(1/2^n, 1/2^{n-1})$ $\ldots (1/2, 1]$

We can now put the line segments in one-to-one correspondence with the natural numbers. If you want to add the point {0} to the collection of pieces, that's OK, too. Adding one element to an infinite set never changes its cardinality. (*Note:* Equally valid arguments can be made without referring to the real number line.)

| 1 | 2 | 3 | 4 | 5 | 6 | ... |
|---|---|---|---|---|---|-----|
| $(1/2, 1]$ | $(1/2^2, 1/2)$ | $(1/2^3, 1/2^2)$ | $(1/2^4, 1/2^3)$ | $(1/2^5, 1/2^4)$ | $(1/2^6, 1/2^5)$ | ... |

**21. Pruning sets.** Removing any elements from a finite set reduces its cardinality because we have reduced the number of things in the set. For infinite sets, we cannot speak of the number of things in the set; instead, we must reinterpret "fewer things" in the sense of cardinalities. For example, if we remove the element 3 from the natural numbers, we are left with a set that has the same cardinality (see Mindscape II.11).

**22. A natural prune.** If we remove all natural numbers greater than 5, the remaining set would consist of the numbers 1 through 5. This new set clearly can't be put into a one-to-one correspondence with the natural numbers; thus, it contains fewer numbers than the set of naturals. It is not enough to remove an infinite set of numbers; you must remove all but a finite set.

**23. Prune growth.** This is clearly not possible for finite sets, because removing things reduces the number of elements in the set, and our notion of larger and smaller sets depends entirely on their numerical size. For infinite sets, we need to define what we mean by "larger" and "smaller" sets. An infinite set $A$ is larger than an infinite set $B$ if (1) there exists no one-to-one correspondence between the two sets, and (2) there exists a one-to-one correspondence between the set $B$ and a subset of the set $A$. Either the collection of remaining things has the same cardinality as the original set or it has a different cardinality. In the first case, the new subset is the *same size*. In the second, it is *smaller*. Fortunately for our intuition, it isn't possible to increase the size of a set by removing elements.

**24. Same cardinality?** We can only conclude that the first set is not larger than the second set. The first set could be the even natural numbers, the second set could be all the naturals, and the pairing could be the natural one,

$$\{2 \Leftrightarrow 2, 4 \Leftrightarrow 4, \ldots\}.$$

In this situation, the two sets (the even natural numbers and all natural numbers) have the same size. However, the second set could be the set of all real numbers, in which case the second set is larger.

378

**25. Still the same?** If the sets had the same cardinality, then there would exist a correspondence that matched every element from one set with every element of the other; something the problem specifically states cannot happen. Therefore, the sets have different cardinalities. (Which set is larger?)

### III. Creating New Ideas

**26. Modest rationals.** 0/1, 1/1, 1/2, 1/3, 2/3, 1/4, 3/4, 1/5, 2/5, 3/5, 4/5, 1/6, 5/6, 1/7, 2/7, . . .
First list all fractions with 1 in the denominator. Then list all new fractions that can be obtained by using 2 in the denominator. Continue increasing the denominator and listing in order all the new fractions that can be created. It is important to point out that each time we increase the denominator, we will create only a finite number of new fractions (so we'll be able to continue increasing the denominator). This method will not work for listing all the rationals greater than 0.

**27. A window of rationals.**

| $\mathbb{N}$ | 1 | 2 | 3 | 4 | 5 | 6 | 7 | 8 | 9 | ... |
|---|---|---|---|---|---|---|---|---|---|---|
| | | | | | | | | | | |
| $\mathbb{R}$ | 0 | 1 | 1/2 | 1/3 | 2/3 | 1/4 | 3/4 | 1/5 | 2/5 | ... |

We haven't found a nice formula, but our description of the method in Mindscape III.26 is sufficient to make the correspondence clear. We now need to show that every rational between 0 and 1 will eventually appear on the list. Every such rational can be expressed as a fraction in lowest terms. Suppose, for example, that the fraction is 5/2347. Because the denominators on the list continue increasing, we'll eventually get to a sequence of fractions with denominator 2347. We'll find our fraction within the first five such numbers. Because this reasoning works for all such rationals, we've shown that the pairing represents a one-to-one correspondence between the two sets.

**28. Bowling ball barrel.** Notice that it's impossible for both barrels to contain only a finite number of bowling balls. So we know that the cardinality of at least one of the barrels is infinite. After reading Chapter 4, you'll want to conclude more. You'll say instead, "The cardinality of one of the barrels must be the same as to the cardinality of the set of *all* the bowling balls."

**29. Not a total loss.** This problem is an abstraction of Mindscapes II.11 and II.12. With subsets of the natural numbers, the following is a useful trick. Pair 1 with the smallest number in the new set. Pair 2 with the second-smallest number in the new set, and so on. For this reason, any infinite subset of the natural numbers can be put in a one-to-one correspondence with the naturals.

**30. Mounds of Mounds.** First it's necessary to find a way of representing the individual candy bars. Either draw a picture, as in Figure 1, or decide upon a naming convention for each bar. For example, 37-right could represent the right candy bar in the 37th package. Now we can list all the candy bars and, in so doing, provide a one-to-one correspondence with the naturals.

1-left, 1-right, 2-left, 2-right, 3-left, 3-right, 4-left, . . .

This should remind you of the proof that the integers have the same cardinality as the natural numbers.

Figure 1   Countability of Mounds candy bars.

**31. Piles of peanuts.** See Figure 2 for a concise picture proof. Alternatively, we could label the peanuts. Let 3(5th row) be the third peanut in the fifth row. Now we can list the peanuts in a manner that is reminiscent of the listing of rationals in Mindscape III.26:
1(1st row), 1(2nd row), 2(2nd row), 1(3rd row), 2(3rd row), 3(3rd row), 1(4th row), . . . .

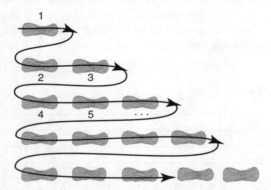
Figure 2   Peanuts in piles that increase in number (zigzag counting)

**32. The big city.** Let's suppose that the streets are laid out as in Figure 3. There is no natural starting point; so pick an intersection, label it 1, and start spiraling outward. As you move outward, label each intersection. Because you will eventually hit any single intersection, you've provided a valid one-to-one correspondence between intersections and natural numbers.

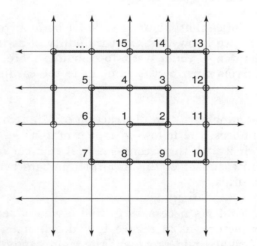
Figure 3   Infinite avenues and streets (spiral counting)

**33. Don't lose your marbles.** To see the similarity between this problem and Mindscape III.32, imagine each avenue represents each box of marbles and a marble is placed at each intersection. Because the set of intersections is in one-to-one correspondence with the naturals, it's likely that the set of marbles will be as well. Suppose that the marbles are laid out as in Figure 4; we could use our standard picture proof to show that this is indeed so.

380

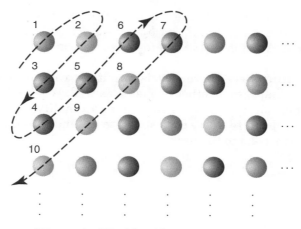

Figure 4   Marbles (diagonal counting)

**34. Make a guess.** In Section 3.3, we'll discover that the set of all real numbers has a cardinality strictly bigger than the natural numbers. Similarly, the set of all points on this page (abstractly speaking) cannot be put into one-to-one correspondence with the naturals.

**35. Coloring.** In base 2, all real numbers can be represented with only 0's and 1's. For example, 0.1 (base 2) represents 1/2, 0.01 (base 2) represents 1/4, 0.001 represents 1/8, and 0.0101010101010101 . . . represents 1/4 + 1/16 + 1/64 + · · · = 1/3. In this way, we can imagine the set of all the colorings as similar to the set of all real numbers between 0 and 1. And, indeed, both sets turn out to have a cardinality strictly greater than the naturals.

### IV. Further Challenges

**36. Ping-Pong balls on parade.** The last question, "Can you name a ball left in the barrel?" gives the solution away. Is ball number 1 in the barrel? No, it was gone after the first step. How about number 2? No, it was gone after the second step. You can't name a single ball in the barrel because they were all eventually taken out. It's unsettling to think that even though the number of balls in the barrel increases after each step, after *all* the steps, there are none left. This is exactly why the expression "infinity minus infinity" is meaningless outside of context.

**37. Naked Ping-Pong balls.** Because the process of adding and removing balls is not explicit, the question, "What's left?" is ill-posed: It doesn't make sense. To see that one can get different outcomes, imagine the following procedure for adding and removing balls. Suppose you first add balls 1–10, then remove the 5th ball. Now add balls 11–20, and remove the 6th ball; add 21–30, and remove the 7th ball. In the end, you're left with balls numbered 1 through 4. In this case, infinity minus infinity is 4. Try finding a way to add and remove balls so that you are left with infinitely many balls.

**38. Primes.** This is a specific instance of Mindscape III.29, the solution of which showed that any infinite subset of the natural numbers can be placed in a one-to-one correspondence with the naturals. Because there are infinitely many primes, we can conclude that there as many primes as natural numbers.

**39. A grand union.** In the proof that there were as many integers as naturals, we took two infinite sets (all the positives and all the negatives) and lumped them together. We'll use the same idea here; but, to get started, we must find a way to represent the elements in the two sets. Call the two sets $A$ and $B$, and let the elements be described with the following notation:

$A = \{a_1, a_2, a_3, \ldots\}$, $B = \{b_1, b_2, b_3, \ldots\}$. The following listing proves that the "huge" set is the same size as the naturals: $A < B = \{a_1, b_1, a_2, b_2, a_3, b_3, \ldots\}$. After duplicate elements (where $a_i = b_j$) have been removed, $A < B$ is still in one-to-one correspondence with the naturals.

**40. Unnoticeable pruning.** Because the set is infinite, it contains elements that we could call $a_1, a_2, a_3, \ldots$ plus, perhaps, some other elements as well. We could remove $a_1$ and $a_2$ from the set, for example, and still have a set of the same cardinality, because we could make $a_1$ correspond to $a_3$, $a_2$ correspond to $a_4$, $a_3$ correspond to $a_5$, and so on, and have the elements that aren't $a_i$'s simply correspond to themselves.

### 3.3 The Missing Member

#### I. Developing Ideas

**1. Shake 'em up.** Georg Cantor was the first mathematician to show that there are more real numbers than natural numbers, demonstrating that there are different "sizes" of infinity.

**2. Detecting digits.** The first digit of the first number is 1. The second digit of the second number is 4. The third digit of the third number is 7.

**3. Delving into digits.** The number 0.12345678910111213141516 . . . is constructed by writing the natural numbers in sequence. The 14th digit is 1. The 25th digit is 7. The 31st digit is 0.

**4. Undercover friend.** Create a number with first digit not 3 or second digit not 8 or third digit not 2 and you have a number not on your friend's list. Some examples are 482, 392, 383, 125.

**5. Underhanded friend.** You're stuck here. Because you friend has shown you no digits of her last number, any number you pick might be the third on her list.

#### II. Solidifying Ideas

**6. Dodge Ball.** Player 2 can win with the following simple strategy: For the first move, focus only on the first letter of Player 1's sequence. If it's an X, place an O in the first box; if it's an O, place an X in the first box. For the second move, focus only on the second letter of the second sequence. Similarly, place a different letter in the second position, and so on. When the game is finished, Player 2's sequence will differ from each of Player 1's sequences in at least one place. It differs from the first sequence in the first position, the second sequence in the second position, and so on.

**7. Don't dodge the connection.** Given any countable list of real numbers (one for each natural number), Cantor described a method for creating a new real number not on the list. Instead of a 6 × 6 Dodge Ball game, Cantor's game had infinitely many rows and infinitely many columns. Because Cantor started with a listing of the reals, as if Player 1 had filled out the entire Dodge Ball chart before Player 2 began. Finally, instead of filling a table full of X's and O's, Cantor created his new number using two digits (say, 3's and 5's), and placed them in such a way that the new number differed from the $n$th number on the list in at least the $n$th decimal place.

**8. Cantor with 3's and 7's.** The proof remains the same.

**9. Cantor with 4's and 8's.** The proof remains the same.

**10. Think positive.** Pair each positive real number with its additive inverse (its negative). Because the negative of any positive number is negative and the negative of any negative number is positive, no numbers are left out. This one-to-one correspondence shows that the two sets have the same cardinality.

**11. Diagonalization.** The essence of this proof lies in the construction of a new decimal number that is not on a countable list of reals. During this construction, Cantor only looked

383

at the first digit after the decimal of the first number, the second digit after the decimal of the second number, the third digit after the decimal of the third number, and so on. If we were to write all the listed numbers with the decimals aligned, then the digits that Cantor used in his construction would form a diagonal line.

**12. Digging through diagonals.** Adjoining all the natural numbers behind a decimal creates the first number. Similarly, adjoining all numbers divisible by 2 creates the second number, and adjoining all numbers divisible by 3 creates the third. By describing this pattern, we've described an entire list of real numbers. Let's create a new number not on the list, using 1's and 2's, with 2's used whenever possible. The diagonal digits in the list above are 1, 4, 9, 2, 5, 2, . . . . So, our new number will begin 0.222121 . . . . (We can't easily describe this number.) By construction, our new number differs from each number on the list in at least one decimal place.

**13. Coloring revisited.** Suppose that the set of all possible colorings was the same size as the naturals. There would necessarily exist a one-to-one correspondence between colorings and the natural numbers; that is, we could list all the possible colorings. Now suppose we have such a list in front of us. Using the diagonalization argument, we can create a new coloring that isn't on the list. If the first circle in the first coloring is red, paint the first circle in the new coloring blue (and vice versa). Continue in this way, changing the color of the $n$th circle in the $n$th coloring. This procedure defines a new coloring that is, by construction, different from all the other colorings.

**14. A penny for their thoughts.** The set of all outcomes has a cardinality greater than that of the naturals. The proof is identical to the proof in Mindscape II.13. In fact, because it's easy to construct a one-to-one correspondence between these two sets, they are necessarily the same size.

**15. The first digit.** Suppose our list was:
0.722222222222222 . . .
0.100000000000000 . . .
0.010000000000000 . . .
0.001000000000000 . . .
0.000100000000000 . . .
0.000010000000000 . . .
$\vdots$

     The first element of the diagonal is a 7, and the rest are all 0's. The newly constructed number $M$ is then 0.72222222222, which is precisely the first number on the list. Because $M$ differs from the $n$th number in the $n$th decimal position (except for $n = 1$), only the first row could possibly match $M$. The moral is that if you don't dot all of your i's and cross all your t's, Player 1 will win, and your proof will fall flat.

### III. Creating New Ideas

**16. Ones and twos.** Note the connection between this problem and Mindscapes II.13 and II.14. You can either use Cantor's diagonalization argument with 1's and 2's or provide a one-to-one correspondence between this set of numbers and the set of all possible red-blue colorings of an infinite set of marbles. Either way, you've proved that this set of numbers is uncountable (that is, has a larger cardinality than the naturals).

**17. Pairs.** This procedure does work—an example is given below:

0.**24**434322234 . . .
0.43**22**3253242 . . .
0.234**24**432133 . . .
0.23435**44**2432 . . .

⋮

$M = 0.22442222$ . . .

Note that each pair of digits in $M$ is determined by a single number on the list. In addition, $M$ is different from each number on the list. For example, it differs from the third number in the third pair, the fourth number in the fourth pair, and so on.

**18. Three missing.** The text describes a method for creating a new number using only 2's and 4's. Modifying the rule so that it uses only 1's and 3's allows for the creation of a second number not on the list. Use 5's and 6's to get the third number. Any pair of numbers (stay away from 0's and 9's) will suffice, but make sure there isn't any overlap. (Why won't using 1's and 2's for the first number and 1's and 3's for the second necessarily give you two different numbers?)

Here's an entirely different solution that uses a single rule. Create the first number in the normal way. Now, insert this new number into the first position of the list. With this *new* list, use the same rule to create the second number. Repeat this process one last time. Throw the second number on top of the new list and apply the rule again. In the end, you've found three numbers that aren't anywhere on the original list, and they're all different from each other.

**19. No Vacancy.** If the set of reals stopped by one person for each real number, the people would not all be able to fit into the motel, even if it were entirely empty. If we *could* fit them all, then we would have found a one-to-one correspondence between the natural numbers and the real numbers. Cantor proved that this is impossible, so the Not Enough Room sign would have to go up.

**20. Just guess.** Read Section 3.4 to see if your guess was correct.

**IV. Further Challenges**

**21. Nines.**
0.300000000000 . . .
0.022222222222 . . .
0.002222222222 . . .
0.000222222222 . . .
0.000022222222 . . .

⋮

The construction described above results in $M = 0.299999999$ . . . $= 0.3$; but 0.3 is the first number on the list. In a similar manner, try concocting a scenario where using 2's and 0's is equally bad.

**22. Missing irrational.** No valid diagonalization procedure can guarantee that the missing real is rational. What if such a procedure were applied to a listing of all rational numbers? If the new number is different from every number on the list, it is necessarily irrational, because all the rationals are already on the list.

We can, however, guarantee that the missing number is irrational. The construction described in this problem will certainly produce a real that is not on the list. The 3's and 5's were added to the mix only to ensure that the number didn't contain any repeating segments. To see this, imagine that the new number is rational. It would necessarily contain a repeating sequence of numbers. For the sake of argument, let's suppose that this repeating sequence is 200 digits long. If the sequence contained only 2's and 4's, then we would only have finitely many 3's and 5's in our number. By construction, our number has infinitely many 3's and 5's, so at least one of the 200 digits is a 3 or a 5. But now we have too many 3's and 5's. Once we pass the first repeating sequence, a 3 or a 5 occurs every 200 digits. In the constructed number, however, 3's and 5's occur with less and less frequency. So the number doesn't have a repeating segment of length 200. The same line of reasoning can be extended to show that the number doesn't have any repeating segments of *any* length.

## 3.4 Travels Toward the Stratosphere of Infinities

### I. Developing Ideas

**1. Which are which?** Given the set $S = \{a, b, c, d\}$, $\{a\}$ is a subset of $S$, $a$ is an element of $S$, $\{\ \}$ is a subset of $S$, $d$ is an element of $S$, and $\{a, b, c, d\}$ is a subset of $S$.

**2. Power play.** The power set of a set $S$ is the set of all subsets of $S$.

**3. Universal emptiness.** The empty set is a subset of every set.

**4. With three or without.** Here's a list of all the subsets of $\{1, 2, 3\}$ that do not contain 3: $\{\ \}$, $\{1\}$, $\{2\}$, $\{1, 2\}$. There are four such subsets. Here's a list of all the subsets that do contain 3: $\{3\}$, $\{1, 3\}$, $\{2, 3\}$, $\{1, 2, 3\}$. There are four of these subsets as well, for a total of eight subsets of $\{1, 2, 3\}$.

**5. Solar power.** The power set of {Earth, Moon, Sun} has eight elements: $\{\ \}$, {Earth}, {Moon}, {Sun}, {Earth, Moon}, {Earth, Sun}, {Sun, Moon}, {Earth, Sun, Moon}.

### II. Solidifying Ideas

**6. All in the family.**
$\{\ \}$ — The no-diner dinner
$\{1\}$, $\{2\}$, $\{3\}$, $\{4\}$ — The lonely dinners
$\{1, 2\}$, $\{1, 3\}$, $\{1, 4\}$, $\{2, 3\}$, $\{2, 4\}$, $\{3, 4\}$ — The one-on-one dinners
$\{1, 2, 3\}$, $\{1, 2, 4\}$, $\{1, 3, 4\}$, $\{2, 3, 4\}$ — The let's-talk-about-the-other-one dinners
$\{1, 2, 3, 4\}$ — The whole family dinner
There are a total of $2^4$, or 16 possible dinners.

**7. Making an agenda.** Each agenda item either appears or does not appear on the final agenda. The number of different agendas is equivalent to the number of different ways that eight boxes could be checked with either Y's (yes's) or N's (no's).

| 1 | 2 | 3 | 4 | 5 | 6 | 7 | 8 |
|---|---|---|---|---|---|---|---|
| N | N | N | N | N | N | N | N |
| N | N | N | N | N | N | N | Y |
| N | N | N | N | N | N | Y | N |
| N | N | N | N | N | N | Y | Y |
| N | N | N | N | N | Y | N | N |
| ... | ... | ... | ... | ... | ... | ... | ... |

Each of the eight items could have a Y or an N under it. So, for each of the two possibilities for item 1, there are two for item 2, two for item 3, and so on, for a total of $2^8$, or 256 ways to fill out the table. (The first row above corresponds to the no-agenda agenda. The second row corresponds to the agenda consisting only of the eighth director's item, and so on.)

**8. The power of sets.** Count the number of left curly brackets associated with each element of the thingie on the right side. The number of left brackets will equal the number of left parentheses in the name of the set.
@ and ! are in $S$.
$\{!, \#, \%\}$ and $\{\#\}$ and $\{!, @\}$ are in $\wp(S)$.
$\{\ \{!\}, \{@\}, \{\#\}, \{\%\}, \{\&\}\ \}$ and $\{\ \{@\}, \{\$, !\}\ \}$ are in $\wp(\wp(S))$.
$\{\ \{\{@, !\}\}, \{\{\$\}\}\ \}$ and $\{\ \{\{@\}\}, \{\{\#, \$\}\}, \{\{!\}, \{\%, \&\}\}\ \}$ and $\{\{\{!\}\}\}$ are in $\wp(\wp(\wp(S)))$.

**9. Powerful words.** The set $\wp(\wp(S))$ is the set of all subsets of the set $\wp(S)$. In other words, the elements of $\wp(\wp(S))$ are collections of subsets of $S$. The set $\wp(\wp(\wp(S)))$ is the set of all possible combinations of collections of subsets of $S$. Realistically, there is no clear way to describe $\wp(\wp(\wp(\wp(S))))$. The simplest is a recursive type definition: $\wp(\wp(\wp(\wp(S))))$ is the set of all subsets of $\wp(\wp(\wp(S)))$.

**10. Identifying the power.** $\{m, a, t, h\}$, $\{\ \}$, and $\{m\}$ are all elements of $\wp(S)$. See Mindscape II.8 for more practice.

**11. Two Cantor.** Let $M$ denote our new subset. Because 1 is not an element of its associated subset, we want $M$ to contain 1 as an element. Because 2 is an element of $\{1, 2\}$, we don't want $M$ to contain 2. Because there are no other elements to consider, we're done. $M = \{1\}$.

**12. Another two.** Again, let $M$ represent the new subset. Because 1 isn't a member of $\{2\}$, we put 1 in the new subset. Similarly, because 2 isn't a member of $\{1\}$, we let 2 be a member of $M$. We've exhausted all the elements of $S$, so we're done. $M = \{1, 2\}$.

**13. Cantor code.** $\{$you, found, it$\}$

**14. Finite Cantor.** The subset $\{\&, \%, \#\}$ is a member of $\wp(S)$ that is not associated with any elements of $S$ in the given correspondence.

**15. One real big set.** The set of all subsets of the real numbers is a set with that cardinality exceeds that of the reals.

## III. Creating New Ideas

**16. The Grand Real Hotel.** Unfortunately, this hotel is not nearly as easy to visualize as the Hotel Cardinality, but we can use our experience with the smaller hotel to help us. The set of real numbers could not fit into the Hotel Cardinality because the cardinality of the reals is greater than that of the natural numbers. To find a set too big for the Grand Real Hotel, it is necessary to construct a set with an even larger cardinality. A natural choice is the power set of the reals.

**17. The Ultra Grand Hotel.** This question is equivalent to "Is there a largest cardinality?" or "Is there a largest set?" The answer to both questions is "no." If there were a largest set, how would it compare in size to its power set? This question is at the heart of the following proof-by-contradiction argument: Assume there is a largest set **S**. Cantor's theorem states that the power set of **S** is even larger, contradicting the assumption that **S** is the largest set. So our assumption is wrong; there is no largest set.

**18. Powerful counting.** The short answer is "a lot." $S$ contains 6 elements. $\wp(S)$ contains $2^6$ elements (see Mindscape II.7). $\wp(\wp(S))$ contains $22^6$ elements, $\wp(\wp(\wp(S)))$ has $2226$ elements, and so on. By working inside out, a pattern emerges; there are as many 2's as there are $\wp$'s. With this powerful insight, we can boldly skip to the end to conclude that the set $\wp(\wp(\wp(\wp(\wp(\wp(S))))))$ has $222222^6$ elements. (This number has too many digits to print. For comparison purposes, $22^6 = 18446744073709551616$.)

**19. Russell's barber's puzzle.** Imagine that this barber is very rigid in his way of life. He classifies all men as either self-shavers or non-self-shavers, and he shaves all the people in the second class and none of the people in the first. Furthermore, imagine that during a midlife crisis, he asks himself where he fits in. If he belongs to the self-shavers group, then he shaves himself, but this would go against his rule of only shaving people in the non-self-shavers group. If he belongs to the non-self-shavers group, then he doesn't shave himself, but this, too, is a problem because now he's in the group of people that he is supposed to shave. He can't classify himself without breaking his rules.

The set *NoWay* has a similar worldview. It classifies sets as either self-included or non-self-included, and it chooses as its elements all the members of the second class. Now assume that *NoWay* is a member of itself. Belonging to the self-included class, it does not meet the criterion to belong to itself, which is a contradiction. Now suppose that *NoWay* isn't a member of itself. Belonging to the non-self-included class meets the membership criterion and so it should be a member of itself. But this is a second contradiction; hence, the paradox.

**20. The number name paradox.** Start small. There are $27 \times 27 = 27^2$ different ways to compose two character sentences, because there are 27 choices for the first character and 27 choices for the second. Similar reasoning shows there are $27^{50}$ different sentences 50 characters in length. This is a large but finite number, and because the set of natural numbers is infinite, so is the set $T$. Because "the smallest number contained in $T$" has fewer than 50 characters, that number must be in $S$ by definition of $S$. On the other hand, every number in $T$ is not in $S$, so the smallest number contained in $T$ cannot be in $S$. We have paradoxically shown that that number is both in $S$ and not in $S$.

## IV. Further Challenges

**21. Adding another.** If the set were the same size as the natural numbers, then this would be easy. We could use the correspondence with the natural numbers to name each element in our set. The first set could have elements {1, 2, 3, 4, . . .}, and the second set could have elements {0, 1, 2, 3, 4, 5, . . .}. (Here we're naming the extra element 0.) Now, it is easy to construct a correspondence to show that these two sets are the same size—pair each integer in the first set with its predecessor.

The subtlety of this problem is that we don't know how big the set is, and therefore we cannot easily label or discuss all the elements. As is often the case, it is only necessary to label *some* of the elements. So we'll divide our original set into two parts, the Countable collection and the Potentially big collection. The new set that has one extra element is similarly divided into two parts: the Countable collection + 0 and the Potentially big collection. The second part of each set is identical, so it is natural to pair these elements with themselves. The first parts are both countable collections, so we can use the ideas in the previous paragraph to complete the one-to-one correspondence.

**22. Ones and twos.** Associate 3 with the pair 11, 4 with the pair 12, 5 with the pair 21, and 6 with the pair 22. A decimal number with 1's and 2's can be associated with a decimal with 3's, 4's, 5's, and 6's by grouping the 1's and 2's in pairs and substituting the appropriate 3, 4, 5, or 6. For example, the decimal number 0.122212111121 . . . after grouping would look like 0.12 22 12 11 11 21 . . . and would correspond to 0.464335 . . . This correspondence is one-to-one.

### 3.5 Straightening Up the Circle

**I. Developing Ideas**

**1. Lining up.** It is not possible to draw a line segment with more points than $L$. You can always construct a one-to-one correspondence between the points in any two line segments.

**2. Reading between the lines.** Pair up the points on $L$ and $M$ as follows. For each point on $L$, draw the line from the top vertex of the triangle through that point of $L$ (see the dashed line segment). Where that line intersects $M$ is the corresponding point. Clearly, every point on $L$ is matched with a point on $M$. Likewise, any point on $M$, together with the top vertex of the triangle, determines a unique point on $L$ for the correspondence.

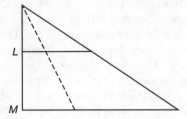

**3. De lines and Descartes.** We verify that $(0, 3)$ satisfies the equation $y = -x + 3$ by substituting $x = 0$ and $y = 3$ to get $3 = -0 + 3 = 0 + 3 = 3$, which is valid. Similarly, because $0 = -3 + 3$, the point $(3, 0)$ also satisfies the equation.

**4. Dashed line rendezvous.** Because $L$ is given by the equation $y = 2$, the dashed line will intersect $L$ at the point where $y = 2$ and $y = -x + 3$. Thus, $2 = -x + 3$, giving $x = 1$. So, the point of intersection is $(1, 2)$.

**5. Rendezvous two.** Because $M$ is given by the equation $y = 1$, the dashed line will intersect $M$ at the point when $y = 1$ and $y = -x + 3$. Thus, $1 = -x + 3$, giving $x = 2$. So, the point of intersection is $(2, 2)$.

**II. Solidifying Ideas**

**6. A circle is a circle.** Place the circles inside one another so that their centers coincide with the point $C$, as in Figure 1. Now imagine all possible rays emanating from $C$ and let points lying on the same ray correspond to each other. That is, for an arbitrary point $S$ on the small circle, draw a ray emanating from $C$ through $S$, continuing so that it intersects the big circle at the point $B$. In this way, the point $B$ corresponds to the point $S$. Because each ray intersects the two circles in exactly one point and because all the points on both circles are hit, the above correspondence is one-to-one.

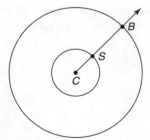

Figure 1   Circle within a circle

**7. A circle is a square.** As in Mindscape II.6, imagine placing the small circle inside the larger square and letting rays emanate from a point inside the circle (see Figure 2). Each ray intersects both the circle and the square and pairs up the two intersection points. Because all points are accounted for, the correspondence is one-to-one, and the two sets of points have the same cardinality, regardless of their sizes.

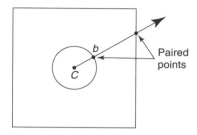

Figure 2   Circle within a square

**8. A circle is a triangle.** See solutions to Mindscapes II.6 and II.7. Place the circle inside the triangle, and pick a point inside the circle as the origin of an infinite number of rays. Each ray intersects the two figures and defines the one-to-one correspondence (see Figure 3).

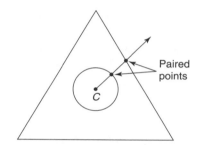

Figure 3   Circle within a triangle

**9. Stereo connections.** All pairings are made via line segments from the top of the circle (see Figure 4).

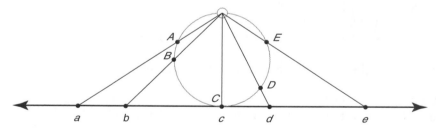

Figure 4   Stereographic projection part 1 (see pictures in text)

**10. More stereo connections.** As in Mindscape II.9, all pairings are made via line segments from the top of the circle (see Figure 5). Remember that the top point of the circle is not associated with any part of the original line segment; thus, it is not included in the correspondence.

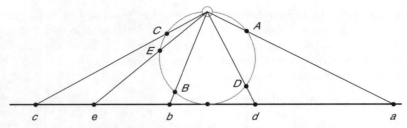

Figure 5    Stereographic projection part 2

**11. Perfect shuffle problems.** Deshuffling the digits in $p = 0.12000100010001\ldots$ gives us the only point in the plane that could possibly correspond to $p$: $(0.10000000000000\ldots,$ $0.20101010101010101\ldots)$. Unfortunately, $p$ remains unpaired because these $(x, y)$ coordinates correspond to a *different* point. Remember that we decided to express all numbers, whenever possible, with trailing 9's instead of trailing 0's. First, rewrite the coordinates as $(0.09999999999999\ldots, 0.20101010101010101\ldots)$, and then shuffle the digits to get $0.029091909190919091909190919091909190\ldots$. This means that the point $p$ doesn't correspond with any point on the plane. In other words, the given correspondence is not one-to-one because some points are left out.

**12. More perfect shuffle problems.** Deshuffling the digits of $p = 0.12001001001001001001$ $\ldots$ gives us the coordinates $(0.101001001001001\ldots, 0.2001001001\ldots)$. Because the coordinates do not have decimal expansions with repeating 0's, the decimal representations do not need to be changed. As a result, shuffling these coordinates gives us the point $p$ back again. No problems.

**13. Grouping digits.**
Write $x$ as 0.6  4  3  4  0004  8     9 7 2  1     1 09  8 0003  4 . . . .
Write $y$ as 0.4  9  8  6       7  4  005  4  6  08  0009   7 07     07 6 . . . .
Shuffling gives 0.644938460004784900574261081000909780700030 7 . . . .

**14. Where it came from.** Group the digits of the point together as
0.9  9  004  08  7  8  0004  04  8 2  02  4  4  0005  01  1  01 9  09  09 . . .
$x = 0.9$ 004  7  0004  8  02  4  01  01  09  09 . . .
$y = 0.9$ 08  8  04  2  4  0005  1  9  09  09  09 . . .
$(x, y) = (0.900470004802401010909\ldots, 0.90880424000519090909\ldots)$.

**15. Group fix.** Write the digits of $p$ as 0.1  2  0001  0001  0001  0001  0001  0001 . . .
$x = 0.1$  0001  0001  0001  0001 . . .
$y = 0.2$  0001  0001  0001  0001 . . .
$p$ gets paired with the point ( $0.1000100010001\ldots, 0.2000100010001\ldots$).

## III. Creating New Ideas

**16. Is there more to a cube?** We need only generalize the proof that a line segment has as many points as a filled-in square. First, to eliminate the ambiguity inherent in any decimal representation of real numbers, express numbers with repeating 9's whenever possible. For an arbitrary point in the solid cube, say $(x, y, z) = (0.50045555\ldots, 0.020202\ldots, 0.00123999\ldots)$, group the digits into clusters of nonzero digits, and then shuffle the clusters.

$$x = 0.\,5\ 004\ 5\ 5\ 5\ldots, \quad y = 0.\,02\ 02\ 02\ 02\ 02\ldots, \quad z = 0.\,001\ 1\ 2\ 3\ 9\ 9\ldots.$$

$(x, y, z) \Leftrightarrow 0.50200100402150225022350295029\ldots$ Clearly, each point in the cube corresponds to a point on the interval $(0, 1)$. Because the process is completely reversible, we know that every decimal in the unit interval is hit. (Because we always deal with clusters of nonzero digits, we never need to worry about constructing a decimal with an infinite trail of 0's.) So, just as with the filled-in square, this procedure determines a one-to-one correspondence between the points on the unit interval and the points on a solid cube.

**17. T and L.** Cut up, the two letters as in Figure 6. Notice that both figures are made up of a single solid line segment (including the two endpoints) and one half-open line segment where one of the endpoints is missing. We can use the methods of this section to show that the two solid line segments have the same cardinality. Similarly we can show that the sets of points on the two half-open segments have the same cardinality, which completes the proof.

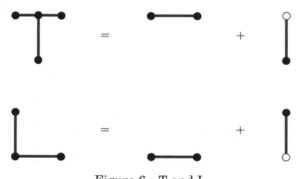

Figure 6   T and L

**18. Infinitely long is long.** Yes; this section demonstrated a one-to-one correspondence between the entire real line and the unit interval. We can use this correspondence to correspond any given set of points on the real line into a subset of the unit interval.

**19. Plugging up the north pole.** Under the stereographic projection, every point on the real line is accounted for; thus, there is no natural point on the real line to associate with the filled-in north pole. However, as with the Hotel Cardinality problems, we could create a vacancy by modifying the correspondence as we did in Mindscape IV.21 from Section 3.4.

After removing the north pole from a circle, the remaining open-ended segment can be flattened to an open-ended segment that has a natural one-to-one correspondence with the interval $(0, 1)$. By pairing the north pole with 1, we complete the one-to-one correspondence (see Figure 7).

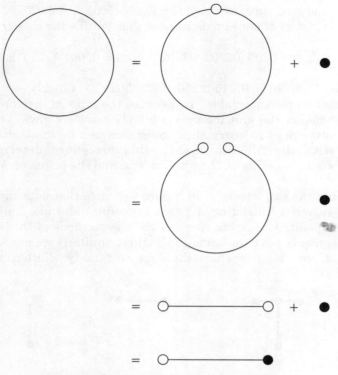

Figure 7   Circle and half-open interval

**20. 3D Stereo.** We need to extend the stereographic projection to three dimensions. Place the punctured sphere on the plane, and identify points on the plane with points on the sphere by drawing lines through the missing north pole. Given an arbitrary point on the sphere, we find its mate by constructing the line from this point to the north pole. Because this line is not parallel to the plane, it will eventually intersect the plane in a single point that is the planar point to which our arbitrary point corresponds. Because we can get all of the points in the plane in this way, our correspondence is one-to-one, proving that the set of points on the unit sphere has the same cardinality as the set of points on the plane.

**IV. Further Challenges**

**21. Stereo images.** Longitudinal lines correspond to straight lines in the plane passing through the south pole. Latitudinal lines correspond to circles centered about the south pole. (See Figure 8.)  In fact, there is a one-to-one correspondence between circles on the sphere and circles on the plane; that is, circles get mapped to circles under stereographic projection.

394

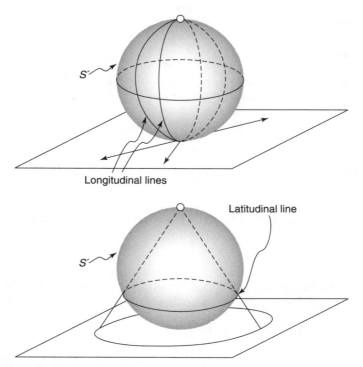

Figure 8   Latitudinal and longitudinal lines under stereo projections

**22. Grouped shuffle.** First note that the mapping is completely reversible. Given an arbitrary decimal, say $p = 0.45201044456001000006981\ldots$, we group the digits as 4  5  2  01  04  4  4  5  6  001  000006  9  8  1 ... . The odd-numbered groups form the decimal for $x$, and the even-numbered groups form the decimal for $y$.

$x = 0.4\quad 2\ 04\ 4\quad 6\ 000006\ 8\ldots = 0.4204460000068\ldots$
$y = 0.5\ 01\quad 4\ 5\ 001\qquad 9\ 1\ldots = 0.5014500191\ldots$

Because $p$ is necessarily written without an infinite tail of 0's, we will have an infinite number of groups; thus, $x$ and $y$ will have an infinite number of nonzero decimals. Unlike the original method for shuffling digits, we will never need to change the decimal representation of any of the numbers. To reverse the process, we once again group $x$ and $y$ into clusters of nonzero digits. Note that this grouping is unique and will be equivalent to our previous grouping:

$x = 0.4204460000068\ldots = 0.4\quad 2\ 04\ 4\quad 6\ 000006\ 8\ldots$
$y = 0.5014500191\ldots\quad = 0.5\ 01\quad 4\ 5\ 001\qquad 9\ 1\ldots$
So, when we deshuffle the clusters, we get $p$ back again,
$p = 0.4\ 5\ 2\ 01\ 04\ 4\ 4\ 5\ 6\ 001\ 000006\ 9\ 8\ 1\ldots$
$\quad = 0.45201044456001000006981\ldots.$

Because the mapping is reversible, we necessarily hit all the points $p$ in the unit interval $(0, 1]$, as well as all the points in the filled in unit-square $(0, 1] \times (0, 1]$.

Where does $p = 1.0$ go? Because $p = 0.9999\ldots$, $p$ gets mapped to the point $(x, y) = (1, 1)$. This means that the open interval $(0, 1)$ gets mapped to the interior of the filled-in square $(0, 1) \times (0, 1)$.

### 4.1   Pythagoras and His Hypotenuse

**I. Developing Ideas**

**1. The main event.** In a right triangle, the square of the length of the hypotenuse is equal to the sum of the squares of the lengths of the other two sides.

**2. Two out of three.** A right triangle with legs of length 1 and 2 has hypotenuse of length $\sqrt{1^2 + 2^2} = \sqrt{5}$. In a right triangle with one leg of length 1 and hypotenuse of length 3, the other leg will have length $x$, where $\sqrt{1^2 + x^2} = 3$. Squaring both sides to solve for $x$ yields $x^2 + 1 = 9$; so, $x = \sqrt{8} = 2\sqrt{2}$.

**3. Hypotenuse hype.** A right triangle with legs of length 1 and $x$ has hypotenuse of length $\sqrt{x^2 + 1}$.

**4. Assessing area.** We know the base of the rectangle; we need to find the height. The diagonal divides the rectangle into two right triangles, each with one leg of length 4 and hypotenuse of length 5. The length of the other leg will be the height of the rectangle. If $x$ denotes the length of this leg, the Pythagorean Theorem tells us that $5^2 = x^2 + 4^2$. So, $x^2 = 25 - 16 = 9$, yielding $x = 3$. Thus, the area of the rectangle is $4 \times 3 = 12$ square inches.

**5. Squares all around.** The figure shows a right triangle. Adjoined to each side of the triangle is a square with area equal to the square of that side of the triangle. The Pythagorean Theorem says that the area of the largest square (the one on the hypotenuse) is equal to the sum of the squares on the other two sides.

**II. Solidifying Ideas**

**6. Operating on the triangle.** See the figure. When the angles are aligned, they will form a straight line, or half of a complete rotation. Because a full rotation corresponds to 360°, the sum of the angles in a triangle is 180°.

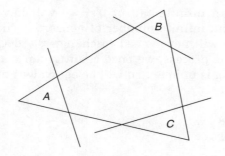 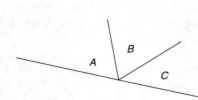

**7. Excite your friends about right triangles.** Describe the proof of the Pythagorean Theorem to someone who has never seen it before.

**8. Easy as 1, 2, 3?** In any right triangle, the side opposite the right angle is the largest. In this case, the length of the hypotenuse is 3. If it is indeed a right triangle, then the Pythagorean Theorem must hold—that is, $1^2 + 2^2 = 3^2$. But 5 isn't 9, so we don't have a right triangle. The only set of consecutive numbers that works here is $3^2 + 4^2 = 5^2$.

**9. Sky high.** "Directly above her" means that the triangle formed by the student, the kite, and the spectator is a right triangle. If the height is $H$, then the Pythagorean Theorem says that $H^2 + 90^2 = 150^2$. (If you make a 3-4-5 right triangle 30 times bigger, you get a 90-120-150 right triangle, which makes solving for $H$ easier!) The kite is flying 120 feet high.

**10. Sand masting.** The tip of the mast, the base of the mast, and the stern of the sailboat form a right triangle. Because the backstay is the longest side of the triangle, we have $H^2 + 50^2 = 130^2$. So, $H = \sqrt{130^2 - 50^2} = 120$. (Another useful triplet is the 5-12-13 right triangle.)

**11. Getting a pole on a bus.** When we measure the length of a box, we don't measure the length of the diagonal; rather, we measure the lengths of the sides—even though the diagonal is larger than either of the sides. So Sarah somehow found a 3-foot × 4-foot box and placed Adam's fishing pole along the diagonal. Once again, the 3-4-5 triplet comes to the rescue.

**12. The scarecrow.** Because the hypotenuse is longer than all the other sides of the triangle, it can't be one of the legs. If $H$ is the length of the hypotenuse, then $3^2 + 3^2 = H^2$, or $H = \sqrt{18} = 3\sqrt{2}$. In general, if the two shorter sides have length $L$, then the hypotenuse will have length $L\sqrt{2}$. Note, this is the 45-45-90 triangle that you get by cutting a square along a diagonal.

**13. Rooting through a spiral.** The first triangle is a special case of Mindscape II.12. It's a 45-45-90 triangle, so the hypotenuse has length $\sqrt{1^2 + 1^2} = \sqrt{2}$. The hypotenuse of this triangle becomes a leg of the second triangle. The longest side of the second triangle is $\sqrt{1^2 + (\sqrt{2})^2} = \sqrt{1+2} = \sqrt{3}$. Similarly, the third triangle's hypotenuse has length $\sqrt{1^2 + (\sqrt{3})^2} = \sqrt{1+3} = \sqrt{4}$, and so on. The hypotenuse of the $N$th triangle has length $\sqrt{N+1}$.

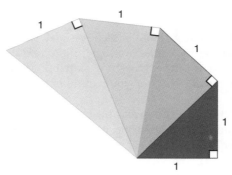

**14. Is it right?** It's close! $(2.6^2 + 8.1^2) = 72.37$, and $(8.6^2) = 73.96$, so we don't have a right triangle. If the third side had length $\sqrt{72.37} = 8.507\ldots$, then we'd have a right triangle. Imagine gradually shrinking the third side from length 8.6 to 8.507. The largest angle is opposite the largest side; when we shrink the side, the opposing angle gets smaller. At the end of our transformation, the largest angle is 90°, so it is intuitive that the largest angle was originally *larger* than 90°. If $a^2 + b^2 < c^2$, then the largest angle is larger than 90°. If the opposite inequality holds, then all angles are less than 90°.

**15. Train trouble.** See the figure. Break the triangle into two separate right triangles and use the Pythagorean Theorem. We have $h^2 + 2640^2 = 2641^2$, or $h = \sqrt{2641^2 - 2640^2} = 72.7$ feet.

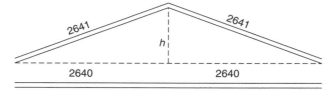

## III. Creating New Ideas

**16. Does everyone have what it takes to be a triangle?** This friend of yours is lying, confused, or both. No side of a triangle can be larger than the other two sides put together. If his longest side were instead 5641, then his triangle would have zero area because it would consist of three sides lying in a straight line.

**17. Getting squared away.** See the figure. Before moving the top triangle, side $a$ represents the length of one side of the alleged square. After the triangle is moved, the same side represents the other side of the alleged square. This proves that the figure is indeed a square. The second square is proved in a similar manner.

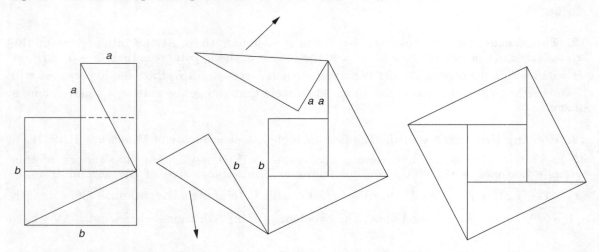

**18. The practical side of Pythagoras.** Make a right triangle out of two sides of your patio. Measure the lengths of the two sides and the diagonal. If they satisfy the Pythagorean Theorem, then your patio is true; otherwise it's skewed. If it is skewed, you can use the results to Mindscape II.14 to decide whether your angle is too large or too small.

**19. Pythagorean pizzas.** Because the area of a circular pizza is proportional to the square of its diameter, we could rephrase the question in terms of square pizzas. This is where right triangles come in. If the area of the medium and small pizzas were equal to the area of the large pizza, then the lengths of the three diameters would form a right triangle. If the largest angle is more than 90°, then you'd want the large pizza. If it's less than 90°, you'd take the two smaller pizzas.

**20. Natural right.** The longest side has length $r^2 + s^2$, so we need to check the following equation: $(2rs)^2 + (r^2 - s^2)^2 = (r^2 + s^2)^2$. Expanding gives $(4r^2s^2) + (r^4 - 2r^2s^2 + s^4) = (r^4 - 2r^2s^2 + s^4)$, which is true. So, yes, there are infinitely many integer valued Pythagorean triplets.

## IV. Further Challenges

**21. Well-rounded shapes.** You cannot "square" the circle with only a ruler and a compass. However, this problem shows that you can "washer" a circle. Construct an annulus with area

398

equal to a given circle. See the figure. The shaded circle has area $\pi b^2$, and the annulus has area $\pi R^2 - \pi r^2$. The right triangle shows that these two areas are the same because $r^2 + b^2 = R^2$ (so that, $R^2 - a^2 = b^2$).

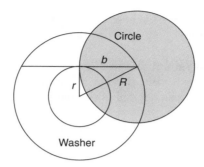

**22. A Pythagorean Theorem for triangles other than right triangles.** See the figure. The two right triangles give $c^2 = (a')^2 + h^2$, $b^2 = (a'')^2 + h^2$. Subtracting and noting that $(a') = a - a''$ gives $c^2 - b^2 = (a - a'')^2 - (a'')^2$. Expanding and simplifying proves the formula $a^2 + b^2 = c^2 + 2aa''$. When the angle between sides $a$ and $b$ approaches 90°, $a'$ gets smaller and smaller. When the angle *is* 90°, $a'' = 0$, and the formula reduces to the Pythagorean Theorem.

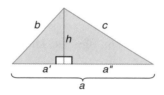

### 4.2   A View of an Art Gallery

#### I.  Developing Ideas

**1. Standing guard.** A gallery with three sides will have a floor plan in the shape of a triangle, requiring only one guard.

**2. Art appreciation.** If a polygonal closed curve in the plane has $v$ vertices, then there are $v/3$ vertices from which it is possible to view every point on the interior. If $v/3$ is not an integer, then the number of vertices needed is the biggest integer less than $v/3$.

**3. Upping the ante.** For a gallery with 12 vertices, you need at most 12/3, or 4 guards. With 13 vertices, you need at most 13/3, or 4.33 . . . guards; that is, you need at most 4 guards. With 11 vertices, you need at most 11/3, or 3.66 . . . guards, that is, at most 3 guards.

**4. Create your own gallery.** Answers will vary.

**5. Keep it safe.** Only 1 camera is needed. Place it at any one of the four inside vertices.

#### II.  Solidifying Ideas

**6. Klee and friends.** Explain the Klee Art Gallery Theorem.

**7. Putting guards in their place.** See the figures. There are several ways to do this.

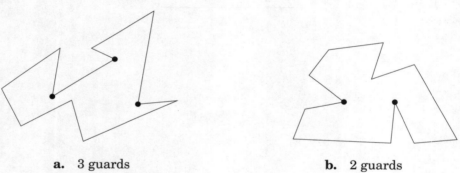

**a.**  3 guards            **b.**  2 guards

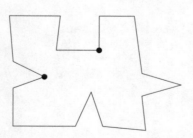

**c.**  2 guards

**8. Guarding the Guggenheim.** See the figures.

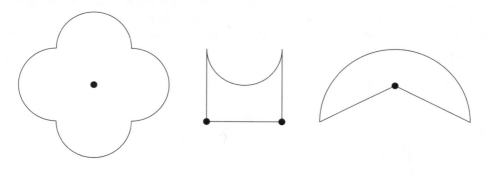

     **a.** 1 guard          **b.** 2 guards          **c.** 1 guard

**9. Triangulating the Louvre.** See the figures. There are numerous correct triangulations.

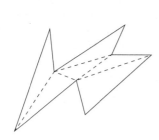 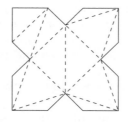 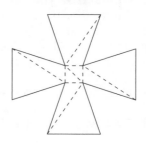

**10. Triangulating the Clark.** The triangulations in the figures are examples of correct triangulations. There are a variety of correct answers.

 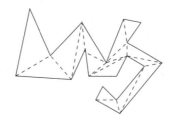 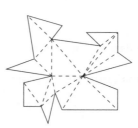

**11. Tricolor me.** You are free to pick the colors of two adjacent vertices, but after that, the colors of the rest of the vertices are determined. See the figure for colorings. All colorings will look like these with the exception of a permutation of the three colors (for example, your blue might be my red).

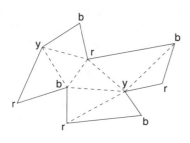 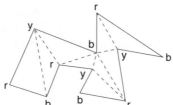 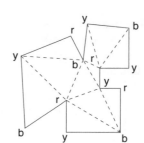

401

**12. Tricolor hue.** Pick a side and arbitrarily label the two vertices with different colors. The other vertices are now determined. Locate the triangle containing the chosen side, and color the remaining vertex. You've colored two new sides. Find any triangles containing these new sides and color the remaining (uncolored) vertex. Repeat this process until you're done. See the figures.

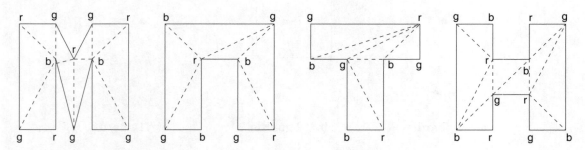

**13. One-third.** The relevance of this Mindscape: Suppose we triangulate and color a six-sided museum. If $R$, $Y$, and $B$ represent the number of red, yellow, and blue vertices, then we have $R + Y + B = 6$. In the text, we proved that one of these numbers must be smaller than or equal to $v/3 = 6/3 = 2$. There are three ways to express 6 as a sum of three numbers: $1 + 1 + 4$, $1 + 2 + 3$, $2 + 2 + 2$. In each case, we have *at least* one number less than or equal to 2.

**14. Easy watch.** Draw a regular hexagon, and put the guard in the center.

**15. Two watches.** See the zigzag museum in the figure.

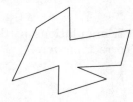

### III. Creating New Ideas

**16. Mirror, mirror on the wall.** Any point will work for any of these figures.

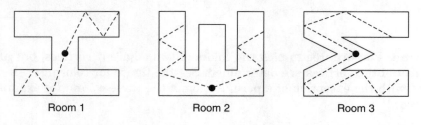

Room 1     Room 2     Room 3

**17. Nine needs three.** See the comb-shaped museum in the figure. It is necessary to have a guard standing inside each of the extended triangular regions. Because these extended triangles don't intersect, there is no way to make do with fewer guards. Try generalizing this shape to draw a 12-sided room that needs at least 4 guards.

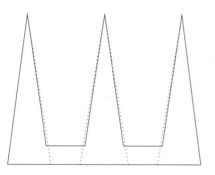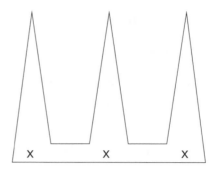

**18. One-third again.** Suppose you have the natural number $N = A + B + C$. If $A < N/3$, $B < N/3$, and $C < N/3$, then $A + B + C < N$.

**19. Square museum.** See the figure for possible guard placements. As always, there are several ways to do this. The museums that look like 3's need at least three guards, whereas the museum that resembles the number 2 needs only two guards.

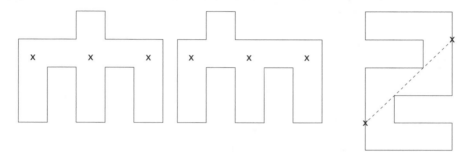

**20. Worst squares.** We'll modify the comb-shaped museum used for Mindscape III.17. See the figure below for the 20-sided museum. Note that this is easily modified to generate the 12- and 16-sided museums.

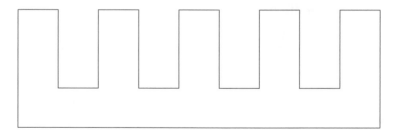

**IV. Further Challenges**

**21. Pie are squared.** This problem isn't directly related to the text, but it *is* neat. The approximating rectangle has width equal to the radius of the circle $r$, and length equal to about half the circumference $\pi r$. The area of the rectangle should be close to $\pi r^2$, which is the familiar formula for the area of a circle. The beauty is that you can visualize the limiting process of dividing the circle into an infinite number of pieces. In the end, you are left with a rectangle with area *exactly* the area of the circle.

**22. I can see the light.** This question is really the same as the House of Mirrors Question (page 226). If a light illuminates the whole room, then a person standing at that point would be able to see every point.

### 4.3 The Sexiest Rectangle

### I. Developing Ideas

**1. Defining gold.** A rectangle is golden if the ratio of the longer side to the short side is exactly $\varphi = \dfrac{1+\sqrt{5}}{2}$, the Golden Ratio.

**2. Approximating gold.** The Golden Ratio is approximately 1.618, so 1.67 is closer than any of the other numbers.

**3. Approximating again.** In decimal form, the four ratios are 5/3 = 1.66 . . . , 11/8.5 = 1.29 . . . , 14/11 = 1.2727 . . . , and 1.5454 . . . . Thus, the first object, the $3 \times 5$ card, has proportions closest to those of a Golden Rectangle.

**4. Same solution.** There are several ways to see that these equations have the same solutions. Inverting both sides of one equation yields the other. Also, cross-multiplying both equations yields the same quadratic: $\varphi^2 - \varphi = 1$.

**5. X marks the unknown.**
**a.** Given $\dfrac{2x}{1} = \dfrac{1}{x-1}$, multiply both sides by $x - 1$ to get $2x^2 - 2x = 1$, or $2x^2 - 2x - 1 = 0$. The quadratic formula gives solutions $x = \dfrac{-(-2) \pm \sqrt{(-2)^2 - 4(2)(-1)}}{2(2)} = \dfrac{2 \pm \sqrt{12}}{4} = \dfrac{2 \pm 2\sqrt{2}}{4} = \dfrac{1 \pm \sqrt{2}}{2}$.

**b.** Given $\dfrac{x}{3} = \dfrac{2}{x-4}$, cross-multiply to get $x^2 - 4x = 6$, or $x^2 - 4x - 6 = 0$. The quadratic formula gives solutions $x = \dfrac{-(-4) \pm \sqrt{(-4)^2 - 4(1)(-6)}}{2(1)} = \dfrac{4 \pm \sqrt{40}}{2} = \dfrac{4 \pm 2\sqrt{10}}{2} = 2 \pm \sqrt{10}$.

**c.** Given $\dfrac{3x}{2} = \dfrac{1}{x+1}$, cross-multiply to get $3x^2 + 3x = 2$, or $3x^2 + 3x - 2 = 0$. The quadratic formula gives solutions $x = \dfrac{3 \pm \sqrt{3^2 - 4(3)(-2)}}{2(3)} = \dfrac{3 \pm 2\sqrt{33}}{6}$.

### II. Solidifying Ideas

**6. In search of gold.** Find three examples of Golden Rectangles.

**7. Golden art.** Several Golden Rectangles can be created using lines in the painting.

**8. A cold tall one?** Yes; Stand a Golden Rectangle on its shorter side, and the base is shorter than the height. Stand it on its longer side, and the height is shorter. To be golden, the ratio of the longest side to the shortest side has to be the Golden Ratio.

**9. Fold the gold.** The dimensions of the new rectangle are exactly half of the original. The ratio between the longer and shorter sides remains unchanged, so it still represents a Golden Rectangle.

**10. Sheets of gold.** If you don't mind a severely folded piece of paper, you can create your Golden Rectangle with only one rectangular sheet of paper without recourse to a straightedge. For a pristine Golden Rectangle, it's necessary to cannibalize the second sheet (see the figure). Using two 8.5 × 11-inch papers, make four folds on one paper to get the correct lengths; mark the lengths on the second sheet, and use the straightedge and scissors to cut out the rectangle.

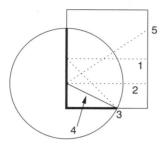 

1. Fold in half lengthwise.
2. Fold again in half (let half of the length represent 1 unit of length).
3. Create the diagonal fold (3) by moving the lower left corner to the first fold (marking the lower thickened edge of length 2 units).
4. Mark the solid line to be used in the next fold ($\sqrt{5}$ units).
5. Make the fourth fold (5) by bringing the solid line in alignment with the left edge of the paper. (This marks the end of the long thickened line.)
6. Transfer the lengths of the thickened edges to the other paper. Use a straightedge (or fold) to complete the Golden Rectangle.

**11. Circular logic?** See the figure. If the original rectangle had lengths 2 and $(1+\sqrt{5})$, then the new rectangles would have dimensions $2 \times 1$ and $2 \times \sqrt{5}$, neither of which is a Golden Rectangle.

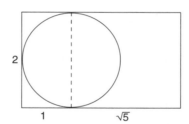

**12. Growing gold.** Note that after we add the square, we have a rectangle with the following property: If we remove the largest square, we are left with a Golden Rectangle. Though not proved in the text, the Golden Rectangle is the *only* rectangle that has this feature; thus, the original rectangle is indeed a Golden Rectangle. If the original sides had length 1 and $\varphi$, then the new rectangle would have sides of length $1 + \varphi$ and $\varphi$. Verify by calculator or by algebra that $(1 + \varphi)/\varphi = \varphi$ to prove that the new rectangle is also golden.

**13. Counterfeit Gold?** In Mindscape II.12, we start with a Golden Rectangle and find that repeatedly adding squares doesn't change the golden nature of the rectangles. If we start with *any* rectangle, the process of adding larger and larger squares makes the resulting rectangles more and more golden. The length of each new edge is the sum of the lengths of the previous two edges. This harks back to the Fibonacci sequences in Chapter 3, where we showed that the limiting ratio was independent of the starting two numbers. (What if we continually *remove* the largest rectangle? Do we get more golden as we go?)

**14. In the grid.** Consecutive Fibonacci numbers (1, 1, 2, 3, 5, 8, 13, . . .) are wonderful approximators of the Golden Ratio. In a 10 × 10 grid, an 8 × 5 rectangle is your best bet. You

405

can convince yourself of this by computing the 100 possible ratios and seeing which is closest to $(1+\sqrt{5})/2$. Surprisingly, it doesn't help to consider nonhorizontal right triangles.

**15. A nest of gold.** The enlarged picture will be identical to the original picture, but its position will be rotated 90°. After removing the largest square and expanding the picture, we'll still have a nested sequence of Golden Rectangles that follow a spiral pattern. Because there is only one way to create such a nested sequence (always remove the square *away* from the center), the two figures are the same.

### III. Creating New Ideas

**16. Comparing areas.** Let $c$ denote the length of the short side of the smaller Golden Rectangle. If $\varphi = (\sqrt{5}+1)/2$, then $h = c\varphi$ and $b = h\varphi = c\varphi^2$. So, Area(G) $= bh = c^2\varphi^3$, whereas Area(G') $= ch = c^2\varphi$. The ratio of areas is $\varphi^2 = (\sqrt{5}+3)/2$. If we enlarge all the dimensions of the small rectangle by a factor of $\varphi$, then the lengths will agree with the larger Golden Rectangle. The corresponding area increases by $\varphi^2$.

**17. Do we get gold?** Let the sides of the square be 3 units long. The right triangle has base 1 and height 3, so its hypotenuse has length $\sqrt{10}$. The resulting rectangle has base $2+\sqrt{10}$ and height 3, for a ratio of $(2+\sqrt{10})/3 = 1.720\ldots$ The smaller rectangle has sides 3 and $\sqrt{10}-1$, for a ratio of $(\sqrt{10}-1)/3 = 1.387\ldots$

**18. Do we get gold this time?** Let the starting rectangle be 2 units wide and 1 unit high. The new longer rectangle has length $1+\sqrt{2}$ and width 1, for a corresponding ratio of $\sqrt{2}+1 = 2.141\ldots$ The smaller rectangle has width $\sqrt{2}-1$ and a ratio of $1/(\sqrt{2}-1) = 2.141\ldots$ It has the same proportions. To see that these two expressions represent the same number, multiply both numerator and denominator by $\sqrt{2}+1$. The dimensions of the starting rectangle do not matter; you'll still end up with two equally proportioned rectangles.

**19. A silver lining?** See the figure. The three triangles in the figure are all similar, each is a factor of $\varphi$ bigger than the other. The part of the diagonal outside the square is the smallest, the part inside the square is larger by a factor of $\varphi$, and the whole diagonal is still larger by the same factor. To see this, remove the largest square from the smaller Golden Rectangle. We now have three Golden Rectangles in the drawing, each a factor of $\varphi$ bigger than the next. Note that the larger leg of the three triangles corresponds to the larger side of the three Golden Rectangles.

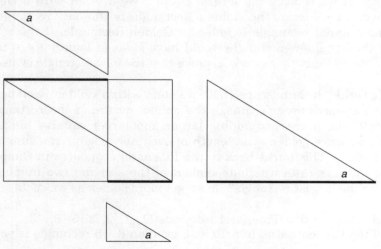

**20. Cutting up triangles.** See the figure. Make a triangle out of the midpoints of the three sides. This new triangle will be one-fourth as large as the original triangle and inverted. The remaining three triangles all have the same dimensions.

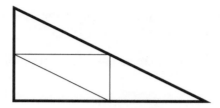

### IV. Further Challenges

**21. Going platinum.** If the original rectangle has height 1 and width $b$, then the new rectangle has height $b - 1$ and width 1. We need to solve the relation $1/(b - 1) = 2 (b/1)$, or $2b^2 - 2b - 1 = 0$, which has solution $b = (\sqrt{3} + 1)/2$.

**22. Golden triangles.** The hypotenuse has length $\sqrt{5}$. Each of the five triangles has a hypotenuse that is 1 unit and angles that are the same as the original triangle; thus, they are all equal and all Golden Triangles.

### 4.4 Soothing Symmetry and Spinning Pinwheels

#### I. Developing Ideas

**1. To tile or not to tile.** The first, second, and fourth shapes can be used to tile the plane. The star and the triangle with a bite out of it cannot. (Try it!)

**2. Shifting into symmetry.** Each pattern can be shifted in the direction of the arrow, moving one B onto the other.

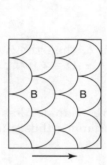 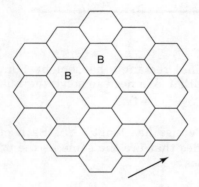 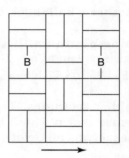

**3. Flipping over symmetry.** All three patterns have a flip symmetry obtained by flipping around a horizontal line drawn through the middle of either B. The center and right-most patterns also have flip symmetries obtained by flipping through a vertical line.

**4. Come on, baby, do the twist.** The first pattern has no rotational symmetry. The middle pattern can be rotated through any multiple of 60°. The right-most pattern has only one rotational symmetry: through 180°.

**5. Symmetric scaling.** The pattern on the left requires four small tiles to create a super-tile (in black) and nine large tiles to create a super-super-tile (in gray). The pattern on the right requires the same. This should not be surprising, because the triangular tiling can be transformed to parallelogram tiling, like the one on the left, merely by joining neighboring triangles appropriately.

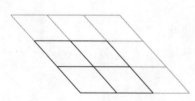 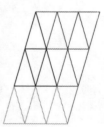

#### II. Solidifying Ideas

**6. Build a super.** Here is a 5-unit super-tile.

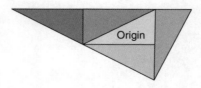

**7. Another angle.** The angle is equivalent to the small angle of the original 1, 2, $\sqrt{5}$ triangle. It has measure arctan(1/2), or 26.56...°. Because this is an irrational number, no nested sequence of super-tiles contains two triangles with the same orientation.

**8. Super-super.** Here is a 25-unit super-tile.

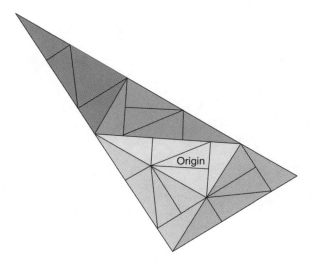

**9. Expand forever.** Fix the interior triangle and continue building super-tiles around it. For each super-tile (and for the interior triangle) draw the largest possible inscribed circle (see the figure). Because the dimensions of each super-tile grow by a factor of $\sqrt{5}$, so do the radii of the circles. Because the circles are nested and their radii tend to infinity, every point in the plane will eventually be covered.

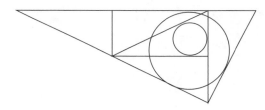

**10. Triangular expansion.** The new equilateral triangle is a larger inverted version of the first. It's rotated 180°.

**11. Expand again.** Unlike the 4-unit super-tile, the 16-unit super-tile has the same orientation as the original (center) equilateral triangle.

**12. One-answer supers.**

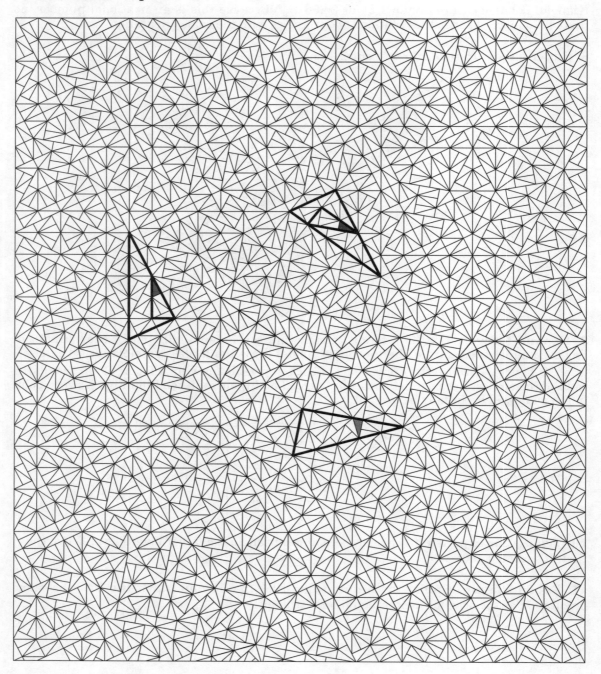

**13. Close to periodicity.** Look at the 5-unit super-tiles containing the tile marked with a star and the one marked with a 1. Because they are in different positions in those 5-unit super-tiles, none of the surrounding tiles will line up.

**14. Golden periodicity.** Yes; but it won't look anything like the Pinwheel Pattern. Put two Golden Triangles together to form a 2 × 1 rectangle. Using this pair as a building block, tile the plane as you would with rectangles. There are several possible periodic patterns.

**15. Many answer supers.** Shifting the super-tiles 1 unit will create different possible super-tiles.

### III. Creating New Ideas

**16. Personal Escher.** See the tilings in the figures. The nine tiles represent the basic tiling pattern. The top and bottom edges are identical, as are the left and right edges. More complicated patterns with one or two tiles can be made by grouping four of them together (see figure) to make one standard tile. Note that the two 4-unit tiles have top and bottom edges equal, and so on.

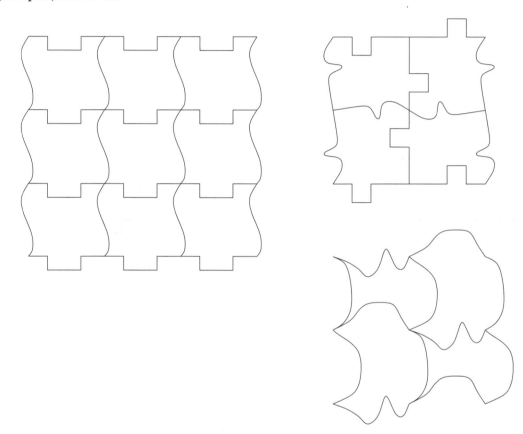

**17. Fill 'er up?** The square, the hexagon, and the triangle can tile the plane, but the octagon can't. If two octagons meet at a side, they form a right angle that prohibits other octagons from filling the empty space.

**18. Friezing.** These friezes are real works of art, so the patterns are not exact. However, the first three have symmetries that shift every point to the right. The second frieze (just the central strip) can be flipped left to right; however, the last two friezes cannot be flipped left to right. The second frieze can be flipped over or flipped and shifted; however, neither of the last two can be flipped.

**19. Wallpapering.** All but the first wallpaper patterns can be shifted to the right. The second wallpaper pattern has a symmetry that reverses the right and the left sides across a vertical central line. The third pattern can be flipped over across a horizontal line that is about 2/3 of the way up from the bottom. After flipping, it can also be shifted. The third pattern can also be rotated through 180° around the center of one of the big flowers.

**20. Commuters?** To see that order matters, it is easier to look at the end results of the modified square pattern of Mindscape III.16. Flipping diagonally and rotating by 90° is equivalent to flipping about a horizontal line. Reversing the steps is equivalent to flipping about a vertical line.

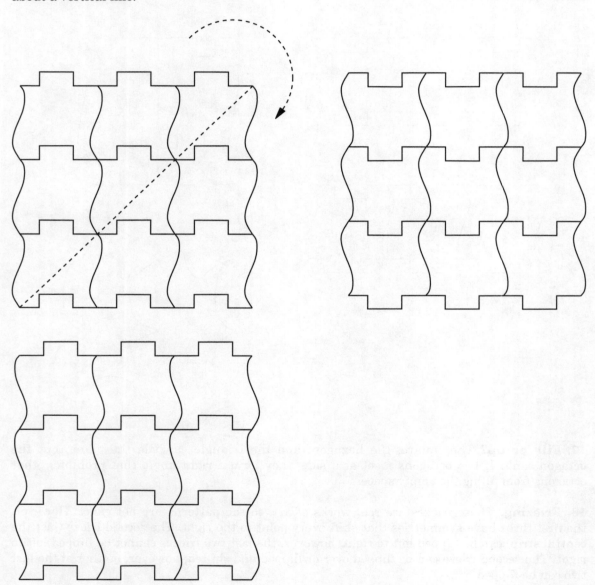

## IV. Further Challenges

**21. Penrose tiles.** Here is the start of a Penrose tiling:

**22. Expand forever.** (See Mindscape II.9.) First, show that you can create arbitrarily large versions of the original shape. Second, argue that if the original shape is glued to the center of a large floor, the process of creating larger and larger shapes will cover the entire floor. (1) If you need 10 tiles to construct a larger version of the original shape, then making 10 separate 10-unit tiles and arranging them in the same pattern allows you to create an even larger version of the original shape. (2) If the new shape covers it on all sides, then each new super-tile will expand away from the center.

413

## 4.5  The Platonic Solids Turn Amorous

### I. Developing Ideas

**1. It's nice to be regular.** A polygon is regular if all sides are the same length and all angles have equal measure. Here are regular polygons with 3, 4, 5, 6, 7, and 8 sides. (Note that the regular 7-gon is particularly hard to draw.)

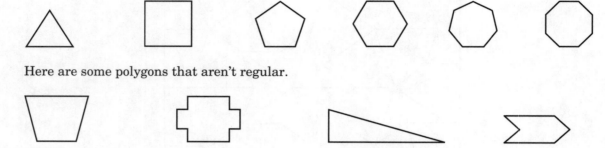

Here are some polygons that aren't regular.

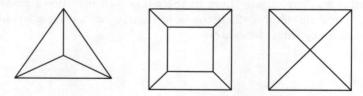

**2. Keeping it Platonic.** A solid is regular if all its faces are identical regular polygons meeting at the same number of edges at each vertex.

**3. Count 'em up.** A cube has 6 faces, 12 edges, and 8 vertices. A tetrahedron has 4 faces, 6 edges, and 4 vertices.

**4. Defending duality.** The cube and the octahedron are duals for the following reason: If you start with a cube and put a vertex in the middle of each face and then join by an edge any two vertices with adjacent faces, you obtain an octahedron. Similarly, start with an octahedron, put a vertex on each face, then join by an edge any two vertices with adjacent faces. The result is a cube.

**5. The eye of the beholder.** The solid on the left is a tetrahedron viewed by looking down on a vertex. The middle solid is a cube viewed by looking straight down at a face (the perspective here is skewed, because the face closer to you should actually look larger than the face farther away.) The third solid is an octahedron viewed by looking directly at a vertex.

### II. Solidifying Ideas

**6. Build them.** Make a complete set of the Platonic solids.

**7. Unfold them.** Pages 274–275 show one way to unfold each of the regular solids. Many other unfoldings are possible.

414

**8. Edgy drawing.**

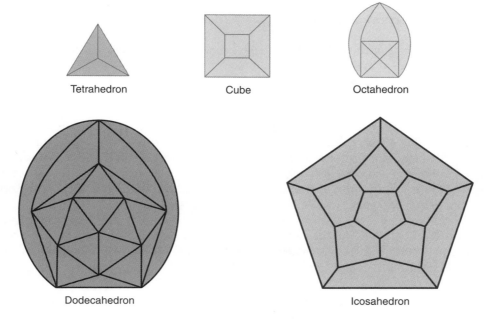

Tetrahedron  Cube  Octahedron

Dodecahedron  Icosahedron

**9. Drawing solids.** See the figures.

 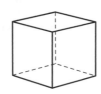  

**10. Life drawing.** Draw the regular solids.

**11. Count.** 2 is the magic number. Tetrahedron: $4 - 6 + 4 = 2$. Cube: $8 - 12 + 6 = 2$. Octagon: $6 - 12 + 8 = 2$. Dodecahedron: $20 - 30 + 12 = 2$. Icosahedron: $12 - 30 + 20 = 2$.

**12. Soccer counts.** View the soccer ball as an icosahedron with its vertices sliced off. The 12 vertices of the icosahedron become the 12 pentagons on the soccer ball, and the 20 triangular faces become hexagons, for a total of 32 faces. The vertices come in groups of five, each corresponding to a vertex of the original icosahedron. So, there are $12 \times 5 = 60$ vertices. There are $12 \times 5 = 60$ edges between the hexagons and the pentagons and $20 \times 3/2$ edges between hexagons (20 hexagons, 3 edges per hexagon, and an overcount by a factor of 2), for a total of 90 edges. Once again, $60 - 90 + 32 = 2$.

**13. Golden rectangles.** Assume that the icosahedron has unit length. Because the base of the first rectangle is an edge, it also has unit length. Focus on the vertices of one of the vertical edges of the rectangle. There is exactly one pentagon formed by the edges of the icosahedron containing these two vertices, and this imaginary vertical edge cuts the pentagon in two. The length of the vertical side is $2 \times \sin(54) = (\sqrt{5} + 1)/2$, which is the Golden Ratio.

**14. A solid slice.** Cubes, tetrahedrons, and dodecahedrons yield triangles when their vertices are sliced. Octagons yield squares, and icosahedrons yield pentagons. The number of sides of the boundary corresponds to the number of faces that meet at a vertex.

**15. Siding on the cube.**
Faces: 6 faces of the cube × 4 glued faces = 24 total faces
Vertices: 8 original vertices of the cube + 1 new vertex for each of the 6 faces of the cube = 14 total vertices
Edges: 12 original edges of the cube + 4 new edges for each of the 6 faces = 36 total edges.
$V - E + F = 14 - 36 + 24 = 2$

## III. Creating New Ideas

**16. Cube slices.** Slicing off a vertex generates a triangle, because the cutting plane intersects three sides of the cube. If we continue making parallel cuts, that triangle will get larger until additional sides intersect the cutting plane. At this point, the slices will look like triangles with one or more vertices cut off. If the cuts generate growing equilateral triangles, then when we are halfway through the cube, the slice will be a hexagon. Depending on where we cut, we can get a wide variety of 3-, 4-, 5-, and 6-sided shapes.

**17. Dual quads.** The edges of the octagon connect the centers of the square faces. Each edge is the hypotenuse of a right triangle with legs of length 1/2. Therefore, the edges of the octagon have length $\sqrt{2}/2$.

**18. Super dual.** Octahedron edges have length $3\sqrt{2}/2$. Draw a vertical line from the vertex of an equilateral triangle to the base of the opposite side. The center of the triangle cuts this vertical line so that the longer piece is twice the length of the shorter (calculate lengths explicitly). This 2:1 proportion is unchanged when we view the triangle from other perspectives. View the surrounding octahedron from a vertex (see the figure). Let $x$ represent the length of the new edges, and use the 45-45-90 triangle to solve for $x$.

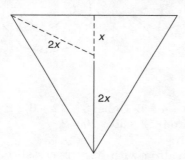
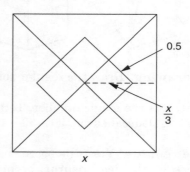

**19. Self-duals.** The edge lengths are one-third of the original tetrahedron. As in Mindscape III.18, it helps to view the tetrahedron from a vertex and to note that the center of an equilateral triangle is one-third as high as the vertex.

**20. Not quite regular.** There are two types of answers. Technically, you could subdivide the faces of an icosahedron into smaller triangles so that most vertices (the new ones) are surrounded by six triangles, while some vertices (the old ones) are surrounded by five. Many dome structures are made this way. This is slightly unsatisfying, because if six equilateral triangles meet at a vertex, you really have only one hexagonal face. As an alternative, arrange things so that more than six triangles surround a vertex. This can be done by progressively replacing equilateral triangles with the top part of a tetrahedron. (One triangle gets replaced with three.) The more triangles added, the more the surface undulates. You lose convexity, but you get real, honest-to-goodness sides.

416

## IV. Further Challenges

### 21. Truncated solids.

| Solid (pretruncated) | Number of Truncated Vertices | Number of Edges | Number of Faces |
|---|---|---|---|
| Tetrahedron | 12 | 18 | 8 |
| Cube | 14 | 36 | 24 |
| Octahedron | 14 | 36 | 24 |
| Dodecahedron | 32 | 90 | 60 |
| Icosahedron | 32 | 90 | 60 |

New $F$ = Old $F$ + Old $V$

New $E$ = Old $E$ + Old $V$ × number of faces that meet at a vertex

New $V$ = Old $V$ × number of faces that meet at a vertex

Each old vertex is cashed in for a new face and as many new vertices as there are faces that meet at a vertex.

### 22. Stellated solids.

| Solid (prestellated) | Number of Stellated Vertices | Number of Edges | Number of Faces |
|---|---|---|---|
| Tetrahedron | 8 | 18 | 12 |
| Cube | 24 | 36 | 14 |
| Octahedron | 24 | 36 | 14 |
| Dodecahedron | 60 | 90 | 32 |
| Icosahedron | 60 | 90 | 32 |

New $V$ = Old $V$ + Old $F$

New $E$ = Old $E$ + Old $F$ × number of sides per face

New $F$ = Old $F$ × number of sides per face

Each old face is cashed in for a new vertex and as many new triangular faces as there were vertices per old face. Note that these solids are the duals of the solids in Mindscape IV.21.

## 4.6 The Shape of Reality?

### I. Developing Ideas

**1. Walking the walk.** Each walk is seven blocks long. If you only travel east or north, you must travel exactly four blocks east and three blocks north to get from $X$ to $Y$, for a total of seven blocks.

**2. Missing angle in action.** Because the angles of any triangle in the plane must sum to $180°$, we find in the first triangle $x = 180 - 70 - 50 = 60$, in the second triangle $x = 180 - 50 - 20 = 110$, and in the third triangle $2x = 180 - 90 = 90$, so $x = 45$.

**3. Slippery X.** We can't determine the size of angle $x$. For a triangle drawn on a sphere, the angles do not have a fixed sum. Triangles that are small compared with the radius of the sphere have angles that sum to just over $180°$, whereas triangles that cover a large portion of the sphere can have angles that sum to $270°$ or more.

**4. A triangular trio.** Triangle $A$ has the largest angle sum. Triangle $B$ has the smallest.

**5. Saddle sores.** The angle $x$ must be less than $90°$, because the angles of this triangle must sum to less than $180°$.

### II. Solidifying Ideas

**6–8. Travel agent.** Iceland is between Austin and Tehran; Alaska is between Austin and Beijing; and a direct Massachusetts–Beijing flight will send you over the north pole.

**9–11. Latitude losers.**

| City Pair | Latitude Distance | Great Circle Distance |
|---|---|---|
| Beijing, China–Chicago, Illinois | 8300 mi | 6600 mi |
| Mexico City, Mexico–Bombay, India | 11100 mi | 9700 mi |
| Sydney, Australia–Santiago, Chile | 7900 mi | 7100 mi |

Use the activity, The Great Circle, from the Interactive Explorations CD-ROM for the great circle distances, and use the latitude-longitude coordinates to compute the entries in the first column. Alternately, you can search the Internet for great circle distances. Beijing and Chicago are on a latitude line $40°$ below the equator. Their longitudes are separated by $171°$. With an earthly radius of 4000 miles, the latitude trip is $(2\pi)(4000 \times \cos(40°))(171/360)$.

**12–16. Triangles on spheres.**
**12.** 41, 37, 104; total = $182°$
**13.** 76, 88, 63; total = $227°$
**14.** 57, 88, 64; total = $209°$
**15.** 4, 179, 4; total = $187°$
**16.** 180, 180, 180; total = $540°$ (The entire path is itself a great circle.)

**17–21. Spider and bug.** See the figure for the optimal unfolding of the boxes. Minimum distance $D$:
**17.** $D = 2$
**18.** $D = \sqrt{1^2 + 2.25^2} = 2.46 \ldots$
**19.** $D = \sqrt{2^2 + 1^2} = 2.236 \ldots$

418

**20.** $D = \sqrt{2^2 + 2^2} = 2.82\ldots$

**21.** $D = \sqrt{2.5^2 + 1.5^2} = 2.19\ldots$

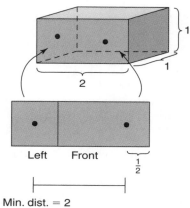

Left  Front  $\frac{1}{2}$

Min. dist. = 2

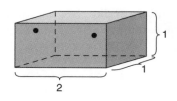

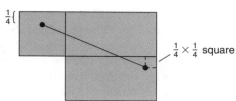

$\frac{1}{4} \times \frac{1}{4}$ square

Min. dist. $= \sqrt{1 + (2 + \frac{1}{4})^2} = 2.46$

$\sqrt{1 + 2.25^2}$

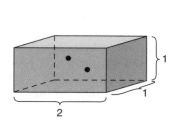

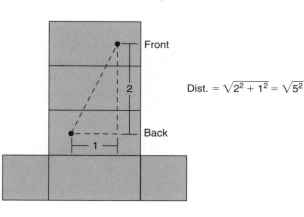

Front

Back

Dist. $= \sqrt{2^2 + 1^2} = \sqrt{5^2}$

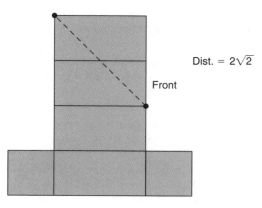

Front

Dist. $= 2\sqrt{2}$

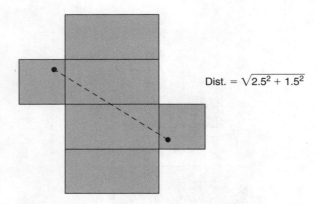

Dist. $= \sqrt{2.5^2 + 1.5^2}$

**22. Becoming hyper.** Follow the instructions to make a hyperbolic plane.

**23. Deficit angles.** The angles will sum to *less* than 180°. The sphere, the plane, and this hyperbolic sheet are examples of surfaces with positive, zero, and negative curvature. On a sphere, the sum of angles adds to *more* than 180°.

**24. Same old.** Because every vertex is surrounded by seven equilateral triangles, there is no way to differentiate one vertex from another. The patterns are the same.

**25. Gauss II.** Earth is locally *very* flat. Suppose you found three buildings so that one pair was separated by 3 miles, another pair by 4 miles, and the third by 5 miles. The sum of these angles would exceed 180° by less than 0.0001°. (See Mindscape II.12 for a method of calculating this.)

**III. Creating New Ideas**

**26. Big angles.** Every triangle divides the sphere into two triangular regions that cover the entire sphere. Suppose we have a tiny triangle whose angles ($a$, $b$, and $c$) sum to just over 180°. Now look at the *other* outside triangle with angles $360 - a$, $360 - b$, and $360 - c$. These angles sum to just under 900°—an excess of almost 720°.

**27. Many angles.** If all three great circles intersect at a point, you get *no* triangles; otherwise, you get eight triangles whose total sum is $8 \times 180 + 720 = 2160$. (The sum of the angles of eight planar triangles plus an excess of 720°.) The simplest example is three perpendicular circles that cut the sphere into eight equilateral triangles, all of whose angles are 90°.

**28. Quads in a plane.** For each quadrilateral, we can find a diagonal line that divides it into two triangles. Because the sum of the angles in a planar triangle is 180, the sum of angles in a planar quadrilateral is $2 \times 180 = 360$.

**29. Quads on the sphere.** Just as the area of a triangle is proportional to its spherical excess (sum of angles – 180), the area of a quadrilateral is proportional to its spherical excess (sum of angles – 360). Area $= 4\pi r^2$(sum of angles in radians – $2\pi$).

**30. Parallel lines.** All lines on a sphere are great circles, and any two of them meet in exactly two points. There are no parallel lines on a sphere.

**31. Floppy parallels.** There are an infinite number of parallel lines that pass through the chosen point. This completes the trilogy: On the plane, there is one such parallel line; on the sphere, none; and in hyperbolic geometry, there are infinitely many.

420

**32. Cubical spheres.** There are eight triangles (one for each vertex), and each triangle has three right angles for a total of 8 triangles × 3 angles per triangle × 90° per angle = 2160°. The spherical excess is 720°.

**33. Tetrahedral spheres.** There is a triangle for each of the four vertices, each with three 120° angles. 4 triangles × 3 angles per triangle × 120° per angle = 1440°. 1440 − (4 × 180) = 720, as expected.

**34. Dodecahedral spheres.** Recall that there are 20 vertices and that three pentagons meet at a vertex. Each pentagonal face will have a newly drawn point in the middle surrounded by five 72° angles. Each drawn triangle corresponds to one vertex so that the sum of all the angles is the product: 20 triangles × 3 angles per triangle × 72° per angle = 4320. Note that 4320 − (20 × 180) = 720.

**35. Total excess.** That the total excess for triangulated spheres is 720° is intimately related to Euler's characteristic formula, $V - E + F = 2$. In the previous Mindscapes, consider a further triangulation that connects the vertices of the drawn triangles with their centers (the old vertices). Here, the total number of vertices is $V + F$, and the total number of triangles is $2E$. The sum of all the angles is found by summing all the angles around each vertex—that is, $360(V + F)$—and the total excess is found by subtracting 180° for each triangle: $360(V + F) - 180(2E) = 360(V - E + F) = 360 \times 2 = 720$.

## IV. Further Challenges

**36–39. Geometry on a cone.**

**36. What is the sum of the three angles?** A triangle not containing the center will have exactly 180° because it would remain a triangle after unrolling the original flat piece of paper.

**37. What is the sum of the angles of your triangle?** The sum of the angles of a triangle *containing* the cone point is $180 + z$, always! (See proof below.)

**38.**

| Angle of Pie | Sum of Angles of Triangles | Difference with 180° |
|---|---|---|
| 30° | 510° | 330 (= 360° − 30°) |
| 60° | 480° | 300 (= 360° − 60°) |
| 90° | 450° | 270 (= 360° − 90°) |
| 180° | 360° | 180 (= 360° − 180°) |

**39.** The figure shows what an arbitrary triangle would look like if the cone were flattened. The angles of the original triangle on the cone are $a$, $b$, and $c$. Flattening the triangle creates a six-sided polygon. An $n$-sided polygon has $(n - 2)180°$, so the polygon in the figure has 720°. Because the line segment $BC$ is straight on the cone, the angles $d$ and $e$ add up to 180°.

$$a + b + c + (d + e) + (f) \quad\quad = 720,$$
$$a + b + c + (180) + (360 - z) = 720$$
$$a + b + c \quad\quad\quad\quad\quad\quad\quad = 180 + z.$$

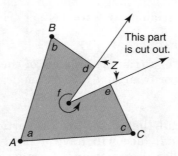

**40. Tetrahedral angles.** Each vertex of a tetrahedron has three equilateral triangles coming together there. Each angle of an equilateral triangle is 60°, so the total angles at each vertex is $3 \times 60° = 180°$. Hence, $360° - 180° = 180°$. There are four vertices of a tetrahedron, and $4 \times 180° = 720°$. We could do a similar computation for other solids; for example, the cube—at each vertex, $360° - (3 \times 90°) = 90°$. There are eight vertices of the cube, and $8 \times 90° = 720°$.

422

## 4.7  The Fourth Dimension

### I.  Developing Ideas

**1. At one with the universe.** Here are the points $x = 3$, $y = -5/2$, and $z = 2.25$.

Because we need only one number to identify a particular point in space, the dimension of the space should be 1.

**2. Are we there yet?** The value $x = 4$ does not specify a point in the plane using Cartesian coordinates, because there are many points (an infinite number) with $x$-coordinate 4. Thus, the dimension of the plane must be greater than 1.

**3. Plain places.** The three points are plotted here. There are no points in the plane that require three numbers to plot.

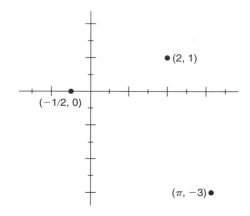

**4. Big stack.** A huge stack of sheets of paper would end up looking like a solid block with dimension 3, even though each sheet of paper has basically only two dimensions.

**5. A bigger stack.** This is analogous to Mindscape I.4. A huge stack of 17-dimensional spaces would look like an 18-dimensional block (in our imaginations, anyway).

### II.  Solidifying Ideas

**6. On the level in two dimensions.** Object 1 resembles a circle or a diamond; Object 2, a triangle; Object 3, a square. The last sequence of points represents cross sections of a line.

**7. On the level in three dimensions.** Object 1: sphere or two cones on top of one another (a diamond spun about its axis). Object 2: Tetrahedron. Object 3: vaselike object (a capped off cylinder with a bulge in the middle). Object 4: a donut.

**8. On the level in four dimensions.** Because we don't run into these objects in ordinary life, there is no common name for them. These are objects generalized from the previous problems. Object 1: a hypersphere. Object 2: a hypercube. Object 3: a line that happens to sit in four dimensions. Object 4: a hypertetrahedron.

**9. Tearible 2's.** The last two barriers require tearing to get the gold; the first two do not.

**10. Dare not to tear?** Take the left line segment (which is wider than the gold) and stretch it into a semicircle that comes out of the page (into the third dimension). If we look only at the plane containing the gold, the container appears to have a hole in it, and we can slide the gold through that hole. To cover our tracks, we would then carefully push the container back from the third dimension to its original configuration.

**11. Unlinking.** Generalizing Mindscape II.10 allows us to unlink the rings. Take a small portion of one of the rings and push it into the fourth dimension. If we look only at the original three dimensions, it appears as if the rope were cut. Take the other ring and slip it past this hole in the rope to unlink the rings. Finally, push the small portion of the rope back into the original 3-dimensional space.

**12. Unknotting.** The pictures left to right have four, five, and six crossings, respectively. At each crossing, one portion of the rope passes under another portion. The 4-dimensional trick described in Mindscape II.11 allows us to have that portion of the rope go *over* instead of under. That is, we can invert any crossings we want. The only challenge in this problem is to invert as few crossings as possible. The first picture requires inverting only one crossing, but the other pictures require inverting two crossings each.

**13. Latitude.** The object is a regular sphere (not a hypersphere). All of her slices are 2-dimensional objects, and if you stack them on top of each other, you'll get a 3-dimensional object (not four).

**14. Edgy hypercubes.** It's an artistic feat to make a 2-dimensional drawing of a 6-dimensional cube. See the figures for the 4D and 5D edge drawings. A 3D cube has $4 + 4 + 4 = 12$ edges. A 4D cube has $12 + 12 + 8 = 32$ edges. A 5D cube has $32 + 32 + 16 = 80$ edges. And a 6D cube has $80 + 80 + 32 = 192$ edges.

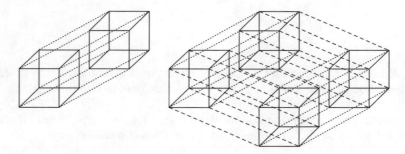

**15. Hypercube computers.** All vertices in an $n$-dimensional cube are connected by a sequence of less than $n$ edges. The longest distance is 2 for a square, 3 for a cube, 4 for a hypercube, and so on. A good way to approach this problem is to represent the vertices by their coordinates and to note that two vertices are connected by an edge if their coordinates differ in only one place.

## III. Creating New Ideas

**16. *N*-dimensional triangles.**

| Triangle's Dimension | Number of Vertices | Number of Edges | Number of 2D Faces | Number of 3D Faces |
|:---:|:---:|:---:|:---:|:---:|
| 1 | 2 | 1 | 0 | 0 |
| 2 | 3 | 3 | 1 | 0 |
| 3 | 4 | 6 | 4 | 1 |
| 4 | 5 | 10 | 10 | 5 |
| 5 | 6 | 15 | 20 | 25 |
| *n* (in general) | $n$ | $n(n-1)/2$ | $n(n-1)(n-2)/6$ | ($n$ choose 4) |

In each case, all the vertices are connected to all the other vertices. To find all the edges, it is enough to ask, "How many ways can we choose two different vertices?" ($n$ choose 2) = $(n!)/(2!(n-2)!)$. To find all the triangles, we ask, "How many ways can I choose three different vertices?" ($n$ choose 3) = $(n!)/(3!(n-3)!)$.

**17. Doughnuts in dimensions.** A circle can be viewed as a circle of points, and a torus is just a circle of circles. Each point on the top circle (at the highest 4D level) is part of a sphere. To see this, imagine taking radial cross sections of each of the slices. A radial cut of the first slice gives a circle. This circle is connected to smaller circles above and below that slice and are reminiscent of the cross sections of a sphere. Just as the points on the top circle are connected smoothly together, so are the corresponding spheres.

**18. Assembly required.** See the figure. There are 24 square faces—6 around the center cube (hidden) and 12 next to the 12 edges of the center cube (see the faces labeled 1, 2, and 3 in the figure). The remaining 6 faces are labeled *A*, *B*, *C*, *D*, *E*, and *F* in the figure.

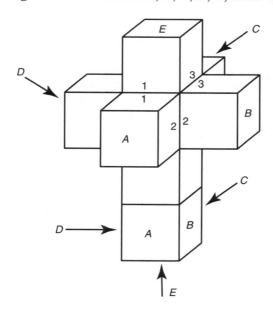

**19. Slicing the cube.** The first slices will intersect only three sides, so they will be triangles. The triangular cross sections will increase in size until the cross section intersects all six sides of the cube. At this point, the tips of the triangular cross sections themselves will be sliced off. Halfway through, you'll see a perfect hexagon and then the cross sections will repeat themselves, as in the figure.

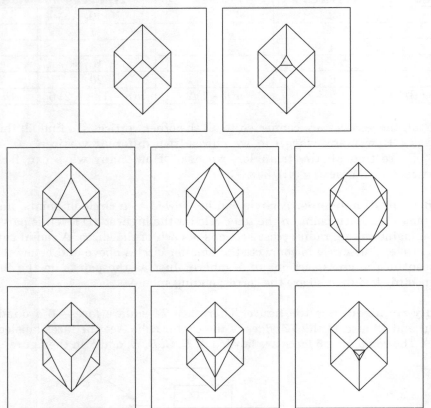

**20. 4D swinger.** The circle and dot swung around produce a circle sitting inside a torus. If we only swing the point, we get two linked circles. If we swing two linked circles in 4D, we get two linked tori. If we spin the center point of a sphere about four dimensions, we will create a circle that is linked to a sphere.

## IV. Further Challenges

**21. Spheres without tears.** See the figure. In the 4-dimensional sphere, let's resort to cross sections to illustrate pushing the inside of the sphere into the fourth dimension. Note also that you can push the insides of the two spheres in opposite directions so that the boundaries can be glued.

426

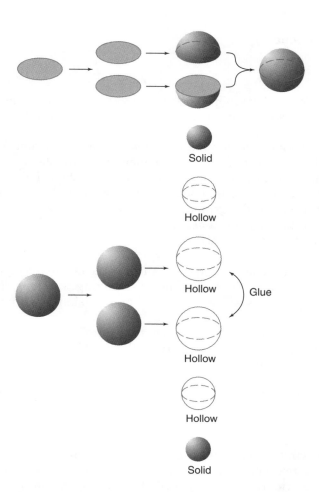

**22. Linking.** Despite the added freedom of moving within four dimensions, the two are still linked. If you pushed the whole circle into the next level, it would appear to be linked to the smaller circle in that level. The top point keeps us from pushing the circle too far into the fourth dimension.

**5.1 Rubber Sheet Geometry**

**I. Developing Ideas**

**1. Describing distortion.** Two objects are equivalent by distortion if we can stretch, shrink, bend, or twist one, without cutting or gluing, and deform it into the other.

**2. Your last sheet.** A toilet paper tube is equivalent by distortion to a CD if we assume that the CD has thickness. Imagine the toilet paper tube is made of very squishy, malleable material. Just flatten it out until it's the diameter of a CD, then shrink the hole until it's the size of the hole in the CD. For the finishing touch, draw out the little tendrils from around the edges of the hole to match the gripper teeth that ring the edge of the center of the CD. Because the material of the original tube was only stretched or shrunk into a CD shape, the tube is equivalent by distortion to the CD.

**3. Rubber polygons.** The triangle, square, and pentagon are all equivalent by distortion because they can all be obtained by stretching or shrinking the same rubber band. All other polygons can also be obtained, as well as circles, ellipses, or shapes with curving boundaries.

**4. Out, outred spot.** Removing the red spot from the X leaves four pieces; removing it from the Y leaves three pieces; removing it from the Z leaves two pieces. Thus, the original letters are not equivalent by distortion.

**5. De-vesting.** The answer here is in the doing!

**II. Solidifying Ideas**

**6. That theta.** Removing the two intersection points leaves the theta curve in three pieces, whereas removing *any* two points on the circle always leaves exactly two pieces. Suppose there were a set of deformations that turned the theta curve into a circle. Mark the two intersection points red and follow the deformation process. At any stage, the removal of the two red points should leave the distorted theta curve in three pieces; yet, when we get to the final stage of the circle, we find that the object falls into just two pieces. This contradiction shows that our assumption is wrong: The theta curve is not equivalent to the circle.

**7. Your ABCs.** Note that the category depends very much on the font. We have picked a simple font without serifs to simplify the categorization. There are nine groups (A, R), (B), (C, G, I, J, L, M, N, S, U, V, W, Z), (D, O), (E, F, T, Y), (H, K), (P), (Q), (X).

**8. Puzzled?** Refer to the figures to see solutions to the kit topology puzzle.

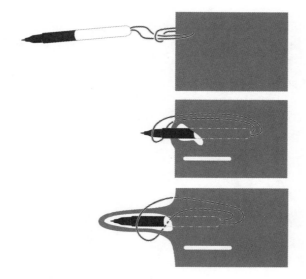

**9. Half dollar and a straw.** They are equivalent. It doesn't matter whether you view the objects as having thickness. If you viewed them as ideal 2-dimensional objects, the straw is a cylinder and the defaced coin is a disk with a hole in it. Deform the straw so that the bottom of it is the same radius as the drilled hole and stretch the top so that it is as wide as the coin. Now squash the straw vertically so that the entire straw lies in a single plane. Voilà; you've deformed the straw into the coin. A similar process works for the thickened 3-dimensional objects.

**10. Drop them.** With rubber undies, yes; with real undies, probably not. Grab hold of the right side and stretch them so that you can pull them around your right leg. Now you're left with long undies that are hanging out of your right pant leg. Because they are no longer around your right leg, you are free to pull out the undies. See the figures.

429

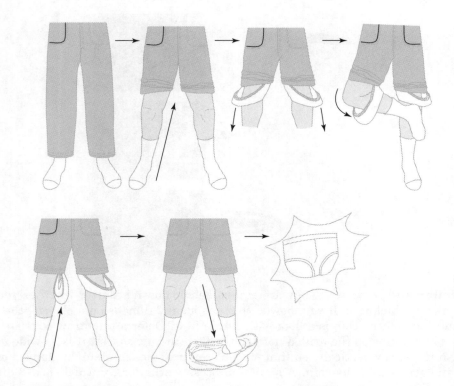

**11. Coffee and doughnuts.** Yes; with the short answer being that they both have a single hole in them. To really show equivalence, it's necessary to describe the distortion that turns one object into the other. For example, take your Gumby-like coffee mug and flatten out the part that holds coffee so that it looks like a flat disk connected to a ring. Now shrink the disk so that the resulting object looks more like a ring with a bump. Smooth out the bump, thicken the ring, and you have a Gumby-like doughnut.

**12. Lasting ties.** Yes, you can unlink yourselves. With your left hand, make a small almost-loop out of your friend's rope and slip it between your right wrist and the handcuff, over your hand, and then back out between your wrist and handcuff.

**13. Will you spill?** Yes. First rotate your hand 180° by bringing the glass under your armpit. Roll your shoulder forward and continue rotating your hand (while straightening out your arm) until you've rotated the palm of your hand 360°. You might feel close to dislocation, but don't stop here. Swing your arm in front of you and bend your elbow so that the glass is just above your head (540°). Continue swinging your arm above your head until the glass is back in its original position. You've rotated the glass 720° with your arm intact!

**14. Grabbing the brass ring.** The square obstacle is irrelevant. Two pieces of the rope go over the ring. Pull them around the ring in opposite directions until they are at the other side of the ring. The rope now slips right off.

**15. Hair care.** Unlike Mindscape II.9, this problem depends on whether you viewed these objects as 2- or 3-dimensional. Viewed as a 3D object, both the hair pin and comb are distortable into solid balls and so, in turn, are distortable into each other. A 2D comb, however, is not distortable into a 2D hair pin. A 2D comb is like an E (from Mindscape II.7) with many more horizontal lines, while a 2D hair pin is equivalent to a C.

**16. Three two-folds.** Cutting on the folds results in papers with two distinct holes; so, the cut sheets are equivalent. Cutting opposite the folds results in four semicircles cut off the boundaries of the original sheet. This cut sheet is actually equivalent to the original uncut sheet.

**17. Equivalent objects.** The two mugs and bottle cap are equivalent. The bowls, plate, the bottle part of the water bottle, and the ice cube tray are equivalent. Other objects are not equivalent to others.

**18. Clips.** By comparing with a circle, the question asks us to view this as a 2-dimensional paper clip. The paper clip can be unfolded to resemble a line segment, so it is not equivalent to a circle. (Removing any single point cuts the paper clip in two but leaves the circle intact.) It is, however, equivalent to a staple, mechanical pencil fillers, the metal spirals that keep notebooks together, and so on.

**19. Pennies plus.** Yes; show the equivalence by giving the penny a quarter turn and squashing the thickened semicircles into thickened rectangles. Both objects are equivalent to a solid sphere.

**20. Starry-eyed.** Suppose the five- and six-pointed stars are made of flexible plastic. Imagine placing a metal cylinder with a small radius inside each star and expanding the radius until both stars are stretched into tightly fitted circles. This shows that both stars are equivalent to a circle, as well as how to distort one star into the other.

**21. Learning the ropes.** Simplify both pictures by shrinking the tail of both ropes into a short straight line. The first rope becomes a circle with a tag pointing inside, and the second rope becomes a circle with a tag pointing outside. If the distortion is limited to two dimensions then the objects are different because the tag is trapped by the circle. With three dimensions, we simply rotate the tag around to show the equivalence. See the figure.

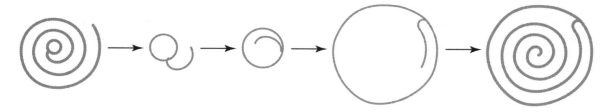

**22. Holy spheres.** Yes; the rules of distortion allow us to move the holes around. Take the top hole and move it around the other three holes.

**23. From sphere to torus.** No; the torus and the sphere are not equivalent, so some of the manipulations implied by the drawing are not allowed. The first three steps are okay, but the last two steps are ambiguous, because they necessarily involve a cutting of the sphere (separating points that used to be touching) or pasting (gluing distant points together), both of which are illegal operations under the rules of distortion.

     Why are they not equivalent? There exists a closed curve on the torus that, when removed, leaves the torus intact. No such curve exists on the sphere. If we could distort the torus into the sphere, the nonseparating curve on the torus would be deformed into a separating curve on the sphere. This contradiction disproves the equivalence.

**24. Half full, half empty.** In one case, there is only one exterior surface; but when the juice is at the top, there are two exterior surfaces (one inside the glass, one outside). So these are not equivalent by distortion.

**25. Male versus female.** There exists one point on the female symbol that cuts it into four pieces, and there exist no points on the male symbol that will cut it into more than three pieces. These symbols are different by the same reasoning that showed the theta curve different from the circle (Mindscape II.6).

### III. Creating New Ideas

**26. Holey tori.** The simplest answer, though not so rigorous, is that they are not equivalent because one object has six holes and the other has five holes. Though it is nontrivial to do so, there is a way to define "hole" so that the number of holes in a surface doesn't change under deformation. An alternative approach: Find six nonseparating curves on the surface of the six-holed object and argue that any six curves on the surface of the five-holed object will cut it in at least two pieces. This last point is just as difficult to make rigorous.

**27. More holey tori.** First count the number of holes in each object. If they are different (as in Mindscape III.26), you can stop; but if they are the same (as in this problem), it is necessary to show explicitly how to distort one object into the other. Once you know that it is possible, it isn't hard to do. Grab the four corners of the large squarish hole and pull them toward the center. See the figures.

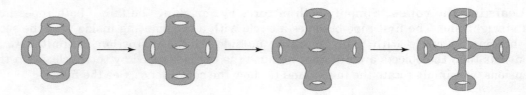

**28. Last holey tori.** Because both objects have five holes, they can be distorted into each other. Grab the tori at the ends and move them around the surface until they are both attached to the center tori. See the figures.

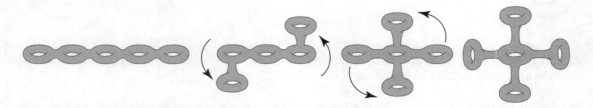

**29. Beyond the holey inner tube.** Convincing yourself that you can invert a punctured sphere and a punctured torus will bring you a long way toward believing that this too can be done. It will also give you ideas on how to perform the inversion. See the figures for one technique. Here, we expand the patch, push the two-holed torus through the opening, and then shrink the patch again.

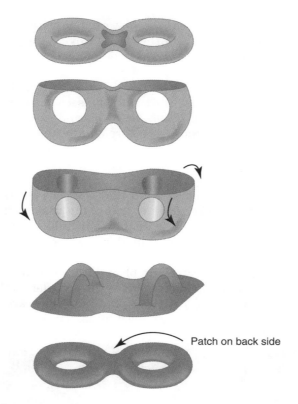

Patch on back side

**30. Heavy metal.** Simply straighten the pieces.

**31. Rings around the ring.** See the figures. Because two linked circles cannot be unlinked in three dimensions, it is comforting that the circles remain linked throughout the deformation.

**32. The disk and the inner tube.** They are equivalent. It is easier to visualize deforming the first object into the punctured inner tube—deform the flat part into an extension of the tube. In the deformation, the boundary of the rubber disk becomes the boundary of the patch.

**33. Building a torus.** Fold the sheet into a cylinder by gluing opposite sides together. Now bend the tube around so that the top circle is right up against what was formerly the bottom circle. Gluing these circles completes the process. Opposite sides are glued together, and the edges match up in the natural way, no twisting necessary. See the figures.

Glue

**34. Lasso that hole.** The first torus can be distorted into the second by grabbing one puncture and dragging it around the hole of the torus. Moving the puncture around the hole a second time will generate the third torus.

**35. Knots in doughnuts.** As in Mindscape III.34, we solve this by dragging and sliding the holes around the surface of the torus. Start with the second torus; as you slide the worm hole around, shrink it so that the knot becomes less and less complicated. Unknotting requires dragging one end of the tube around and through the hole of the torus.

### IV. Further Challenges

**36. From knots to glasses.** Yes; keep in mind that the tube is just as deformable as the thickened knot. Slide part of the thickened knot across the tube so that it appears as if the thickened knot is now looped to itself. *Note:* If the solid tube were colored as in the first picture, the final deformed object wouldn't be colored in the same way as the second picture.

**37. More Jell-O.** See the figures. Stretch the block into a U shape so that the inner tube becomes straight. Thicken the thin outside portion of the cylinder so that it matches up with the rest of the figure. Now twist the figure so that the holes are pointing in the same direction to get our standard picture of the two-holed torus.

**38. Fixed spheres.** Yes; If the sphere were distortable, this would be easy—shrink the sphere and thread it through the knot. We could just as easily expand the knot and move the entire knot around the sphere instead. It's a matter of perspective. For example, when you are on a roller coaster, it appears as if you are moving and the track is staying put. But could it not be that you are standing still while the track and the rest of the world are moving around you?

**39. Holes.**
**a.** There exist two nonseparating loops on the two-holed torus (one loop through each of the holes).
**b.** Removing any two loops on a torus will leave at least two pieces (removing a nonseparating loop on the torus leaves a cylinder, and removing any loop from a cylinder will chop it in two).
**c.** *If* the two loops were equivalent, then the two-holed torus could be deformed into the torus. This same deformation would deform the two nonseparating loops of the two-holed torus into two nonseparating loops of the torus. Because no such loops exist, this represents a contradiction, proving our first assumption wrong. They are, therefore, not equivalent.

**40. More holes.** As in Mindscape IV.39, the important facts are (a) There exist three nonseparating loops on the three-holed torus, and (b) there do not exist three such curves on the two-holed torus. If we assume that the three-holed torus can be deformed into a two-holed torus, then we could use the deformation to find three nonseparating loops on the two-holed torus. This contradicts the facts above and proves that these are inherently different objects. In general, tori with different numbers of holes are not equivalent.

## 5.2 The Band That Wouldn't Stop Playing

### I. Developing Ideas

**1. One side to every story.** A Möbius band is a surface with one side and one edge modeled as follows: Take a strip of paper, give it a half-twist, then tape the two ends together.

**2. Maybe Möbius.** One way to determine if a loop of paper is a Möbius band is to place a pencil in the middle of a side and, without lifting the pencil, draw a line along the middle, all the way around until you return to where you started. If the resulting penciled circle appears on "both sides" of the band, then the band is a Möbius band because it has only one side.

**3. Singin' the blues.** Once you construct the band, you should see a white edge meeting a blue edge at the place where you taped the edges together. This Möbius band has one side with a blue portion and a white portion.

**4. Who's blue now?** This construction does not yield a Möbius band. When the edges are taped together, the white edges will meet on one side and the blue edges will meet on the other, preserving the two-sided quality of the loop.

**5. Twisted sister.** Your sister will get a plain loop with no twists. The second half-twist in the same direction will simply undo the first.

### II. Solidifying Ideas

**6. Record reactions.** Just do it.

**7. The unending proof.** Read the band.

**8. Two twists.** There are two edges and two sides before cutting. The cutting process will cut each side in half, for a total of four sides with two additional edges. This is consistent with two strips of paper with two edges each. Before cutting, you can see that the two edges form two linked circles. After cutting, the two strips will still be linked. In addition, the two strips also have two half-twists.

**9. Two twists again.** This is exactly the same as Mindscape II.8. You get a thin and thick copy of the original strip linked together.

**10. Three twists.** Because the original strip has only one side and one edge, cutting along the center line will produce two sides and leave two edges that are as long as the single edge in the original strip. The resulting strip is a knotted circle—it is a trefoil knot.

**11. Möbius length.** With the left and right ends identified appropriately, a rectangular strip of paper of length $L$ represents a Möbius band with boundary edge of length $2L$. The centerline of the Möbius band still has length $L$. Cutting along this centerline produces a longer two-sided strip with edges still of length $2L$, but the centerline of this new strip is $2L$ as well.

**12. Möbius lengths.** The center strip is a thinner version of the original, but the additional two-sided loop is twice as long.

**13. Squash and cut.** You do get two pieces, but surprisingly, one of the pieces is a Möbius band.

**14. Two at once.** Because the two strips are connected to each other, there is only one two-sided strip (with two full twists). After cutting along the core, you'll be left with two linked copies of the original strip. If you make a circle out of one strip, the other strip will loop around it twice.

**15. Parallel Möbius.** It isn't possible. Consider the analogous problem of placing two ordinary two-sided strips side by side. It is easy to imagine gluing the two strips together along one side, creating, in effect, a doubly thick strip. The other two sides remain glue-free. The problem with gluing two Möbius bands together is that there is only one side, so no part of either strip is without glue. If you glued the two sides together, nothing would remain. There isn't enough room in three dimensions to perform this delicate feat.

**16. Puzzling.** If the bump and groove are symmetrical so the pieces fit together even if you turn one over, then they can be made into a Möbius band. With the pictured pieces, however, you cannot.

**17. Möbius triangle.** The centerline of the Möbius band forms a triangle. The outer edges of a flattened 4.5-inch Möbius band form a perfect hexagon. The shortest Möbius band could be made starting with a trapezoid of height 1 consisting of three equilateral triangles. Folding these triangles on top of one another and taping the angled edges of the trapezoid creates a Möbius band covering an equilateral triangle three times. We can make the trapezoid into a rectangle by cutting off a right triangle from one side and inverting it and taping it to the other side. That rectangle has length $\sqrt{3}$, because it would be equivalent to one-and-a-half sides of an equilateral triangle of altitude 1.

**18. Thickened Möbius.** The Möbius band that we create with paper really is a thickened Möbius band, because paper has thickness. The two edges can be seen by drawing a line on the strip far from the centerline. Though the strip has only one side, the drawn line actually has two sides, and these two sides correspond to the two edges of the thickened Möbius band.

**19. Thickened faces.** The one edge of the thin Möbius band becomes the second face of the thickened Möbius band. Locally, the thickened Möbius band resembles a squarish rectangular beam, so intuition leads us to expect four faces. Just as with the edges of a Möbius band, the opposite faces are really connected and there are only two faces.

**20. Thick then thin.** The thickened Möbius band of Mindscapes II.18 and II.19 has two faces and two edges, and the object is completely symmetrical with respect to these edges and faces. Whether you shrink the top (and bottom) faces or the left (and right) faces, you'll end up with a Möbius band.

**21. Drawing the band.** The loop formed by the edge of the Möbius band and the loop formed by the centerline are linked together. As you pull in the drawstring, the outer edge will become a smaller and smaller circular loop. But because this loop is linked with the centerline loop, you will always have extra cloth to contend with, preventing the drawstring from closing up neatly.

**22. Tubing.** It still has only one side, but it has two edges.

**23. Bug out.** Have the bug walk over the top of the handle part of the Klein bottle and keep going around. Soon the ladybug is "inside."

**24. Open cider.** By rotating the Klein bottle, the liquid will go up and around the thin part and come out.

**25. Rubber Klein.** Compare with Mindscape III.33 from Section 5.1. Glue the left and right sides to make a cylinder. Now bend the cylinder into a U shape, pass one end through the side of the cylinder (not possible in three dimensions), and then join the two ends together. See the figures.

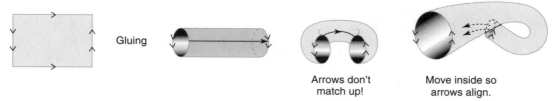

Gluing

Arrows don't match up!

Move inside so arrows align.

### III. Creating New Ideas

**26. One edge.** With a red marker, start drawing a line from the middle of the bottom edge to the right end of the bottom edge. Lift the marker and move it to the top left corner of the rectangle (it represents the same point after twisting and pasting). Continue moving left to right until you get to the top right corner. Again, because this corner is connected to the lower left corner, continue marking left to right from the lower left corner. Eventually you will get back to the middle of the bottom edge where you started. You've covered all the boundary edges in the band, so you've proved that there is only one edge.

**27. Twist of fate.** The half-twist of the Möbius band arises from identifying two lines with arrows pointing in different directions. The centerline of the Möbius band crosses this identification line one time—corresponding to the one-half twist. The centerline of the cut Möbius band is twice as long and crosses the identification band *two* times in the same direction.

**28. Linked together.** The pieces are interlocked because the centerlines of the two objects form linked rings. Equivalently, the edge of the Möbius band links the centerline. To see this, wrap the strip around to form a cylinder with the centerline forming a perfect circle. Identify the plane containing this circle. As you give the strip a half-twist to form the Möbius band, you'll find that the edge of the Möbius band intersects the plane once inside and once outside the circle, showing that the two loops are indeed linked.

**29. Count twists.** See Mindscape III.27. Traveling along the center band, you cross the identified lines one time, corresponding to a single half-twist (just like the original Möbius band). Traveling along the outer band, you cross the identified lines twice, corresponding to a strip with two half-twists.

**30. Don't cross.** Suppose you could draw such a curve. Given this hypothetical curve, label the intersection points on the left edge of the identification diagram as $A$, $B$, and $C$ ($A$ on top, $C$ on bottom). Label the intersection points on the right side as $a$, $b$, and $c$, respectively (again $a$ on top, $c$ on bottom). The way edges are identified, it must be the case that $A$ is identified with $c$ and $C$ is identified with $a$, leaving the middle piece disconnected from the rest. This contradiction proves that a triple crossing is impossible.

**31. Twisted up.** Start drawing lengthwise anywhere in the strip and continue until you get back to where you started. When you finish, you will have completely described one side of the strip (the side with the drawn line). If there is a blank side, then you have an even number of twists; if there is no other side, then you have an odd number of twists.

**32. Klein cut.** The two cuts separate the Klein bottle into two cylindrical pieces. Consider cutting the edge identification diagram *before* pasting the edges together. (See the figures.) Nevertheless, it is good practice to visualize where the cuts are located on the Klein bottle.

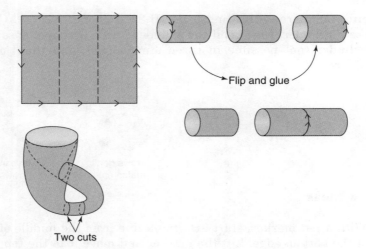

Flip and glue

Two cuts

**33. Find a band.** Look at the solution to Mindscape II.25, which shows how a cylinder passing through itself can be glued together to make a Klein bottle. If we cut the cylinder lengthwise, we will be left with two long strips of paper. Watch what happens to one of the strips after the gluing. Both strips become Möbius bands. See also Mindscape III.35.

**34. Holy Klein.** Take a picture of a Klein bottle, and widen the hole to see that the remaining part is as pictured. Alternatively, use the model of a Klein bottle as a square with opposite edges identified in pairs, with one pair reversed. On it, draw one Möbius band straight across and another twisted strip, as shown. Note that the remaining part of the Klein bottle is just a disk in the hole.

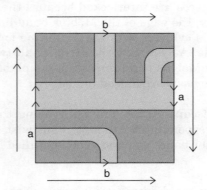

**35. Möbius Möbius.** Instead of making two vertical cuts, as in Mindscape III.32, make two horizontal cuts symmetric about the centerline, as in the figure (these two lines represent a single looped curve on the Klein bottle). It is easier to find the Möbius bands by making the cuts on the diagram *before* gluing the associated edges.

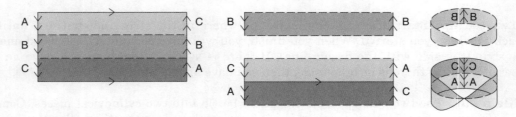

## IV. Further Challenges

**36. Attaching tubes.** Slide one end of the tube around the band, as shown.

**37. Möbius map.** See the figures for two maps, one of which has six countries all bordering one another, which shows that you need at least six colors to color this map.

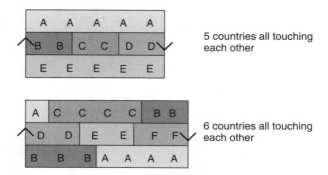

5 countries all touching each other

6 countries all touching each other

**38. Thick slices.** Because there are two edges (see Mindscape II.19), cutting off one corner edge will leave two pieces, each containing one of the edges. The small piece will be twice as long as the larger and will wrap around the larger one two times so that the two pieces are still inseparable.

**39. Bagel slices.** You have to try this! The bagel is still in one piece, though it now resembles a coiled ring with a circumference twice that of the original bagel.

**40. Gluing and cutting.** If we cut the rubber sheet and then identify the appropriate edges, we will get two objects: a Möbius strip and a disk (see figures). The original object can be re-created by gluing the disk back onto the Möbius strip.

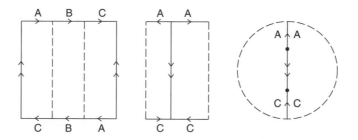

## 5.3 Feeling Edgy?

### I. Developing Ideas

**1. What a character!** The Euler Characteristic is given by $V - E + F = 2$.

**2. Count, then verify.** For the given graph, $V = 5$, $E = 6$, and $F = 3$; thus, $V - E + F = 5 - 6 + 3 = 2$.

**3. Sneeze, then verify.** A rectangular box has 6 faces, 12 edges, and 8 vertices; thus, $V - E + F = 8 - 12 + 6 = 2$.

**4. Blow, then verify.** For the peace symbol, $V = 5$, $E = 8$, and $F = 10$; thus, $V - E + F = 5 - 8 + 10 = 2$.

**5. Add one.** The original graph has $V = 3$, $E = 3$, and $F = 2$. Adding a vertex in the middle of one edge yields new values of $V' = 4$, $E' = 4$, and $F' = 2$, for which $V' - E' + F' = 4 - 4 + 2 = 2$.

### II. Solidifying Ideas

**6. Bowling.** The surface of a bowling ball is topologically equivalent to a sphere, because what appear to be holes are just indentations of the surface. Therefore, the Euler Characteristic is 2.

**7. Making change.** The original graph has seven vertices, eight edges, and three faces ($7 - 8 + 3 = 2$). Figure a adds 1 edge and 1 face. Figure b adds 1 edge and 1 vertex. Figure c adds 1 edge and 1 face. Figure d adds 1 vertex and 1 edge. In any case, we get a new edge and a new vertex or a new edge and a new face. In both cases, the formula $V - E + F = 2$ remains valid.

**8. Making a point.** The total number of vertices goes up by 1, the total number of edges goes up by 1, and the number of faces remains unchanged. The new edge cancels the new vertex in the computation so that the result remains unchanged. $(V + 1) - (E + 1) + (F) = V - E + F$.

**9. On the edge.** You cannot reduce the number of faces in a graph by adding edges. If the new edge is created by connecting two previous vertices, then one new region is added. If the new edge is created by adding one vertex and connecting it to an old vertex, then no new regions are created. In either case, the formula $V - E + F$ remains unchanged. There is no other way to add an edge to a graph.

**10. Soap films.** In all cases, $V - E + F = 2$. In the second picture, $V = 6$, $E = 9$, and $F = 5$. In the third picture, $V = 5$, $E = 8$, and $F = 5$.

**11. Dualing.** The regular solids and their duals have the same number of edges. The number of vertices in the dual equals the number of faces of the original solid, and vice versa. As a result, the computation $V - E + F$ is the same for both objects. Because all regular solids are topologically equivalent to a sphere, their Euler Characteristic is 2.

**12. No separation.** Connected graphs on the plane satisfy the formula $V - E + F = 2$, where $F$ represents the number of pieces of the plane after removing the edges. If the plane isn't separated into different pieces, we have $F = 1$, so that $V - E = 1$.

**13. Lots of separation.** As in Mindscape II.12, the formula $V - E + F = 2$ applies with $F = 231$. We can rewrite this as $E = V + 229$, showing that there are 229 more edges than vertices.

**14. Regions.** Substituting $E = 151$ into the formula $V - E + F = 2$ gives $V + F = 153$. If $F = 153$, then $V = 0$, which is not possible, because every edge must have a vertex at both ends. A loop is the special situation where the same vertex serves for both ends. The maximum number of faces corresponds to having the minimum number of vertices, or 1. With $V = 1$, $F = 152$. So you get the maximum number of faces with 1 vertex. 151 loops around this vertex chops the plane into 152 regions.

**15. Psychic readings.** Use $V - E + F = 2$. If $E = 36$ and $F = 18$, then $V = 2 + E - F = 20$. There are 20 vertices.

**16. An odd graph.** If $F$ is odd, and $V$ and $E$ are both even, then $V - E + F = 2$ won't work. $V - E$ is even, $F$ is odd, and an even plus an odd is odd; but 2 isn't odd. This contradiction shows that no such connected graph exists.

**17. Random coloring.** If your squiggle happens to retrace itself, then you could have conceivably drawn any connected map in existence. In this case, we can use an important theorem in mathematics called The Four-Color Theorem to show that you need no more than four colors. It is unlikely that a random squiggle will involve any retracing. If your squiggle crosses itself only in isolated points, then you need only two colors. We won't give a proof here, but the procedure for coloring is straightforward. Pick any region and color it red. Color neighboring regions blue, and their neighbors red, and so on, until all regions are colored.

**18. Coloring America.** It's easy to find three states, each of which borders the other two (for example, California, Oregon, and Nevada). This proves that we need at least three colors. Can we get by with three? No, Nevada and its five neighboring states can't all be colored with only three colors (because 5 is odd).

**19. Circles on a sphere.** There are three faces, two edges, and two vertices: $V - E + F = 3$, which is not the expected number 2. The Euler Characteristic Theorem makes statements about *connected* graphs on the sphere, and the graph described in this problem is not connected.

**20. More circles.** There are three faces, three edges, and two vertices: $V - E + F = 2$. Unlike Mindscape II.19, this graph is connected; thus, the Euler Characteristic Theorem applies.

**21. In the rough.** There are 13 vertices, 24 edges, and 13 faces implied by this 2-dimensional drawing. $V - E + F = 13 - 24 + 13 = 2$. If we looked at this as a 2D graph, we have instead $V = 12$, $E = 18$, and $F = 8$ (count the outside, too). Again, $V - E + F = 2$.

**22. Cutting corners.** Truncated tetrahedron: $V = 12$, $E = 18$, $F = 8$. Truncated cube and truncated octahedron: $V = 14$, $E = 36$, $F = 24$. Truncated dodecahedron and truncated icosahedron (also known as soccer ball): $V = 32$, $E = 90$, $F = 60$. In all cases, $V - E + F = 2$. Compare with Mindscape II.23.

**23. Stellar.** Stellated tetrahedron: $V = 8$, $E = 18$, $F = 12$. Stellated cube and stellated octahedron: $V = 24$, $E = 36$, $F = 14$. Stellated dodecahedron and stellated icosahedron (also known as soccer ball): $V = 60$, $E = 90$, $F = 32$. In all cases, $V - E + F = 2$. Compare with Mindscape II.22.

**24. A torus graph.** The intersection of the two circles gives one vertex, two edges, and one face: $V - E + F = 1 - 2 + 1 = 0$. So long as the graph goes both ways around the torus (so that all regions can be flattened out), $V - E + F = 0$.

**25. Regular unfolding.** Look at the number of vertices. 4 vertices = tetrahedron; 6 vertices = octahedron; 8 vertices = cube; 12 vertices = icosahedron; 20 vertices = dodecahedron.

### III. Creating New Ideas

**26. A tale of two graphs.** The left piece has eight vertices, nine edges, and three faces (counting the outside face). Note, $8 - 9 + 3 = 2$. The right piece has six vertices, seven edges, and three faces. Again, we have $6 - 7 + 3 = 2$. The composite graph has $8 + 6 = 14$ vertices, $9 + 7 = 16$ edges, and $3 + 3 - 1 = 5$ faces (because we counted the outside face twice). So, $V - E + F = 14 - 16 + 5 = 3$, the magic number for two-component graphs.

**27. Two graph conjectures.** Let $V_1$, $E_1$, $F_1$ represent the vertices, edges, and faces of the first piece and $V_2$, $E_2$, $F_2$ represent the analogous components of the second piece. The total number of vertices is $V = V_1 + V_2$; edges, $E = E_1 + E_2$; faces, $F = F_1 + F_2 - 1$, where the $-1$ counteracts the double counting of the outside face. In all, we have $V - E + F = (V_1 + V_2) - (E_1 + E_2) + (F_1 + F_2 - 1) = (V_1 - E_1 + F_1) + (V_2 - E_2 + F_2) - 1 = 2 + 2 - 1 = 3$. The simplest example of such a graph is two dots: $V - E + F = 2 - 0 + 1 = 3$.

**28. Lots of graphs conjecture.** In the same spirit as Mindscape III.27, we can represent the vertices, edges, and faces of the graph in terms of the vertices, faces, and edges of the individual pieces. $V = V_1 + V_2 + \cdots + V_n$, $E = E_1 + E_2 + \cdots E_n$, and $F = F_1 + F_2 + \cdots F_n - (n - 1)$, because we overcounted the outside face $n - 1$ times. $V - E + F = (V_1 + \cdots + V_n) - (E_1 + \cdots + E_n) + (F_1 + \cdots + F_n + 1 - n) = (V_1 - E_1 + F_1) + \cdots + (V_n - E_n + F_n) + (1 - n) = 2n + (1 - n) = n + 1$. The simplest example is a graph with $n$ points: $V = n$, $E = 0$, $F = 1$, giving $V - E + F = n + 1$.

**29. Torus count.** This polygonal torus has 9 vertices, 18 edges, and 9 faces. $V - E + F = 9 - 18 + 9 = 0$, the magic number for polygonal tori.

**30. Torus two count.** Each torus has 9 vertices, 18 edges, and 9 faces. When pasting the two rectangular faces, we identify 4 vertices, identify 4 edges, and lose 1 face entirely. So, $V = 9 + 9 - 4 = 14$, $E = 18 + 18 - 4 = 32$, and $F = 9 + 9 - 2 = 16$. The Euler Characteristic is then $V - E + F = -2$. In general, the Euler Characteristic of a connected sum is the sum of the individual Euler Characteristics less 2.

**31. Torus many count.** Let $V_1 = 9$, $E_1 = 18$, and $F_1 = 9$ represent the one-holed torus of Mindscape III.29; and let $V_2 = 14$, $E_2 = 32$, and $F_2 = 16$ represent the two-holed torus of Mindscape III.30. We can connect the two just as we connected the tori in the previous problem—remove an outside rectangular face from both tori and glue their boundaries together. The newly created three-holed torus has $V_1 + V_2 - 4$ vertices, $E_1 + E_2 - 4$ edges, and $F_2 + F_2 - 2$ faces, so that the Euler Characteristic is $V - E + F = (V_1 + V_2 - 4) - (E_1 + E_2 - 4) + (F_1 + F_2 - 2) = (V_1 - E_2 + F_1) + (V_2 - E_2 + F_2) - 2 = (0) + (-2) - 2 = -4$. We can make an $h$-holed torus by adding $(h - 1)$ tori to a single torus, one by one. Each time we add a torus, the characteristic goes down by 2, so that an $h$-hole torus has characteristic $-2(h - 1)$.

442

**32. Torus tours.** There are two edges, one face, and one vertex. $V - E + F = 1 - 2 + 1 = 0$. After identifying one edge, two of the vertices are identified, leaving at most two different vertices. After the next identification, these two vertices are identified, leaving just one vertex.

**33. Tell the truth.** The object has 60 faces and 30 vertices. Before gluing, there are $60 \times 3 = 180$ edges; after gluing, each edge is associated with two triangles, so that there are 90 edges total. $V - E + F = 30 - 90 + 60 = 0$. She made a one-holed torus (or miscounted the number of vertices). If she made a two-holed torus, we would need $V - E + F = -2$.

**34. No sphere.** There are 60 triangular faces, $F = 60$. We apply glue to the 180 different sides. Then, one by one, we marry together two sides for a total of 90 marriages, $E = 90$. If $n$ triangles meet at a vertex, then the 180 original vertices are put into groups of $n$, leaving $180/n$ total vertices ($n$ vertices before gluing become 1 vertex after gluing). Because the Euler Characteristic for a sphere is 2, we have $(180/n) - 90 + 60 = 2$. But no integer value of $n$ satisfies this formula, which is a contradiction. However, we might have made a torus ($n = 6$) or a six-holed torus ($n = 9$).

**35. Soccer ball.** Let $P$ and $H$ represent the number of pentagonal and hexagonal faces, respectively, so that the total number of faces is $F = P + H$. Because edges are shared by two faces, we have $E = (5P + 6H)/2$, and because three faces meet at a vertex, we have $V = (5P + 6H)/3$. Substituting these quantities into the Euler formula gives $[(5/3)P + 2H] - [(5/2)P + 3H] + [P + H] = 2$. The $H$'s cancel, so $P = 12$. If we assume there are no flat regions (where three hexagons meet at a vertex), then each vertex is connected to exactly two hexagons and one pentagon. Thus, each vertex is associated with a single pentagon, and the total number of vertices is $5P = 60$. Solving for $H$ in the vertex formula above, $60 = (5P + 6H)/3$, gives $H = 20$. Thus, $F = 32$, $E = 90$, and $V = 60$.

## IV. Further Challenges

**36. Klein bottle.** We start with four edges, four vertices and one face. After one pair of edges are glued together (and so, two pairs of points), we are left with three edges, two vertices, and one face. The final identification leaves two edges, one vertex, and one face. $V - E + F = 1 - 2 + 1 = 0$. The Euler Characteristic for the Klein bottle is 0. Compare with Mindscape III.32.

**37. Not many neighbors.** Because all the countries are connected to each other, the resulting graph is connected, allowing us to use Euler's formula; hence, $V - E + F = 2$, or $F = 2 + E - V$. Because $E \geq 3V$, we have $F \geq 2 + 3V - V$, or $F \geq 2 + 2V$. Each face has a minimum of three edges because there are no loops and a maximum of one edge goes between two vertices. Because each edge corresponds to two different faces (or two different edges of the same face), we have that there are at least $3F/2$ edges ($E \geq 3F/2$). Because $F \geq 2 + 2V$, we have $E \geq 3V + 3$. To summarize: We started with the fact that $E \geq 3V$ and used the Euler Characteristic to show that $E \geq 3V + 3$. We could then repeat this argument with this new bound to show that $E > 3V + 7$, which is the basis of an inductive argument proving that the number of edges is infinite, which is a contradiction.

Alternatively, because $E \geq 3V$, let $E = 3V + X$, where $X$ is positive. Euler's formula gives $F = 2 + (3V + X) - V$, or $F = 2 + 2V + X$. Because $E \geq 3F/2$, $2E \geq 3F$; thus $2(3V + X) \geq 3(2 + 2V + X)$. This simplifies to $2X \geq 3X + 6$, which isn't satisfied by any positive integer $X$. This contradiction shows that our original assumption is wrong. Not every country is surrounded by six others.

**38. Infinite edges.** Here, $V - E + F = 7 - 9 + 3 = 1$. One can view the planar graph as the stereographic projection of a graph on a sphere that has a vertex at the north pole. Faces next to the north pole become unbounded faces in the plane, and edges connected to the north pole become infinite rays. If we add an extra point at infinity (to represent the north pole), then the Euler Characteristic formula works as usual, $V - E + F = 2$.

**39. Connecting the dots.** $V = 5$. Count the edges by counting 4 edges for each vertex, for a total of 20. Every edge is overcounted exactly twice, so there are exactly 10 edges, $E = 10$. Euler's formula demands that $F = 7$. By a similar reasoning, we have at least 21 edges by counting 3 edges to a face; because every edge is a part of two different faces, there are at least 21/2 edges. Because the number of doubly counted edges is even, we must have more than 22 doubly counted edges and, hence, more than 11 *real* edges. This contradicts the fact that $E = 10$. (Alternatively, note that there must exist a subgraph of four vertices connected to each other. There is only one way to draw four vertices connected to each other, and the result is that all faces are triangular. A fifth point would lie in one of these triangular regions, making it impossible to connect to all four points.)

**40. Gas, water, electric.** $V = 6$, $E = 9$. Because the graph is connected, Euler's formula gives $F = 2 - V + E = 5$. There are no loops (no one-sided faces), and there is no more than one edge between any two vertices (no two-sided faces). A triangular face would imply that there was an edge between two utilities or two houses; thus, there are no three-sided faces. Each face has a minimum of four edges. Because each edge is a part of two faces (or two edges of the same face), we have $E > 4F/2$, or $E > 10$. This contradiction proves the impossibility of the alleged drawing.

### 5.4 Knots and Links

#### I. Developing Ideas

**1. Knotty start.** All the figures represent mathematical knots except the one on the far right, which is not a knot because it has loose ends and, thus, is not a closed curve.

**2. The not knot.** The unknot is a closed curve that can be untangled to resemble a circle.

**3. Crossing count.** From left to right, the knots have three, zero, and five crossings.

**4. Tangled up.** The given figure is a link because it contains two distinct closed curves.

**5. Ringing endorsement.** The Borromean rings are a link of three rings with the property that removing any one ring causes the remaining two to become unlinked.

#### II. Solidifying Ideas

**6. Human knots.** The five of you will be tangled in both situations. To get the unknot, you need to switch two crossings.

**7. Human trefoil.** You only need two people who are willing to get very close. Jill stands with arms extended out to the side. Jack stands directly behind Jill, puts his left arm below her left arm, and his right arm above her right arm. Wrapping his right arm under his left arm, Jack grabs Jill's hands and completes the trefoil knot.

**8. Human figure eight.** Again two people are enough. Have Jack and Jill stand facing one another with Jack's right hand holding Jill's left. Have Jill reach her right hand over their other clasped hands. Now Jack can weave his left arm around Jill's right and twist his left hand in such a way that his left hand can just reach Jill's right.

**9. Stick number.** Six sticks; see the figure.

**10. More Möbius.** After cutting, a trefoil knot twice as long as the original band remains. This is because the edge of the original triply twisted Möbius band forms a trefoil knot.

**11. Slinky.** It can be unwound to form a perfect circle, so it is not a knot. To see this, imagine pulling one coil at a time through the Slinky. In the end, you are left with a big untangled loop.

**12. More slink.** It is not knotted; the best way to see this is to take your shoelaces, make the knot, and try to untangle it.

**13. Make it.** It is the unknot. Shoelaces, a cut rubber band, or a really long necklace would also do the job.

**14. Knotted.** No, it isn't a knot. It only seems like a knot when you hold the ends. As long as you don't cut the rope, an unknot will remain an unknot because you can always reverse your steps.

**15. Slip.** The slip knot is useful for attaching the rope to objects, but without anything to cling to, the slip knot untangles itself when pulled.

**16. Dollar link.** The paper clips pass through each other, the dollar straightens out, and the paper clips fall to the floor. Surprisingly, the paper clips are linked!

**17. Knotted loop.** You can always untangle the knot.

**18. Borromean knot.** No; the Borromean rings are three separate loops, whereas the trefoil knot is a single loop. Changing the crossings keeps the loops intact and so can't be useful for turning three loops into one.

**19. Unknotting knots.** Switching any crossing in either figure will work.

**20. Alternating.** All the knots are alternating except for the square knot.

**21. Making it alternating.** Start anywhere on the rope and alternate the crossings as you go; first go under, then over, then under, and so on. Surprisingly, when you meet each crossing again, you won't need to change it. In Section 5.3, we pointed out that any such map can be colored with two colors so that regions sharing an edge had different colors. Color the map and imagine driving along the knot. Notice that before each undercrossing, the left region is a certain color and before each overcrossing the left region has the opposite color.

**22. Alternating unknot.** Hold a rubber band between two pens so that the rubber band resembles a circle. Rotate one of the pens by 180°. For every half-twist, the rubber band creates another crossing, despite the fact that the rubber band is still unknotted. See the figure.

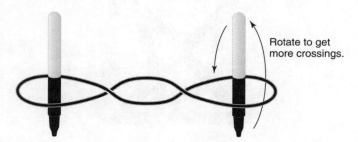

Rotate to get
more crossings.

**23. One cross.** This is easier to see than to prove rigorously. To see it, imagine lifting the top crossing of the loop with your fingers; the rest of the curve will follow along so that you have an unknotted loop dangling from your fingers. More rigorously: The loop is divided into two rings that meet at the crossing. Either one loop contains the other or they are separated. In either case, a simple twisting removes the crossing.

**24. Two loops.** Yes, three crossings are possible (see the figure). However, if the two loops are painted red and blue, the number of red-blue crossings will always be *even*. (At least one of the two loops cannot have a self-crossing; assume that is the blue loop. Pick a point on the red loop that is outside the blue loop. As you travel along the red loop, you will go in and out

446

of the blue loop. After an odd number of red-blue crossings, you will be inside the blue loop; after an even number of red-blue crossings, you'll be outside the loop. To get back to where you started, you must have traveled through an even number of red-blue crossings.)

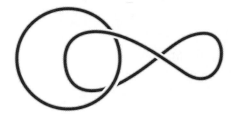

**25. Hold the phone.** Yes; simply thread the phone through all the entanglements, one by one.

### III. Creating New Ideas

**26. More unknotting knots.** For the knot that looks like two trefoil knots spliced together, switching two crossings is enough; switch one of the top three crossings and one of the bottom three crossings. For the jumbled knot, perhaps three are needed (upper-right, upper-left, and lower-left crossings).

**27. Unknotting pictures.** Pick any point and start moving along the knot. Flip crossings when necessary so that it is as if you are constantly traveling under crossings. Stop when you visit a crossing for the second time. This crossing is part of a loop that lies under the rest of the knot. (Imagine shrinking this part of the loop so that all the corresponding crossings vanish. This leaves you with a smaller problem that you can simplify in the same manner.) Now ignore the crossings that are a part of this loop; let them be invisible for the next steps. Continue as before, and repeat the process.

**28. Twisted.** If there is an odd number of crossings, then you have a knotted loop. Otherwise, you are looking at two linked loops.

**29. More alternating.** In both cases, this could be done with string. For the first knot, move the vertical strand that originally goes down the middle to the right, as shown. For the second one, move the segment of the knot between the dots to the new position, as shown.

**30. Crossing numbers.** The unknot is the simplest example. It can be drawn with as many crossings as we want, as we saw in Mindscape II.22.

**31. Lots of crossings.** Grab a piece of the knot and start twisting. For every half-twist, you gain an extra crossing, showing that you can get as many crossings as you want.

**32. Torus knots.** Yes; see the figure. Solid and dashed line segments indicate the knot. If we removed the torus from the picture, we would be left with the standard trefoil knot.

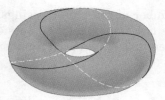

**33. Two crosses.** The main idea is to systematically and exhaustively eliminate all possibilities. The beginning of one such approach: Show that you can deform any two-crossing knot so that the crossings appear as in the figure. Could $a$ connect to $d$? No, because then either $b$ or $c$ would be trapped. Could $a$ connect to $b$? No, because then the $a$-$b$ loop could be untwisted to leave one crossing, and we showed that all one-crossing knots are the unknot. Continue in this way to eliminate the various possibilities.

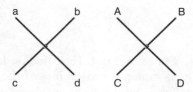

**34. Hoop it up.** Lay the knot on the table so that you can see each crossing clearly. At each crossing, replace the top part of the knot with a semicircular arc that comes up off the table.

**35. The switcheroo.** Shrink the rightmost knot and slide it along the curve until it comes out the other end. Think of the knot as being made of super rubber that can stretch and shrink at will. As you slide the small knot along the curve, the front part gets continually compressed, while the part behind it gets continually expanded.

## IV. Further Challenges

**36. 4D washout.** With four dimensions, there is enough room to switch crossings without cutting the knot. Push a small piece of the knot into the fourth dimension so that the knot appears to be cut, pass the knot through the new opening, and pull the small piece back from the fourth dimension. Because any knot can be untangled if you switch enough crossings (Mindscape III.27), all knots are unknots in 4-dimensional space. The same applies for links.

**37. Brunnian links.** The figure shows five Brunnian links. Note that this picture is easily generalized to an arbitrary number of links.

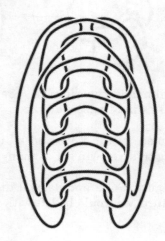

**38. Fire drill.** Start linking the ropes together as we did in Mindscape III.37. When you think you have five floors' worth of rope, put the post through the *last* loop that you used. Throw the rope out the window and climb down.

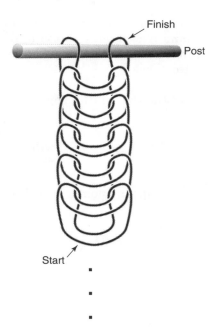

**39. Bing links.** Leaving the pole and bottom two rubber bands fixed, rotate the top two rubber bands around until their two clasping places, which were originally in the right and left sectors, are both moved into the top sector. Do the same kind of rotation with the bottom pair to move their clasp places into the bottom sector.

**40. Fixed spheres again.** The additional lower rope keeps the upper rope knotted. Draw a loop along the rope that also goes along the surface of both spheres. Note that the resulting loop is a trefoil knot. Because trefoil knots can't be undone in three dimensions, there is no hope of unknotting both ropes.

### 5.5 Fixed Points, Hot Loops, and Rainy Days

### I. Developing Ideas

**1. Fixed things first.** Start with two identical disks placed one on top of the other. If the top disk is crumpled in any way and placed back on top of the bottom disk so that no part hangs over the edge, then there must be at least one point on the top disk that lies directly over its original position on the bottom disk.

**2. Say cheese.** The center point of the cheese (the point where the diagonals meet) will be in its original position.

**3. Fixed flap jacks.** The points on the "flipping axis" will be in their original positions after the flip.

**4. Pointing out opposites.** The antipodal points are drawn with bold outlines in the diagram.

**5. Loop around.** A circle of variably heated wire will always have a pair of opposite points at which the temperatures are exactly the same.

### II. Solidifying Ideas

**6. Fixed on a square.** Yes; a filled-in square is equivalent by distortion to a filled-in disk. If you could find a clever way to move a red square sheet so that no red points matched up with their blue mates, we could then derive a contradiction to the Brouwer Fixed Point Theorem. Take the circular blue and red disks, distort them into square sheets, use the clever way to move all the red points, and then distort the blue square back into a circle. The blue points are back in their original positions, yet the red points have all moved.

**7. Fixed on a circle.** No; rotate the red circle by any amount, and all the red points move. With a filled-in disk, rotations fix only the center point; with the center point gone, the Brouwer Fixed Point Theorem doesn't hold.

**8. Winding curves.** For each figure: On the green circle, place a compass that continually points to the corresponding point on the yellow curve. Now move the compass clockwise around the green circle (one time) and notice how many times the compass hand spins around (and in what direction). The winding numbers for the three figures are 1 (clockwise), 2 (clockwise), and 0, respectively. *Warning:* In this case, the winding number agrees with the number of loops the yellow curve makes around the center, but this is not always true.

**9. Winding arrows.** As you travel westerly (or clockwise) around the equator, the hands spin around one time in a counterclockwise fashion.

**10. Under pressure.** Yes, for exactly the same reason given in the text as to why there exist two antipodal points with the same temperature. Let Diff(point $p$) equal the difference between the pressure at point $p$ and the pressure at $p$'s antipodal cousin. If $A$ and $B$ are antipodal points, then Diff($A$) = –Diff($B$). Because Diff($p$) varies continuously from a positive number to a negative number, there exists a point $C$ (between $A$ and $B$) where Diff($C$) is 0. Therefore, $C$ and its antipodal point have the same pressure.

**11. Not the equator.** Yes; the exact same argument applies as for any circular loop, large *or* small.

**12. Home heating.** Draw a circle with a 5-foot diameter somewhere in your room (wall, floor, in the air, etc.). Note that the distance between every antipodal pair of points on the circle is 5 feet. Now use the Hot Loop Theorem to show that there exists a pair of two antipodal points with the same temperature.

**13. Polar populations.** Suppose that at some point, one of the circles cuts 3 people in half. Does this count as 1.5 people or 0 people? If points can be associated with fractions of people, then the number of people changes smoothly from one point to the next, just as the temperature and pressure does. In this case, the answer is clearly yes. Take any great circular loop and apply the Hot Loop Theorem.

**14. Lighten up.** To the extent that lightness varies continuously from point to point, the answer is yes. Draw any great circle and apply the Hot Loop Theorem, replacing the word "temperature" with "lightness." Hairsplitters might argue that light is, in fact, a discrete quantity measured in photons, but we'll not pursue that possibility here.

**15. Shot disk.** No; as in Mindscape II.7, we could rotate the red disk about the center point by 90°.

### III. Creating New Ideas

**16. Lining up.** Yes; measure how far each red point is moved left or right. The right end is either fixed or moves left. The left end is either fixed or moves right. Suppose the red line is the unit interval [0, 1] and $f(p)$ gives the distance that point $p$ moves. If the endpoints aren't fixed, then $f(0) > 0$ and $f(1) < 0$; so, somewhere $f(x) = 0$.

Another reason is that the solution to the Rubber Break-Up Challenge in this section shows that you can't remain on the blue square and distort a red square so that all the points move. If you could distort the red line so that all the points moved, then you could use the same technique to distort the red square so that all the points moved (the red square is just an infinite set of vertical lines). Because the latter is not possible, neither is the former. Brouwer's Fixed Point Theorem holds for the line as well. (There is a variety of different proofs.)

**17. A nice temp.** No; all we know is that *if* the temperature distribution changes, it changes smoothly with no sudden jumps. But we know nothing about the actual temperatures, only that some set of antipodal points has the *same* temperature. We can't determine the exact temperature. For example, the whole Earth could be a constant 32°F.

**18. Off center.** No; move each red point in a straight line toward the missing point. Stop halfway so that the distance between the red point and the missing point is now half of what it was originally. This represents a distortion that moves every red point. For example, if the center point were removed, the disk would be uniformly shrunk by a factor of 2.

**19. Diet drill.** This person may have lost all the weight in the last 10 days (or the first 10 days), so there is no way to determine the exact weight at any specific time during the three-month period. The person's weight changes smoothly over time; so at *some* time, the person's weight was 154.5 pounds, but we can't say exactly when.

**20. Speedy.** You drove 140 miles in 2 hours, for an average speed of 70 miles per hour. At some point in time, you must have been traveling at least 70 miles per hour (otherwise your average speed would be lower). So you will get a ticket for driving at least 5 miles per hour over the speed limit sometime while you were on the tollway.

## IV. Further Challenges

**21. The cut core.** Yes; we could ignore all but one of the removed points and use the same technique that solves Mindscape III.18. But we have more freedom here, so be creative! For example, rotate the outside disk and make the inside diameter thinner. With this extra space, translate the small disk in one direction. Every point has moved, yet we haven't broken any rules.

**22. Fixed without boundary.** No; in Mindscape II.18, we found that removing a single point was sufficient to invalidate the Brouwer Fixed Point Theorem. So, pick one point on the boundary, call it $X$, and ignore the rest. Move each red point in a straight line toward $X$, stopping halfway so that the distance between the red point and $X$ is now half the original distance. It is as if the missing point were exerting a gravitational force on all the red points. The disk compresses itself toward $X$. Because every point moves, there are no fixed points.

**23. Take a hike.** This figure represents a hypothetical path the hiker might have taken. The solid line represents the vertical ascent, and the dashed line describes his trip down. The solid line separates the square into two pieces. Because the dashed line must start at the summit at 8:00 a.m. and end at the base at 5:00 p.m., the dashed line must cross the solid line *somewhere*. We can't say specifically when or where the watches and locations both agreed, but we can say for certain that such an event took place.

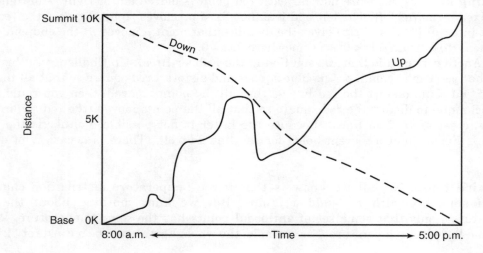

452

### 6.1 Images

#### I. Seeing Things

**1. The incredible shrinking duck.** The picture of the Quacked Wheat box within the Quacked Wheat box is a copy, roughly one-third in size, of the original picture.

**2. Multiplicity.** There is a variety of answers. The largest such sub-figure is half the size of the original picture. Note that the original picture can be subdivided into three of these sub-figures. There are also sub-figures that are 1/4, 1/8, 1/16, . . . as large. (Any power of 1/2 will do.)

**3. Different sizes.** See figures on page 441 of the text for the regions that delineate reduced copies. Note that some of the regions are tilted or turned over as well as reduced.

**4. Blooming broccoli.** It depends on how hungry you are.

**5. Not quite cloned.** The last picture looks strikingly similar to parts of the original Mandelbrot set, although it is easily 10,000 times smaller!

**6. Julia's descendants.** In this case, each half is the whole (except reduced).

**7. Maybe moon.** The craters and shadows appear real.

**8. Exposing forgeries.** Perhaps some of the craters are too similar to one another.

**9. Nature's way.** Branches of trees can resemble miniature trees. The spiral of a conch shell looks the same as you zoom in on its center. Here is an orbiting hierarchy: moons orbiting planets; planets orbiting the sun; stars orbiting the center of the galaxy; galaxies orbiting . . . .

**10. Do it yourself.** Figures like the one in A Tight Weave (see Chapter 1) would work. Many more examples appear in Section 6.3.

## 6.2   The Dynamics of Change

### I. Developing Ideas

**1. Any interest?** At the end of one year, you'll have $(1.03)(\$1000) = \$1030$. After two years, you'll have $(1.03)^2(\$1000) = \$1060.90$. After five years, you'll have $(1.03)^5(\$1000) = \$1159.27$.

**2. Urban expansion.** In 2008, the population of Starburg is expected to be $(1.07)(10,000) = 10,700$.

**3. Pre-sushi.** Population density is the ratio of the actual population to the maximum sustainable population. When there are 3000 fish, the population density is 0.5. When there are 4800 fish, the density is 0.8. When there are 7500 fish, the density is 1.25.

**4. A booby trap.** If $P$ denotes the actual population of boobies, then $P/50,000 = 0.675$. Thus, $P = 33,750$.

**5. Too many.** The average number of birds per square yard is $50,000/17,000 = 2.94$, or about 3 birds per square yard.

### II. Solidifying Ideas

**6. Call your shots.** As the figures show, all destinations are side pockets.

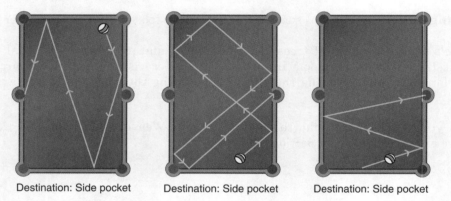

Destination: Side pocket    Destination: Side pocket    Destination: Side pocket

**7. Getting cornered.** See the figure.

Want to sink
the ball here

**8. Double your money.** By the end of the 14th year, you will have \$1980 in the bank, almost double the original balance. A rule of thumb: The number of years times the annual percentage is about 72.

454

**9. Too many.** Let's estimate first: The rule of thumb in Mindscape II.8 says that the population will double roughly every 70 years. So, in 700 years, the population will double 10 times: $2^{10} = 1024$, or about 1000. So, in 700 years, our population will grow by about a factor of 1000, to about $5.7 \times 10^{12}$. In another 700 years, we get another factor of 1000, for a total of $5.7 \times 10^{15}$, which already exceeds the total square yardage. So, in far less than 1400 years, the population growth will decline. The precise solution to $(5.7 \times 10^9) \, 1.01^y = (2.5 \times 10^{15})$ is $y = \ln(2.5 \times 10^{15}/(5.7 \times 10^9))/\ln(1.01) = 1305.6 \ldots$ years.

**10. Rice bowl.** Don't focus on the total number of grains of rice on the checkerboard. Instead, simply estimate the total number of grains on the last square. Because each square has twice the number of grains as the previous one, the last square will have $2^{64}$ grains of rice. That's more than $1.8 \times 10^{19}$ grains of rice. (For comparison, Earth has a volume of only $10^{21}$ cubic millimeters.)

**11. Nature's way.** Hackers work tirelessly to break the hardest codes. Once broken, the codes themselves become harder still, causing hackers to create increasingly powerful code-breaking algorithms.

**12–15. The Game of Life.** See the figures.

**12.**

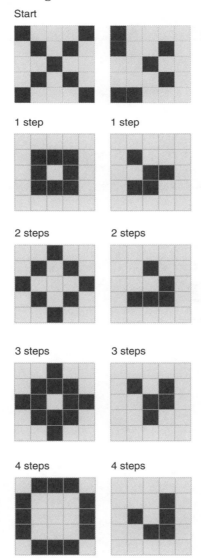

**13.**

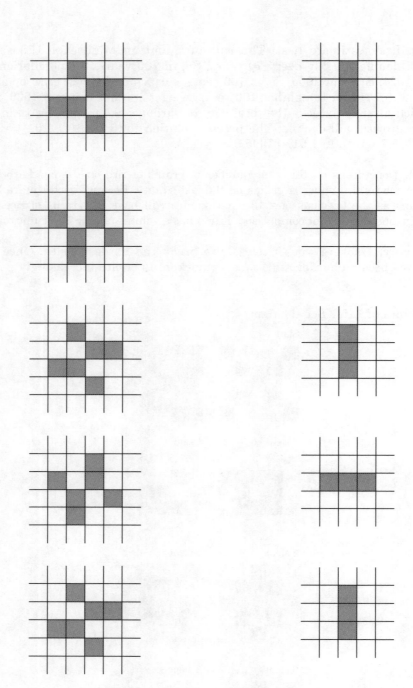

456

**14.**

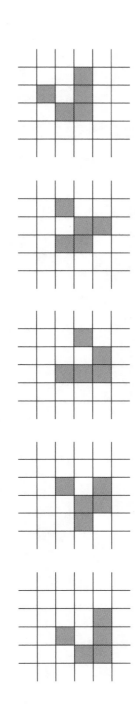

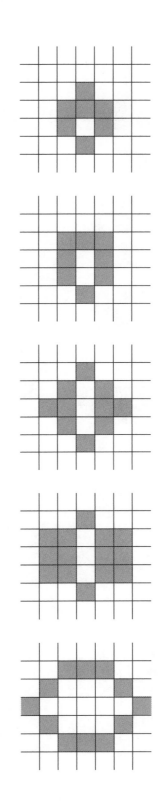

**15.**

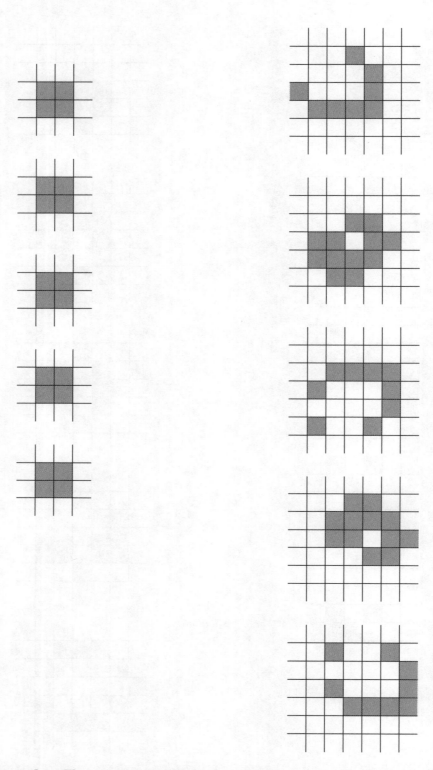

**16–19. Life cycles.** The square in Mindscape II.15 is stable. The symmetric figure in Mindscape II.14 with seven dots grows and then stabilizes. The three asymmetric designs (Mindscapes II.12 and II.14) are migratory and both go down and to the right. The rest eventually become periodic with period 2. None explodes.

458

**20. Life on the Web.** Visit www.heartofmath.com or experiment with the activity, Conway's Game of Life, on the Interactive Explorations CD-ROM.

**21. Explosion.** The figure shows a small population that grows seemingly forever.

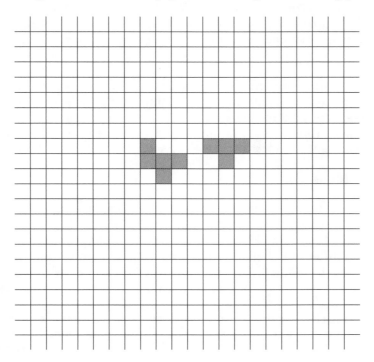

**22. Extinction.** The figure shows a population that will die after a few generations. Notice that the difference between starting populations in this and the previous problem is just one dot.

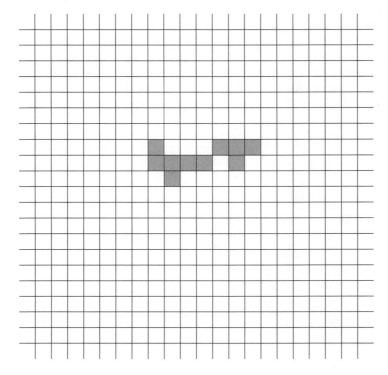

**23. Periodic population.** Combining one figure with period 2 and one with period 3 is an easy way to generate a pattern that repeats after only six generations!

**24. Programmed population.** The two columns agree to within two decimal places until the 3rd repetition and to within one decimal until the 7th repetition. The columns appear to behave independently until the 19th repetition, where by chance the columns line up again. But this correspondence disappears after a few more iterations. The 25th iteration gives $0.6694456060e - 1$ and $0.6245924332$.

**25. Programmed population: the next generation.** In contrast to Mindscape II.24, the 25th iterations yield $0.3832692350$ and $0.3884892781$. The columns are nearly identical, differing by no more than 0.02 at any time. These columns illustrate the notion of a stable iterative process where the outcome is not very sensitive to the initial data, whereas the previous problem represents a process that is very sensitive to its initial conditions.

### III. Creating New Ideas

**26. How many now?** When $n = 1$, the equation becomes $1/2 = 1/4 + c(1/4)(1 - 1/4)$, which when solved for $c$, yields $c = 4/3$. $P_3 = 1/2 + (4/3)(1/2)(1/2) = 5/6$.

**27. Fibonacci.** The difference between the two columns grows exponentially. After 20 iterations, they differ by 13,530, but the ratio between the two columns converges to $2.236 \ldots$. So, the second column is always just a little more than twice the first.

**28. Fibonacci again.** The limiting ratio in both columns is the Golden Mean. In fact, this limiting ratio is independent of the starting seeds. To convince yourself, you might try wildly different starting seeds: Try negative numbers, fractional numbers, irrational numbers, and so on.

**29. Alien antenna.** After four generations, you have $2^4 = 16$ endpoints; after $n$ generations, $2^n$ endpoints. If you reduce $V$ by a factor of $1/2$ each time, what remains is something like the Sierpinski Triangle with all the horizontal segments taken out.

**30. Cobweb plot.** The function is $f(x) = 2.7x(1 - x)$. So the first five iterates of 0.25 are 0.5063, 0.6748, 0.5925, 0.6519, 0.6127. On the graph of $f(x)$, start at 0.25. Go up to the curve. Go over to the diagonal. You should hit the diagonal at the point (0.5063, 0.5063). Go to the curve and back to the diagonal. You should hit the diagonal at (0.6748, 0.6748), and so on.

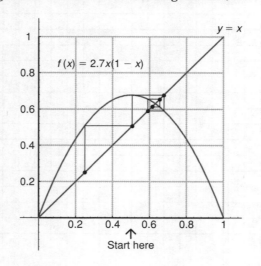

**31. More spiders.** It is periodic. The $x$-coordinate oscillates between 0.79 and 0.52.

**32. Arachnids.** The iterates stabilize at 0.75: $0.25 \to 0.75 \to 0.75 \to 0.75 \to 0.75 \to \ldots$. This equation is identical to the one of Mindscape II.25, where the starting points were 0.245 and 0.246, both of which resulted in a seemingly random sequence of numbers. 0.25 is one of the rare numbers that doesn't produce a random sequence.

**33–35. Making dough.** Let $x$ represent the distance from the left end. For $x < 1/2$, $x \to 2x$; for $x > 1/2$, $x \to 2 - 2x$.

**33. Red.** The only stationary points are $x = 0$ and $x = 2/3$. Either $2x = x$ (solution $x = 0$) or $2 - 2x = x$ (solution $x = 2/3$).

**34. White.** $x = 2/5$ and $x = 4/5$ are the only points with period 2. To find these points, assume $x$ first moves right (so $x \to 2x$) and then that result moves back left ($2x \to 2 - 2(2x)$). Then solve the equation $x = 2 - 2(2x)$. The solution is 2/5. Note that 2/5 goes to 4/5. Because $2 - 2(4/5) = 2/5$, the point returns. If you start at 4/5, you go to 2/5, then return to 4/5.

**35. Blue.** One set of triply periodic points is $x = 2/9$, 4/9, and 8/9. Assume that the general description of the period 3 point is $< 1/2$, $< 1/2$, $> 1/2$. Start with an arbitrary point $x < 1/2$. After one step, $x \to 2x$. Because we are assuming that $2x < 1/2$, we apply $2x$ again for the second step to get $4x$, which we assume is $> 1/2$. Finally, we apply $2 - 2x$ for the third step to give us the equation $x = 2 - 2(4x)$, whose solution is $x = 2/9$. The other set (2/7, 4/7, 6/7) is found by assuming the pattern is $< 1/2$, $> 1/2$, $> 1/2$, which yields $x \to 2x \to (2 - 4x) \to (2 - 2(2 - 4x)) = x$. Solving for $x$ yields $x = 2/7$. See Mindscape II.13.

## IV. Further Challenges

**36. More cobwebs.** Traditional cobweb iteration takes the form "go vertically to the function and horizontally to the diagonal, repeat." Reverse this to find the points that will eventually hit (1/3, 2/3). Notice that there are two possible horizontal moves at each step. For example, both intervals (1/9, 2/9) and (7/9, 8/9) get mapped to (1/3, 2/3), and so they represent all the points that leave after two iterations. Similarly, there are two intervals of length 1/27 that get mapped to (1/9, 2/9) and two intervals of the same length that get mapped to (7/9, 8/9). (So there are four intervals of points that leave after three iterations).

   The sets of intervals that are eliminated at each stage are exactly the sets removed when forming the Cantor Set (see Mindscape IV.38). Once you guess that the Cantor Set is involved, it is easy to show that the set of points that *never* leave the unit square is identical to the Cantor Set.

**37. Yet more cobwebs.** Find a starting point by trial and error with pictures such that the cobweb path eventually intersects the diagonal within every 1/6th of the diagonal; that is, the cobweb path at some time intersects the diagonal somewhere between (0, 0) and (1/6, 1/6); at some time it intersects between (1/6, 1/6) and (2/6, 2/6); and so on. See the figure.

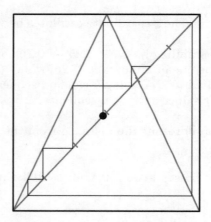

**38. Cantor's cuts.** The endpoints at each stage remain in all future stages; thus, they are in the Cantor Set. After the $n$th stage, the right endpoint of the first interval is $(1/3^n)$ and represents a number in the Cantor Set. The set $\{1, 1/3, 1/9, 1/27, \ldots\}$ is one example of an infinite set within the Cantor Set.

**39. How much is gone?** We removed an interval of length $1/3$ after the first step, two intervals of length $1/3^2$ after the second step, four intervals of length $1/3^3$ after the third step, and $2^{n-1}$ intervals of length $1/3^n$ after the $n$th step. The length of the removed sets is $L = 1/3 + 2/9 + 4/27 + \cdots + 2^{n-1}/3^n + \cdots$. Notice that $(2/3)L = 2/9 + 4/27 + \cdots + 2^{n-1}/3^n + \cdots$, so $L - (2/3)L = 1/3$. (Everything cancels.) Solving this for $L$ gives $L = 1$. A set of length 1 was removed, so the collection of leftover points (the Cantor Set) has no length, or measure 0.

**40. How much remains?** Each number in the Cantor Set has only one expansion in terms of 2's and 0's. (When you rewrite $0.2020022222\ldots$ with leading 0's, you get $0.20201000\ldots$ and invariably introduce a 1 into the representation, which isn't allowed. You encounter a similar problem when replacing a trail of 0's with a trail of 2's.) This allows us to proceed with the typical diagonalization argument: Assume the Cantor Set is the size of the naturals, list them, and go along the diagonal generating a new number not in the set.

In base 2, every real number is written with 0's and 1's. A natural map from the Cantor Set to the reals is the map that replaces all 2's with 1's and changes the base from 3 to 2. So, $20/27 = 0.202000\ldots$ (base 3) goes to $0.101000\ldots$ (base 2) $= 1/2 + 1/8 = 5/8$. This map is not one-to-one, but it does hit *every* real number, some even twice.

## 6.3 The Infinitely Detailed Beauty of Fractals

### I. Developing Ideas

**1. A search for self.** A picture or object exhibits self-similarity if parts of it look identical to larger parts but at a different scale.

**2. Desperately seeking similarity.** Broccoli, coastlines, and ferns often display self-similarity.

**3. Too many triangles?** At stage 2, the Sierpinski Triangle has 9 filled-in triangles. At stage 3, there are 27. At each stage, the number of triangles increases by a factor of 3. At stage 4, there are 81 triangles. These values suggest that at stage $n$ there will be $3^n$ triangles.

**4. Counting Koch.** At stage 2, the Koch curve has 16 line segments. At stage 3, it has 64. At stage $n$, there will be $4^n$ segments.

**5. Argyle art.** The third stage of the collage is shown here.

### II. Solidifying Ideas

**6. Nature's way.** We saw a thin, straight-line ice crystal growing on the surface of a car windshield. Upon closer inspection, there were small perpendicular spikes evenly spaced, that, in turn, had smaller perpendicular spikes as well. This could be related to the self-similar nature of crack growth. A straight crack branches in two, and each branch then branches in two, and so on.

**7. Who's the fairest?** Form an equilateral triangle by placing the three mirrors together. If you drew the path of a light beam inside the triangle, it would hit all three mirrors infinitely many times. So, in theory, if your eye is at that point, you would see each mirror infinitely many times.

**8. Billiards and mirrors.** You would probably see several billiard balls. You wouldn't see infinitely many because the billiard ball itself blocks the one and only infinitely long light path. If the ball is placed correctly, place your eye in line with the center of the nearest mirror and aim for the ball in the *other* mirror.

**9. MTV.** Do it.

**10. Photo op.** Represent the camera as a point light source and find all the paths that eventually come back to the camera. A line perpendicular to the parallel mirror will come back after one reflection. A light placed a little to the right will be a path that hits both mirrors before coming back to the camera. Thus, the cameras will appear smaller as you move to the right.

**11. How many me's?** Light can travel $186 \times 5280$ feet in 1/1000th of a second; it can reflect about $186 \times 1056 = 196{,}416$ times before the shutter closes. There will be one path that takes 196,416 reflections, one path that takes 196,415 reflections, and so on. You'll see roughly 196,416 cameras, each appearing 5 feet farther in the distance than the previous one.

**12. Quacker, Quacker, Quacker.** In the picture, you are holding a Quacked Wheat box honoring a sports star that is 1/10th the size of the picture itself. Make a photocopy 10% of the original size, cut out the tiny picture and tape it over the box in the picture. Now repeat the process. After three iterations, the sports star has been shrunk to 1/1000th his original size and won't be noticeable; the cereal box will be invisible, too.

**13. Sierpinski hexed.** The edge of every triangle at each stage is eventually replaced with a scaled version of the standard Cantor Set. In fact, each point of the Sierpinski Dust can be associated with a pair of numbers in the Cantor Set on the base of the triangle. Specifically, each point of the Sierpinski Dust above the $x$ axis is the apex of an equilateral triangle whose other two vertices are points of the Cantor Set on the base of the original triangle. See the figure.

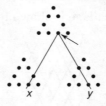

**14. The Kinks.** At the $n$th stage there are $4^{n-1}$ sides, with a bend between each one, for a total of $4^{n-1} - 1$ bends.

**15. Four times.** Your third stage will look like 64 dots (each dot representing a tiny copy of the original picture) arranged in four groups of four groups of four. Note that the edges from the earlier stages will disappear. The pictures will reduce to dots in the limit, and you'll be left with an infinite set of points clustered at the corners of a big square, with each corner being clusters of points near the corners of smaller squares, and so on.

**16. Burger heaven.** The third stage looks like a flock of birds flying in V formation (see the figure). The burgers vanish, leaving instead three copies of the entire picture at different scales.

**17. Ice cream cones.** The infinite stage doesn't depend on the original picture. The final picture will be identical to the final picture in Mindscape II.16—it is a result of the process. When the cone is shrunk to the size of a dot, you won't be able to decipher whether the dot was a burger or an ice cream cone.

**18. Sierpinski boundary.** The limiting picture is the Sierpinski Triangle. As in Mindscapes II.16 and II.17, it doesn't matter what picture you start with. The final result is a function of the process not the input pattern.

**19. Catching Z's.** The length of the second stage is $(9 \times 1/6)24 = 36$ centimeters long. Because each Z in the third stage is again replaced by nine smaller Z's, each 1/6 as long, the length grows by another factor of 9/6. The third stage is $(3/2)^2\, 24 = 54$ centimeters.

**20. Replacement pinwheel.** You get the pinwheel super-tiles described in Section 4.4, "Soothing Symmetry and Spinning Pinwheels."

**21. Koch Stool.** See the figure.

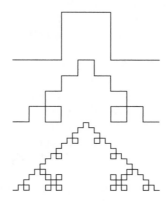

**22. Koch collage stool.** Take any picture, reduce it by a factor of 3, make five copies of it, and overlap the pictures so that their centerlines form the previous figure. Make sure that the second picture from the left is rotated 90° counterclockwise and that the second picture from the right is rotated 90° clockwise.

**23. Sierpinski shooting.** See the figure.

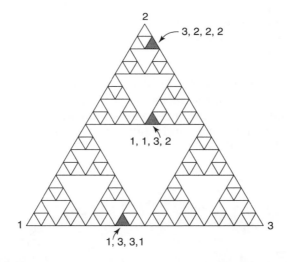

**24. Sierpinski target practice.** Any three consecutive 2's will land you in the triangle labeled $a$. The sequence 2, 1, 3 puts you in triangle $b$, and the sequence 1, 2, 1 puts you in triangle $c$. Work backward; to get to triangle $b$, you must have been in the top corner of the left triangle and then rolled a 3. To get to the top corner of the left triangle, you must have been in the top triangle and then rolled a 1. To get to the top triangle, you need only roll a 2.

**25. Cantor Set.** You are building the Cantor Set by the collage method.

465

## III. Creating New Ideas

**26. Cantor luck.** Moving 2/3 the way toward 0 is equivalent to "divide your current position by 3." You can move all the points in the Cantor Set at once by reducing the set by a factor of 3 and lining it up with the origin. Notice that the Cantor Set neatly lines up with itself, because the left and right thirds are each 1/3 replicas of the entire set. Moving toward 1 is essentially the same.

**27. Cantor Square.** Like the Sierpinski Dust in Mindscape II.13, the edges at each stage will be replaced with Cantor Sets in the limit. The coordinates of each point in the Cantor Square have the form $(x, y)$, where both $x$ and $y$ are members of the Cantor Set.

**28. Cantor Square shrunk.** Let $S_n$ represent the collage after $n$ stages. Because each stage of the collage-making process is a subset of the previous stage ($S_n$ is contained in $S_{n-1}$), we can define this limiting process by saying that the Cantor Square, $C$, is the intersection of all the collage stages: $C = \cap \{S_1, S_2, S_3, \ldots\}$. Similarly, we can apply the collage-making process (CMP) to the Cantor Square by instead applying the process to each of the $S_1, S_2, \ldots$ and then taking their intersection. Notationally this becomes:
$$\text{CMP}[C] = \cap(\text{CMP}[S_1], \text{CMP}[S_2], \text{CMP}[S_3], \ldots) = \cap \{S_1, S_2, S_3, \ldots\} = C.$$

**29. Cantor Squared.** We can move all the points in the square 2/3 toward one of the vertices by shrinking the entire Cantor Square by a factor of 3 and shifting it in the appropriate direction. Because the Cantor Square is made of four such scaled versions of itself, each move will take points in the Cantor Square to other points in the Cantor Square.

**30. Hexed again.** As in Mindscape III.29, you can move each point 2/3 of the way toward a vertex by (a) reducing the entire triangle by a factor of 3 and (b) shifting the triangle to meet up with the appropriate vertex. Because this set is formed by three scaled versions of itself, the reducing and shifting moves points within the Sierpinski Dust to other points within the same set. For every 1 you roll, you'll cover a point in the top triangle; for every pair 2, 3, you'll cover a point in the top corner of the right triangle; and so on. Odds are that you'll hit any given triangle some time.

**31. Pinwheel spun.** In Section 4.4, we fixed a single tile and tiled the entire plane with tiles of the same size. Here we take the opposite approach. We fix a single tile and then tile it with progressively smaller tiles. After three steps, you will get a 125-unit super-tile. The infinite limit is a solid triangle. We would have a more typical (and interesting) fractal pattern if we did not divide the interior triangle.

**32. Antoine's necklace.**

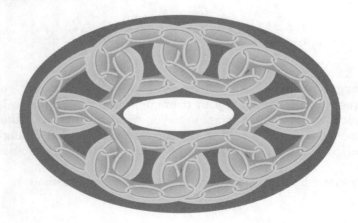

466

**33. Menger jacks.**

**34. A tighter weave.** The third stage is a square with the central $1/5 \times 1/5$ square missing and twenty-four $1/25 \times 1/25$ squares missing.

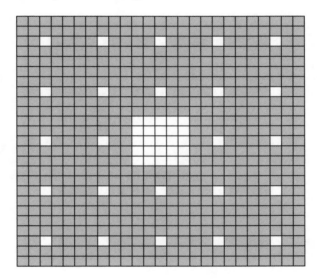

**35. A looser weave.** In both Mindscapes III.34 and III.35, the remaining purple has zero total area. See Mindscape IV.39.

**IV. Further Challenges**

**36. From where?** Reduce a solid square vertically by a factor of 3. Make two copies of the thin rectangle, and position them at the top and bottom portions of the original square so that it is as if the middle third of the square were removed.

**37. Treed.** Modify the collage instructions for the fern. Modify the proportions, eliminate the skew factors for simplicity, and play with the positioning of the four reduced copies of the original.

**38. Flaky.** Start with any picture. Reduce it by 50%, make six copies, and arrange the centerlines of the six pictures to form a hexagon. Experiment with rotating and/or slightly shifting the positions of the six reduced copies.

**39. How big a hole?** In Mindscape III.34, we first remove 1/25 of the square. At the next stage, we remove 24 squares, each 1/25 smaller than the corresponding square in the previous stage. Adding up all the squares that we remove: $1/25 + 24(1/25)^2 + 24^2(1/25)^3 + \cdots = 1/25(1 + (24/25) + (24/25)^2 + \cdots) = 1/25(25) = 1$. So everything is removed. The total area of the purple is 0. In Mindscape III.35, the analogous computation becomes: $4(1/16) + 4(12)(1/16)^2 + 4(12^2)(1/16)^3 + \cdots = 1/4(1 + 12/16 + (12/16)^2 + \cdots) = 1/4(1 + 3/4 + (3/4)^2 + \cdots) = 1/4(4) = 1$.

**40. 4D fractal.** Dividing each dimension into three intervals gives $3 \times 3 \times 3 \times 3 = 81$ sub-hypercubes. Remove the center hypercube and all the hypercubes that share a 3D face with it (there are two such hypercubes in each direction). In all, you are removing 9 of the 81 sub-hypercubes.

**6.4  The Mysterious Art of Imaginary Fractals**

**I. Developing Ideas**

**1. Not Raul.** Gaston Julia was a French mathematician who, in the 1920s, did some of the earliest work on sets that exhibit chaotic behavior. These sets are called Julia Sets, in his honor; computer-rendered drawings of some Julia Sets are among the most recognized images of fractals.

**2. Use your imagination.** The number $i$ is an imaginary number that equals $\sqrt{-1}$. $i^2 = -1$, $(2i)^2 = -4$, $(2i)(3i) = -6$.

**3. Complex plots.** The new points are shown in gray.

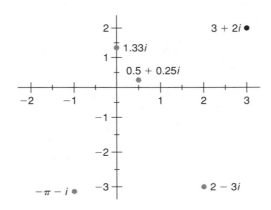

**4. Innie or outie?** If the point $a + bi$ lies inside the Mandelbrot Set, the corresponding Julia Set will have one piece.

**5. Outie or innie?** If the $(a + bi)$-Julia Set has more than one piece, the point $a + bi$ lies outside the Mandelbrot Set.

**II. Solidifying Ideas**

**6. Arithmetic.** $(2 + 3i) + (3 - 7i) = (5 - 4i)$; $(-2 + 5i) + (4 + 6i) = (2 + 11i)$; $(1 - 2i)(2 + 3i) = (8 - i)$; $(-i)(4 - 2i) = (-2 - 4i)$; $(3 - 2i)^2 = (5 - 12i)$.

**7. More arithmetic.** $(1/2 + 3i) + (5/2 - i) = 3 + 2i$; $(-1 + i) + (3 + 2i) = 2 + 3i$; $(1 - \sqrt{2}i)(3 + \sqrt{2}i)$ $= 5 - 2\sqrt{2}i$; $(3 - 2i)(3 + 2i) = 13$; $(2 - \sqrt{5}i)^2 = -1 - 4\sqrt{5}i$.

**8. Quick draw.** $(1 + 2i) + (3 + i) = (4 + 3i)$; $(1 + 2i)(3 + i) = 1 + 7i$; $(-2 + 3i) + (-1 - 2i) = (-3 + i)$; $(-2 + 3i)(-1 - 2i) = 8 + i$. Adding the angles is easy to do visually; multiplying is more difficult. Length of $(1 + 2i)$ is $\sqrt{5}$. Length of $(3 + i)$ is $\sqrt{10}$. Length of $(-2 + 3i)$ is $\sqrt{13}$. Length of $(-1 - 2i)$ is $\sqrt{5}$.

**9. Quick draw II.** $(2 + 2i) + (3 - i) = (5 + i)$; $(2 + 2i)(3 - i) = (8 + 4i)$; $(2 + i) + (-2 - i) = (0 + 0i) = 0$; $(2 + i)(-2 - i) = (-3 - 4i)$. Lengths: $|(2 + 2i)| = 2\sqrt{2}, |(3 - i)| = \sqrt{10}, |(2 + i)| = \sqrt{5}, |(-2 - i)| = \sqrt{5}$.

**10. Be square.** $1/2^2 = 1/4$; $(2 + 2i)^2 = 8i$; $(-1 + i)^2 = -2i$ $(2 + i)^2 = 3 + 4i$.

469

**11. Squarer.** $3 + 4i$: length 5. $(3 + 4i)^2 = -7 + 24i$: length 25.
$3/7 - (4/7)i$: length 5/7. $(3/7 - (4/7)i)^2 = -1/7 - (24/49)i$: length 25/49.

**12–15. Ill iterate I.**
**12.** $0 \ldots i \ldots -1 + i \ldots -i$ (periodic with period 2)
**13.** $i \ldots -1 + i \ldots -i \ldots -1 + i$
**14.** $2 + 3i \ldots -5 + 13i \ldots -144 - 129i \ldots 4095 + 37153i \ldots$ (color green)
**15.** $1 - i \ldots -i \ldots -1 + i \ldots -i$

**16–19. Ill iterate II.** Surprisingly, these *all* tend to infinity eventually, even though they don't appear to after the first few iterations.
**16.** $0 \ldots 0.27 \ldots 0.3429 \ldots 0.38758041 \ldots$
**17.** $0.5 + 0.5i \ldots 0.27 + 0.5i \ldots 0.0929 + 0.27i \ldots 0.2057 + 0.0502i \ldots$
**18.** $2 + 3i \ldots -4.73 + 12i \ldots -121.4 - 113.5i \ldots 1841.0 + 27552.9i \ldots$ (color green)
**19.** $i \ldots -0.73 \ldots 0.8029 \ldots 0.9146 \ldots$

**20–23. Ill iterate III.** Once again, each sequence eventually blows up ($0.5i$ takes 30 iterations before it becomes clear).
**20.** $0 \ldots 0.11 - 0.67i \ldots -0.3268 - 0.8174i \ldots -0.4513 - 0.1357i \ldots$
**21.** $0.5 + 0.5i \ldots 0.11 - 0.17i \ldots 0.0932 - 0.7074i \ldots -0.382 - 0.8019i \ldots$
**22.** $1 \ldots 1.11 - 0.67i \ldots 0.8932 - 2.1574i \ldots -3.747 - 4.524i \ldots$ (color green)
**23.** $0.5i \ldots -0.14 - 0.67i \ldots -0.3193 - 0.4824i \ldots -0.0207 - 0.3619i \ldots$

**24. Orange Julias.** Color the regions between the contour lines.

**25. Julia Webbed.** Visit www.heartofmath.com or use the Interactive Explorations CD-ROM.

### III. Creating New Ideas

**26–28. Great escape?**
**26.** $-1 + i$ grows without bound immediately.
**27.** 0 doesn't grow without bound until after 70 iterations.
**28.** $0.25 + 0.25i$ appears to bounce around inside the unit circle forever. If it does escape, it does so after 200 iterations.

**29–30. Mandelbrot or not?** Numbers corresponding to connected Julia Sets will lie *inside* the Mandelbrot Set; otherwise, the point lies outside the Mandelbrot Set.

**31–34. Zero in.** If the Julia Set is connected, then 0 will lie inside the Julia Set.

**35. Mandelbrot origins.** If $c$ is a point in the Mandelbrot Set, then the corresponding Julia Set will be connected. The previous Mindscapes mention that this is equivalent to saying that 0 is in the Julia Set, which means if we start with 0, the iteration, $z \to z^2 + c$ will remain bounded. Therefore, the Mandelbrot Set is the set of all points $c$ such that the iteration $z^2 + c$ doesn't send 0 off to infinity.

### IV. Further Challenges

**36. The great escape.** Consider the geometric method of squaring and adding complex numbers. If $z_1$ has length greater than 2, then $z_1^2$ has length greater than 4 and $z_1^2 + a + bi$ has length greater than 3, so that $z_2 = z_1^2 + a + bi$ lies outside a circle of radius 3. Similarly, $z_3 = z_2^2 + a + bi$ has length greater than $3^2 - 1 \geq 8$; $z_4 > 8^2 - 1 \geq 63$; and so on.

**37. Bounded Julia.** Suppose $c$ corresponded to an infinitely large Julia Set. There would necessarily exist arbitrarily large points $z_1$ that remained bounded under iterations of $z \to z^2 + c$. Now generalize Mindscape IV.36 to find the contradiction. If $|z_1| > \sqrt{|c|} + 2$, then

$$|z_n| > |z_{n-1}|^2 - |c| = \left(|z_{n-1}| - \sqrt{|c|}\right)\left(|z_{n-1}| + \sqrt{|c|}\right) \geq 2\left(|z_{n-1}| + \sqrt{|c|}\right) \geq 2|z_{n-1}|.$$ This shows that the Julia

Set for the point $c$ is contained within a circle of radius $\left(\sqrt{|c|} + 2\right)$.

**38. Prisoner.** There are only two fixed points, $i$ and $1 - i$. A general approach is to expand the equations and equate imaginary and real parts; this leads to a system of quadratic equations that we don't recommend trying to solve. A better approach is to use the quadratic formula to solve $z^2 + c = z$, where $c = 1 + i$. The quadratic formula gives $z = 1/2 \pm 1/2\sqrt{-3 - 4i}$. Now the goal is to find the number whose square is $-3 - 4i$. This is generally an easier problem: $(\pm(2 + i))^2 = -3 - 4i$.

**39. Always a prisoner.** If $z$ and $c$ represent $a + bi$ and $c + di$, respectively, then our goal is to solve $z^2 + c = z$. The quadratic formula works for complex numbers as well and gives the solution $z = 1/2 \pm 1/2\sqrt{1 - 4c}$, where $\sqrt{1 - 4c}$ represents the square root of a complex number. Though this is difficult to find numerically, we can use the geometric notion of arithmetic to show it exists. To find the square root, take the square root of the length, and divide the angle by 2.

**40. Mandelbrot connections.** No one knows how to solve this question. A new idea is needed.

### 6.5 Predetermined Chaos

#### I. Developing Ideas

**1. Does this thing come with a warranty?** Using two different calculators to do the same calculations does *not* guarantee exactly the same answer. If the calculators hold numbers with different decimal-place accuracy, for example, answers may differ.

**2. Root repeater.** Starting with 0.999 and repeatedly pressing the square root key will give values that get closer and closer to 1. Because the calculator can't hold an ever growing number of decimal places showing 0.99999 . . . , it will eventually display the result as 1. Starting with 0.9999 requires fewer repetitions of the square root key to reach 1.

**3. Transforming experience.** If $x = 0.5$, then $y = 4(0.5)(1 - 0.5) = 1$. If $x = 0.437$, then $y = 4(0.437)(0.563) = 0.984124$.

**4. Up and over.** Starting at (0.4, 0.4) on the diagonal and traveling vertically upward, you hit the point (0.4, 0.8). Starting at (0.4, 0.8) on the inverted **V** and traveling horizontally toward the diagonal, you hit the point (0.8, 0.8).

**5. Over and up.** The points are marked in the figure.

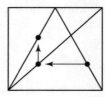

#### II. Solidifying Ideas

**6. Go.** If $x = 0.4$, then the first four iterates represent a square on the cobweb plot. See the figure.

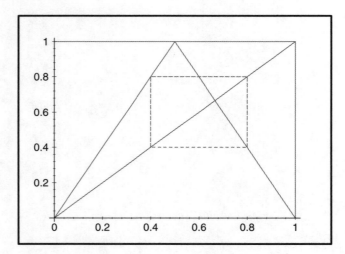

472

**7. Staircase.** The cobweb resembles a staircase with steps that get larger and larger. See the figure.

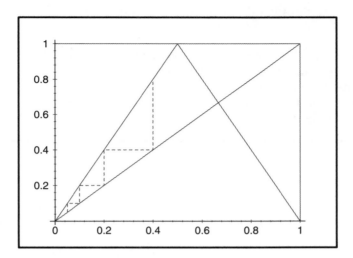

**8. How far?** After four iterations, the point remains stationary. A single step comprises a vertical line *and* a horizontal line.

**9. Points that quit.** Work backward from one of the two fixed points (0 and 2/3). What goes to 0? 0 and 1. What goes to 1? 1/2. What goes to 1/2? 1/4 and 3/4. Therefore, both 1/4 and 3/4 are two starting points that will get fixed at 0 after three iterations. For the fixed point 2/3, we have: 1/3 goes to 2/3; both 1/6 and 5/6 go to 1/3; both 1/12 and 11/12 go to 1/6; and both 5/12 and 7/12 go to 5/6. So, 1/12, 11/12, 5/12, and 7/12 all get fixed at 2/3 after three iterations.

**10. Ups and downs.** See the figure.

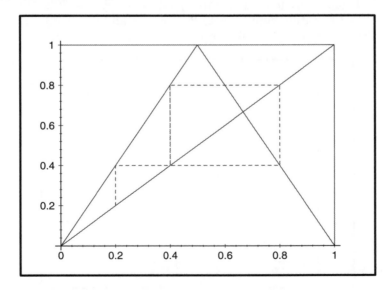

**11. Don't move.** The diagonal line intersects the inverted **V** at two points: 0 and 2/3. The equation of the right line is $y = 2 - 2x$, and the equation of the diagonal is $y = x$. Solving this system gives $x = 2/3$.

**12. Two step.** $x = 0.4$. This is the cobweb plot from Mindscape II.6. In addition to guessing and trial and error, you can find the solution via algebra. Let's assume that $x$ goes up to the left curve ($y = 2x$), over, and then down to the right curve ($y = 2 - 2x$). So $x \rightarrow 2x \rightarrow 2 - 2(2x)$. Solving the equation $2 - 2(2x) = x$ gives the answer.

**13. Three step.** $x = 4/9$. (This problem gives one of the six solutions to Mindscapes III.33–35 in Section 6.2.) The left and right lines are given by $y = 2x$ and $y = 2 - 2x$, respectively. Our point goes left-right-left and then repeats itself. So $x$ goes to $2x$, then $2x$ goes to $2 - 2(2x)$, and finally $2 - 2(2x)$ goes to $2(2 - 2(2x))$. We want this last expression to be $x$, so we solve $2(2 - 2(2x)) = x$. See the figure.

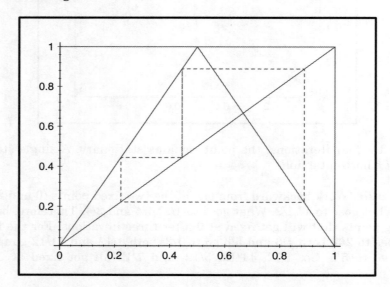

**14. Four step.** Using the reasoning and notation of the previous two Mindscapes, let's try to find a point that goes, left-left-left-right-repeat. $x \rightarrow 2x \rightarrow 2(2x) \rightarrow 2(2(2x)) \rightarrow 2 - 2(2(2(2x))) = x$. This last equation simplifies to $2 - 16x = x$ and has solution $x = 2/17$. The path (2/17, 4/17, 8/17, 16/17) is shown in the figure.

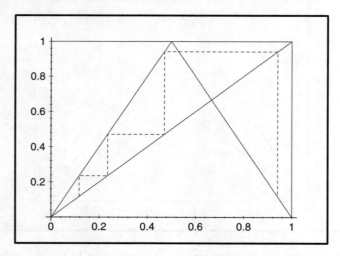

**15. Grow up.** Because 0 goes to 0 and 0.1 goes to 0.2, all the points in the interval [0, 0.1] will get mapped to the interval [0, 0.2].

**16. Grow up again.** As in Mindscape II.15, it is enough to iterate the endpoints of the interval. $0 \to 0 \to 0$ and $0.1 \to 0.2 \to 0.4$, so that $[0, 0.1]$ is mapped to $[0, 0.4]$. See the figure.

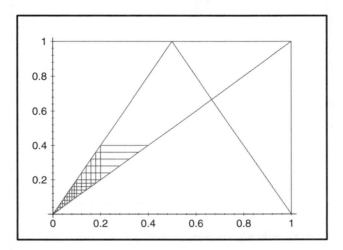

**17. A different patch.** Seven steps. The interval $[0, 0.5]$ maps to the whole diagonal after one step (because 0.5 goes to 1), so the next question is, how many steps until the interval covers $[0, 0.5]$? $0.01 \to 0.02 \to 0.04 \to 0.08 \to 0.16 \to 0.32 \to 0.64$, only six steps.

**18. Target practice.** As we saw in Mindscape II.17, 0.01 goes to 0.32 after five steps, which means that after five steps, the interval $[0, 0.01]$ maps to the interval $[0, 0.32]$. If the interval $[0, 0.01]$ were made of rubber, then this map shows how to distort and stretch the interval to align with $[0, 0.32]$. Therefore, something must hit all the points in between.

**19. Too high.** Both points take only two steps before leaving the unit square.

**20. More gone.** Only three steps. See the figure.

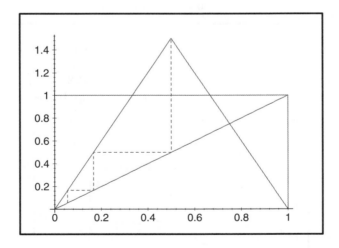

**21. Too short.** The first 50 iterations of 1/3 and 2/3 are shown, giving a visual indication that the points are trapped within the interval $[1/3, 2/3]$. Because neither 1/3 nor 2/3 ever hits the interval $[0, 0.25]$, no other points in the interval will either.

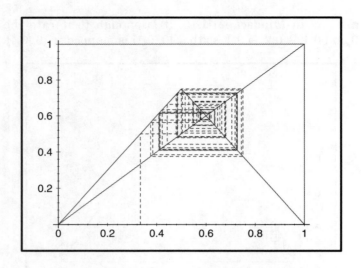

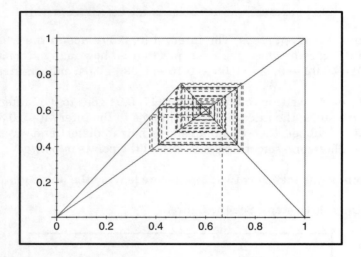

**22. Where to?** The figure shows a cobweb plot of these two iterative processes. Note that the lines corresponding to 0.437 and 0.438 get even closer after each iteration. The 30th iterate of 0.437 is 0.3828197816 and the 30th iterate of 0.438 is 0.3828198948. They differ only in the eighth decimal place because the two paths converge to the same limiting cycle.

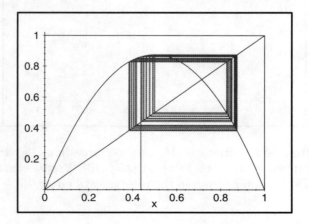

**23. Calculator slips.** The first 30 iterations are shown in graphical form in the figure. The 30th iterates of 0.437 and 0.438 are 0.9247 . . . and 0.0007 . . . , respectively. A graph of the first 5 iterations better explains the chaotic nature of this transformation (see the second figure). Note that the two lines get farther apart after each iteration.

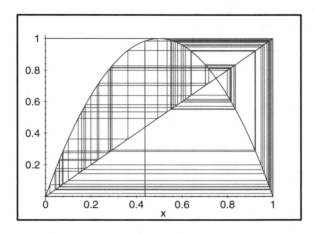

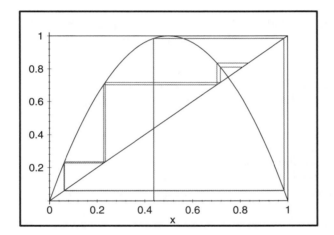

**24. Just missed.** The movie *Sliding Doors* chronicles two parallel lives of Gwyneth Paltrow's character. In one life, she caught the train home; in the other, she missed it and had to wait a few minutes for the next one. This small difference created radical differences in her life and argues that our lives are sensitive to tiny differences at certain points.

**25. Take stock.** It is likely that the three graphs, if scaled, would appear as if they were generated by the same random process (instead of three different ones). The amount that each needs to be scaled will be related to the volatility of the stock, which is used to measure and quantify how quickly a stock's prices fluctuate.

### III. Creating New Ideas

**26. Repulsive.** Under iterations, points greater than 1 tend to infinity while points less than 1 tend to 0.

**27. Attractive.** Continually dividing a number by 2 will bring you closer and closer to 0. For numbers near 0 (say $|x| < 1/2$), $x^2 < 1/2|x|$, so squaring brings you even closer to 0 than does dividing by 2.

**28. Sierpinski attractor.** It doesn't matter what image you start with. In the end, all images are shrunk to the size of a point, and the resulting image is the Sierpinski Triangle, which is actually *defined* to be the attractor of this iterative process.

**29. Two step.** $1/2 \to (1+\sqrt{5})/4 \to 1/2 \to \ldots$ (see the figure.)
Numerically, $0.5 \to 0.8090169945 \to 0.5 \to \ldots$.

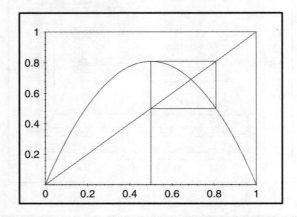

**30. Periodic attraction.** The iterates are 0.48, 0.8077225673, 0.5025834321, 0.8089953967, 0.5000431939, 0.8090169884, ..., which get closer and closer to the repeated values starting with 1/2.

**31. Periodic attraction.** The iterates get closer to the two-periodic cycle in Mindscape III.29. They are $0.8 \to 0.5177708765 \to 0.8079950310 \to 0.5020405471 \to 0.8090035202 \to 0.5000269478 \to 0.8090169922 \to 0.5000000043 \to 0.8090169945 \to 0.4999999998 \to 0.8090169945 \to \ldots$.

**32. Four-peat.** After 12 iterations, the iterates are 0.500884225, 0.874997263, 0.382819683, 0.826940706, 0.500884209, .... Notice that the first and last numbers above differ only in the eighth decimal place. So, we are looking at a limiting cycle of period 4.

**33. Nearly fourly.** The 12 values are tending to the 4 cycle given in Mindscape III.32.

**34. Tent attraction?** No; see the two paths in the figure.

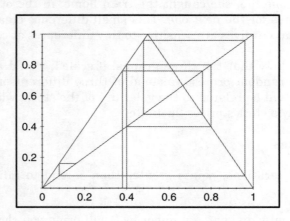

478

**35. Becoming periodic.** (See Mindscapes II.9 and III.34.) 0.4 → 0.8 → 0.4 → .... Let's find a point that gets sent to 0.4 after two iterations. What gets mapped to 0.4? 0.8 and 0.2. What gets mapped to 0.2? 0.9 and 0.1. So, 0.1 → 0.2 → 0.4 → 0.8 → 0.4 → .... Both 0.1 and 0.9 become two-periodic after two steps.

### IV. Further Challenges

**36. The Earth moved.** The circle of points describing the equator remains a circle of points after each step, but at a latitude closer to the north pole. At each step, the ring of points spins and shrinks, getting asymptotically closer to the pole (but never reaching it after a finite number of steps).

**37. Poles apart.** The north pole is an attractor because *all* points near it converge to it, and the south pole is a repeller because points close by move far away, even though the south pole stays fixed.

### 38–40. Logistic cobwebs

**38. $r = 2$.** See the figure. 0.0 and 0.5 are the only fixed points. 0.0 is a repeller, and 0.5 is an attractor, as can be seen in the figure. Each iteration brings the point closer and closer to 0.5.

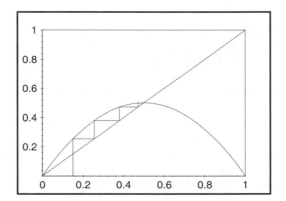

**39. $r = 4$.** We can get a precise answer by numerically solving the eighth-degree equation, $F(F(F(x))) = x$, where $F(x) = 4x(1 - x)$. There are eight real roots corresponding to the two fixed points and two three-periodic cycles. Those two cycles are illustrated in the figure with starting points 0.116977778 and 0.188255099.

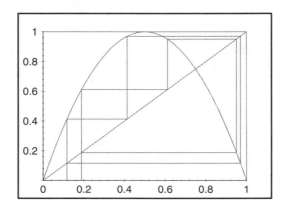

**40. $r = 4$.** Neither three-periodic cycle of Mindscape IV.39 is an attracting cycle. Even a starting point that differs by 0.0001 will lead to a path that quickly diverges.

### 6.6 Between Dimensions

**I. Developing Ideas**

**1. Parallel grams.** You need four parallelograms to build a new one that is twice as long (or wide) as the original. You need nine copies for a new parallelogram that is three times as large.

**2. Moving on up.** The middle version has a scaling factor of two; the right version has a scaling factor of three.

**3. Bigger rug.** You would need eight copies of the carpet to build a new one three times as wide.

**4. Dimension connection.** The dimension $d$ of an object is related to the number of copies $N$ needed to construct a larger version scaled by a factor $S$, by the equation $S^d = N$.

**5. Divining dimension.** The dimension of the object is the value of $d$ for which $2^d = 3$. So, $d = \ln 3/\ln 2 = 1.58496\ldots$.

**II. Solidifying Ideas**

**6. Stay inbounds.** (1) Because the Mandelbrot Set contains a solid 2D square, its dimension is greater than 2; because it lies completely within a 2D plane, its dimension is less than 2. So, this set has fractal dimension 2. (2) The Menger Sponge contains a 2D square but sits inside 3D space; thus, it has fractal dimension between 2 and 3. (3) The Cantor Set contains a 0D point and lies within a 1D space (a line) and so has fractal dimension between 0 and 1.

**7. Regular things.** For the left figure, four copies of the rectangle can make a rectangle twice as large. $S = 2$ (scale), and $N = 4$ (number of copies). $2^d = 4$ implies $d = 2$, which agrees with our expectation. For the right figure, eight copies of the solid brick make a new brick twice as large. $S^d = N$ becomes $2^d = 8$, implying that $d = 3$.

**8. More regular things.** As in Mindscape II.7, four copies of the parallelogram make up a larger one twice as large. $2^d = 4$, so $d = 2$.

**9. Any right triangle.** Four triangles ($N = 4$) put together form a triangle twice as large ($S = 2$). $S^d = N$ becomes $2^d = 4$, implying $d = 2$. So, right triangles have dimension 2.

**10. Sierpinski carpet.** The eight squares surrounding the center are a scaled version of the larger square. Therefore, eight copies make a square three times as large. $S^d = N$ becomes $3^d = 8$, implying a fractal dimension of $d = \ln 8/\ln 3 = 1.89\ldots$ (the base of the logarithm doesn't matter as long as you are consistent.)

**11. Koch Stool.** The collage process for the Koch Stool replaces each straight line with five lines, one-third as large. Applying the collage-making process to the Koch Stool doesn't change anything, which proves that the five copies of the stool can be put together to form a stool three times as large. $3^d = 5$, so that the fractal dimension is $\ln 5/\ln 3 = 1.46\ldots$.

**12. Cantor Set.** As in Mindscape II.11, the relevant numbers can be found in the collage-making instructions. Two copies can be arranged to make a Cantor Set three times as large. $3^d = 2$ has solution $d = \ln 2/\ln 3 = 0.63\ldots$.

**13. Cantor reduced.** Here, two copies ($N = 2$) can make a new set four times as large ($S = 4$). $4^d = 2$, implying $d = 0.5$. This set has fractal dimension 1/2, halfway between that of a point and a line.

**14. Long Koch.** In the second stage, each segment is again replaced by five segments with length four-thirds larger than the original length. Each stage is four-thirds longer than the previous stage. The $n$th stage has length $(4/3)^{n-1}$. Because this tends to infinity as $n$ gets large, the final Koch Curve has infinite length.

**15. Plus.** $N = 3$ is the number of copies needed. $S = 2$ is the increase in size. The fractal dimension is therefore $d = \ln 3/\ln 2 = 1.58\ldots$.

### III. Creating New Ideas

**16. Tinier triangles.** Suppose you make something similar to the Sierpinski Triangle, but this time you make three copies each reduced so that the sides are only one-third of the original side lengths. You then position the reduced triangles at the three corners.

The collage-making instructions imply that three copies of this object (Sierpinski Dust) can be formed to make a new object three times larger. $3^d = 3$ implies that $d = 1$. We have a totally disconnected set of points with fractal dimension 1.

**17. Menger Sponge.** Twenty copies can make a Menger Sponge three times as large, implying that the fractal dimension satisfies $3^d = 20$. $d = \ln 20/\ln 3 = 2.72\ldots$.

**18. Thinning.** The infinite process creates the diagonal. It is easy to verify that the diagonal is a fixed point of the collage-making process, which means that it *is* the limiting fractal.

**19. Not much.** Two copies of the diagonal can be arranged to make one twice as large. $2^d = 2$, implying that the fractal dimension is 1, which agrees with our intuitive notion that lines have dimension 1.

**20. Koched.** There are many answers, but they all involve replacing each segment with five segments that are one-fourth as large. The figure shows one such fractal.

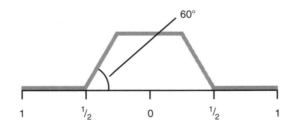

## IV. Further Challenges

**21. Find a fractal.** The Sierpinski Carpet has this dimension. Another possibility: Generalize Mindscape III.16 to three dimensions. Take a cube, reduce it by a factor of 3, make eight copies, and stick them in the eight corners. This process describes a fractal of dimension $\ln 8/\ln 3$, as well.

**22. Find a 1.5 fractal.** First look for integers satisfying $S^{3/2} = N$. Because of the 2 in the denominator, $S$ needs to be a perfect square. The first few solutions are $4^{3/2} = 8$, $9^{3/2} = 27$, $16^{3/2} = 64$, and so on. (a) Reduce a cube by a factor of 4, make eight copies, and stick them in the corners. (b) Cut a square into 16 pieces and remove the center four and the four corners. Repeat the process with the remaining eight squares.

## 7.1 Chance Surprises

### I. Developing Ideas

**1. Taking a spin.** It's very likely that more than half will land tails up.

**2. Birthday surprise.** You need only 23 people to ensure that the chances of two (or more) sharing a birthday is greater than 50%.

**3. Opposite of heads.** Because 47 out of 100 tosses were tails, the answer is 47%.

**4. Penny percent.** Tails came up 22 times, for a fraction of 0.44. Thus, 44% of the tosses came up tails.

**5. Party time.** 3/12, or 1/4, of the children have birthdays in December, for a percentage of 25%. No one has a birthday in February, for a percentage of 0%. Five of the children have a birthday in the same month as another child (December or October). Because 5/12 is approximately 0.42, the percentage is about 42%.

### II. Solidifying Ideas

**6. Edgy Lincoln.** In general, more than 50% will land heads up. Note that balancing a single penny on its edge 100 times will give results different from balancing 100 *different* pennies on their edges. Some pennies have less of a bevel than others.

**7. Spinning Lincoln.** Less than half will land heads up. Again, spinning the same penny 100 times will yield results different from spinning 100 *different* pennies. Use as many different pennies as possible.

**8. Flipping Lincoln.** This will fall closer to 50% because the process is in no way related to the edges of the penny. However, if you are careful, you can possibly stack the odds by trying to flip the penny in exactly the same way every time. Flipping it *high* in the air should minimize this bias.

**9. A card deal stick.** You'll choose the king correctly one-third of the time. Note, these odds would be the same regardless of whether the dealer showed you one of the aces.

**10. A card deal switch.** Surprisingly, you will choose the king correctly two-thirds of the time. Your new choice is based partly on new information obtained from the dealer. This is precisely the Monty Hall problem, which is discussed at length later in Chapter 7.

**11. A card reunion—black first.** After seeing one black card, there are 25 other black cards in the remaining deck of 51. So, the true odds are 25/51 = 0.490 . . . . You should find that roughly half of the black-first pairs had a second black card.

**12. A card reunion.** You should find roughly one-third of the at-least-one-black pairs were both black. There are four equally likely ways to pick two cards: BB, BR, RB, RR. Because we are only concerned with pairs containing a black card, we only have three equally likely outcomes: BB, BR, and RB. Only one pair of the three *equally likely* cases has two black cards, so the odds are 1/3. (The exact probability is 25/77.)

**13. Birthday bash.** The next time you are in a room with 40 people or so, ask each person for his or her birthdays to see whether there is a common birthday.

**14. Presidential birthdays.** James Polk (11th president, 1795) and Warren Harding (29th president, 1865) were both born on January 20. From Washington to Clinton, there is only one match.

**15. Vice-presidential birthdays.** Again, there is only one match. Hannibal Hamlin (Lincoln) and Charles Dawes (Coolidge) were both born on August 27.

## 7.2 Predicting the Future in an Uncertain World

### I. Developing Ideas

**1. Black or white?** The probability that he wears the black pair on a Wednesday is 1 minus the probability that he wears the white pair on a Wednesday, or 2/5.

**2. Eleven cents.** The outcomes are HH, HT, TH, and TT, where the first letter denotes the dime and the second denotes the penny. The event $E$ is the set of outcomes HH, HT, and TH. The probability that $E$ occurs is 3/4.

**3. Yumm.** The probability of getting green is (number of green candies)/(total number of candies) = 2/14, or 1/7. The probability of getting blue is 5/14.

**4. Rubber duckies.** There are eleven numbers (60, 61, ... , 70) out of 75 that result in a stuffed duck; so, the probability of winning a stuffed duck is 11/75 = 0.14666 . . . . There are 59 numbers (1, 2, ... , 59) that yield a consolation prize; so, the probability of neither stuffed duck nor banana is 59/75 = 0.78666 . . . .

**5. Legally large.** If an experiment is repeated a large number of times, then the relative frequency of a particular outcome will tend to be close to the probability of that particular outcome.

### II. Solidifying Ideas

**6. Lincoln takes a hit.** See the figure. The bevel causes the center of gravity to be closer to the heads side than the tails side. In the banging experiment, the penny stands with the beveled edge parallel to the floor (see figure). If the penny lands heads, then the center of gravity takes the left path; for tails, the center of gravity has to take the right path. Because the center of gravity is closer to the left edge, it takes less energy to follow the left path.

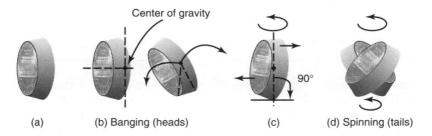

Center of gravity

(a)  (b) Banging (heads)  (c)  90°  (d) Spinning (tails)

The spinning experiment: When it's spinning *really* fast, there is no precession, and the center of gravity lies directly above the edge touching the floor (the head edge). When spinning pennies (or other objects) slow down, they start to precess so that the center of gravity moves in a circular path about the axis of rotation. The slower the spin, the larger the precess, causing the penny to fall so that tails is showing.

**7. Giving orders.** Completely fulfilling all promises (0.00001), Picking the queen of hearts (1/52), Seeing a full moon (1/30), Selecting an ace (1/13), Rolling a 6 (1/6), Picking a black card (1/2), Flying safely (0.9999999). Parentheses indicate nonrigorous guesses at the probabilities.

**8. Two heads are better.** About one-third of the time, two heads should appear.

**9. Tacky probabilities.** In our experiments, about half of the tacks landed on their heads. It seemed that the shorter the fall, the greater the chance they would land on their heads.

**10. BURGER AND STARBIRD.** 17 letters, 4 R's, 2 B's, 9 in the first half (A–M), and 5 vowels. There is a 4/17 chance of getting an R, a 2/17 chance of getting a B, a 9/17 chance of pulling a letter from the first half of the alphabet, and a 5/17 chance of pulling a vowel.

**11. Monty Hall.** The experiments should reflect the analysis of the text. Sticking wins one-third of the time, and switching wins two-thirds of the time. The probabilities of sticking and winning should add to 1.

**12. 7 or 11.** Answer: 2/9. There are six ways to roll a seven, (1-6, 2-5, 3-4, 4-3, 5-2, 6-1) and two ways to roll an eleven (5-6, 6-5). Because there are 36 equally likely ordered rolls, the chances of rolling 7 or 11 is 8/36, or 2/9. If these numbers come up on your first roll in the casino game craps, you automatically win.

**13. D and D.**

| H | H | H | H | H | H | T | T | T | T | T | T |
|---|---|---|---|---|---|---|---|---|---|---|---|
| 1 | 2 | 3 | 4 | 5 | 6 | 1 | 2 | 3 | 4 | 5 | 6 |

Of the 12 equally likely outcomes, there is only one way to get a head and a 4 (H, 4), so the probability is 1/12. Once you know that a 2 is showing, there are only two possible (equally likely) outcomes (H, 2) and (T, 2). So, the chances that the dime is showing heads is 1/2.

**14. The top 10.** There is one 3, four prime numbers (2, 3, 5, 7), five even numbers, and zero numbers evenly divisible by 13. So, the corresponding probabilities are 1/10, 4/10 = 2/5, 5/10 = 1/2, and 0/10 = 0, respectively.

**15. One five and dime.**

| Penny  | H | H | H | H | T | T | T | T |
|--------|---|---|---|---|---|---|---|---|
| Nickel | H | H | T | T | H | H | T | T |
| Dime   | H | T | H | T | H | T | H | T |

Three presidents: 1/8. Exactly two presidents: 3/8. If you only know that a head is showing, then you can rule out the (T, T, T) possibility, so that there are really only seven equally likely outcomes. Now the probability of three presidents is 1/7. Knowing that Lincoln is showing eliminates four of the equally likely outcomes. Given this, the probability of seeing three heads is reduced to 1/4.

**16. Five flip.** Given this new information (assuming this person is trustworthy), the probability of five heads is 1/1. Without the information, the probability of such a feat is rare indeed, 1/32.

**17. Flipped out.** If the coin were normal, then it would land on heads roughly half the time. In fact, the standard deviation in this case is less than 1600 flips, so there is definitely something going on with this coin. Take it to a physicist, take it to a séance, but don't spend it!

**18. Spinning wheel.** The chance of a 13 is 1/38. There are 18 red spaces, 18 black spaces, and 2 green spaces. So, the chance of a ball landing on a red space 18/38 = 9/19.

**19. December 9.** Let's call the two people Tip and Kirk. The chance that Tip was born on December 9 is 1/365 (forget leap year). The chance that Kirk was born on this day is also

1/365. The chance that both Tip *and* Kirk were born on this day is (1/365)(1/365). So, the answer is $(1/365)^2$.

**20. High roller.** Answer: 5/6. It is easier to find the probability that the sum *doesn't* exceed 4. The only such sums are (1-1), (1-2), (2-1), (1-3), (3-1), (2-2). So, the chance of not exceeding 4 is 6/36 = 1/6. As soon as we talk about the 36 equally likely outcomes, we are treating 2-1 and 1-2 differently and 3-1 and 1-3 differently, which is why it is necessary to count them separately in the analysis.

**21. Double dice.** Of the 36 equally likely ordered rolls, there are six ways to get doubles. So, the chance is 6/36 = 1/6. Alternatively, we could think of this as a birthday problem in which there were only six calendar days to the year. It doesn't matter what the first die rolls. The chance that the second die matches it is 1/6.

**22. Silly puzzle.** We hope you didn't spend too long on this one. She knew that there were twins in the class.

**23. Just do it.** Do a birthday survey as described.

**24. No matches.** The chance that the second person doesn't match the first is 365/366. Two birthdays are used up, so the chance that the third doesn't match either of the first two is 364/366. The probability that the first doesn't match *and* the second doesn't match is (365/366)(364/366). Continuing in this way, the probability that the first 40 people don't match is $(365/366)(364/366) \times \cdots \times (327/366) = (365!/326!)/366^{39} = 0.1087681902\ldots$.

**25. Spinner winner.** Four of the 10 equally wide places are numbered 6 or higher. Therefore, the probability of winning is 4/10, or 0.2.

### III. Creating New Ideas

**26. Flip side.** Probabilities are easiest to calculate when we start with a list of equally likely outcomes. The simplest way to do this is to view the coins as *different* and then order the outcomes. With this in mind, there are four equally likely outcomes that contain at least two heads: (T, H, H), (H, T, H), (H, H, T), and (H, H, H). So, the chance of three heads is 1/4.

**27. Other flip side.** Number the coins 1, 2, and 3, and consider the eight equally likely ordered outcomes. The person has only eliminated one of the eight outcomes, which leaves a probability of 1/7 that three heads came up.

**28. Blackjack.** The remaining deck has 50 cards, 12 of which are face cards. So, the probability of getting a face card is 12/50, or 6/25.

**29. Be rational.** No, the probability is 1/3. Half the numbers have no 2's in their prime factorization. Of the half that *are* divisible by two, half of them are divisible by four and half are not. Together this means that 1/4 of the numbers have exactly one 2 in their prime factorization. Classify each number by the number of 2's in its prime factorization.

| Number of 2's in Factorization | 0 | 1 | 2 | 3 | . . . | $N$ | . . . |
|---|---|---|---|---|---|---|---|
| Probability | 1/2 | 1/4 | 1/8 | 1/16 | | $1/2^{N+1}$ | . . . |

For the reduced fraction to contain only odd numbers, both the numerator and the denominator must have originally had the same number of 2's in their prime factorization. Ask instead, "What is the probability that they both had no 2's in their factorization *or* both

487

had one 2 in their factorization *or* both had two 2's in their factorization, and so on?" This translates into $P = (1/2)^2 + (1/4)^2 + (1/8)^2 + \cdots + (1/2^n)^2 + \cdots = 1/4 + (1/4)^2 + \cdots + (1/4)^n + \cdots = 1/3$.

**30. Well red.** This is nonintuitive. The probability that the other side is blue is 1/3. Remember that our notions of probability stem from manipulating sets of *equally likely* outcomes. Red-red, red-blue, blue-red, and blue-blue are not all equally likely outcomes. Assume that all the cards and colors are different; for example, (pink, red), (blue, navy), (ruby-red, sky-blue). You are shown a side of a card with one of the red hues. It might be the pink side, the red side, or the ruby-red side. Among these three equally likely possibilities, in two of them the other side is another red hue and in only one is the other side a blue hue. So the probability that the other side is blue is 1/3.

| Pink | Red | Ruby-red |
|------|-----|----------|
| Red | Pink | Sky-blue |

**31. Regular dice.** The five dice have 4, 6, 8, 12, and 20 sides. The five possible ways to total a 6 are (1, 1, 1, 1, 2), (1, 1, 1, 2, 1), (1, 1, 2, 1, 1), (1, 2, 1, 1, 1), and (2, 1, 1, 1, 1). Because the total number of different equally likely outcomes (where order counts) is $4 \times 6 \times 8 \times 12 \times 20 = 46080$, the probability of totaling 6 is $5/46080 = 0.0001 \ldots$.

**32. Take your seat.** The 40 rows are arranged in 4 groups of 10, and you are equally likely to be in any of the groups (because each row has the same number of seats). The probability that you are in the first group is therefore 1/4. Of the 40 seats, there are exactly 10 window seats. So, the probability of a window seat is $10/40 = 1/4$ as well.

**33. Eight flips.** Better to ask, "What is the probability of seeing *zero* heads?" The probability that the first one is tails *and* the second one is tails *and* ... is $1/2 \times 1/2 \times \cdots \times 1/2 = 1/2^8 = 1/256$. The probability that this doesn't happen is 255/256.

**34. Lottery.** Matching at least one number is the opposite of not matching any numbers, an event whose probability is easier to measure. The chance that the first number isn't a match with 2 or 9 is 8/10, while the chance that the second number also isn't a match is 7/9 because of the nine remaining balls—only seven are different from 2 and 9. The probability of neither matching is $(8/10)(7/9) = 56/90 = 28/45$. So, the probability that you have at least one match is $1 - 28/45 = 17/45$. There are 10 possible balls selected first and 9 selected second, for a total of $(10)(9)$ total ways of selecting two balls in order. Of those 90 possibilities, only 2-9 or 9-2 will match both numbers. So, your probability of matching both numbers is $2/90 = 1/45$.

**35. Making the grade.** The probabilities 0.2 and 0.9 don't add to 1.0. This question is a joke. Because the person missed this probability question, chances are he or she is not a 4.0 student.

### IV. Further Challenges

**36. Cool dice.** Order them according to their lowest numbered side: (two blanks, four 4's) → (three 1's, three 5's) → (four 2's, two 6's) → (all 3's) → (two blanks, four 4's). Each die after an arrow beats the previous die with probability 2/3.

**37. Don't squeeze.** The complementary question, "What is the probability that you didn't locate any defective Charmins?" is easier to answer. The probability that each one is OK is 9/10; so, the probability that all five are OK is $(9/10)^5 = 0.59 \ldots$. So, the probability that you will find a defective Charmin is $0.41 \ldots$.

**38. Birthday cards.** The probability that you don't match on the second draw is 51/52. The chance of seeing three different cards is (51/52)(50/52). The chance of seeing 10 different cards is (51/52)(50/52) . . . (43/52) = 0.397 . . . .

**39. Too many boys.** The decree would have no effect. Each birth has an equal chance of resulting in a boy or a girl. So we would expect 50% of the children to be boys and 50% girls, no matter who is having the children. (Of course, this assumes that there is no genetic predisposition to having male or female children.)

**40. Three paradox.** Implicit in this fallacious reasoning is that there are four equally likely outcomes (3 heads, 0 tails), (2 heads, 1 tail), (1 head, 2 tails), and (0 heads, 3 tails). In two of the four cases, all three coins show the same face. *Warning:* These four cases are not equally likely. They occur with probabilities 1/8, 3/8, 3/8, and 1/8, respectively. It is better to use the eight equally likely outcomes (HHH), (HHT), (HTH), (HTT), (THH), (THT), (TTH), and (TTT). Here, only two of the eight cases have all the coins showing the same face.

An extreme example of this kind of fallacious reasoning: The chance of my winning the lottery tomorrow is 1/2 because I either win or I don't. Clearly the two outcomes are not both equally likely.

## 7.3 Random Thoughts

### I. Developing Ideas

**1. Daily deaths.** Assuming 365 days in a year, an average of 52,000,000/365, or about 142,466, people die each day.

**2. Wake-up call.** The probability the event will *not* happen is $1 - 1/1000 = 999/1000$, or 0.999. The probability the events do not occur for two days is $(0.999)(0.999) = 0.998001$. The probability of no anticipated coincidences in an entire year is $(0.999)^{365} = 0.6940698 \ldots$

**3. More than 12 monkeys.** If we put a very large number of monkeys at word processors, eventually one of them will type out the script for *Hamlet*.

**4. Get shorty.** Buffon was an 18th century French scientist who estimated the value of $\pi$ by tossing breadsticks onto a tiled floor and counting the number of times the breadsticks crossed a line between the tiles.

**5. Nothing but heads.** Because there are eight outcomes (HHH, HHT, HTH, HTT, THH, THT, TTH, TTT), and only one has all heads, the probability of getting all heads is 1/8, or 0.125.

### II. Solidifying Ideas

**6. Pick a number.** Ideally, each number is picked with probability 1/4, so that, on average, $50/4 = 12.5$ people will pick each number. However, in practice, 3 is selected much more often than probability would dictate.

**7. Personal coincidences.** My grandfather played the Michigan lottery (pick 6 out of 44) for the first time after seeing five of six lottery numbers in a dream. He told us of the dream *before* buying the ticket (making additional side bets that he would win), and ended up getting five of the six lottery numbers. (True story.)

**8. No way.** It isn't so surprising. First, you probably have a lot of friends, and most of them are traveling back to school on the last day of spring break. Second, Chicago is one of the major hubs of many airlines, so many students will have connections through Chicago.

**9. Enquiring minds.** No; a psychic's predictions are not better than any other person's predictions.

**10. Milestones.** Two or three famous people appear in the obituaries of such magazines each week. This number may be small due to space restrictions in the magazine; nevertheless, this corresponds to roughly $2.5 \times 52 = 130$ famous deaths each year.

**11. Local mortality.** Sampling a county of 6000, there were 16 deaths over the course of 10 weeks, which corresponds to an estimate of 83 deaths per year, for a 1.3% death rate.

**12. Unlucky numbers.** What is the probability that *none* of them die? From Mindscape II.11, the odds that a random person lives are $1 - 83/6000 = 5917/6000$. The probability that all 1000 live is roughly $(5917/6000)^{1000} = 0.00000089$. It's *very* unlikely.

**13. A bad block.** Suppose only 39 people died during every two-week period. This accounts for only $39 \times 26 = 1014$ deaths. So, 39 deaths per two weeks is not enough.

**14. Eerie.** The probability is the same as for any two people, 1/365. The probability that two people have the same birthday *and* will die in the same year is much lower. In this Mindscape, however, we were already told that the two celebrities died in the same year, so it doesn't enter into the probability.

**15. Murphy's Law.** Something can go drastically wrong every minute of our lives: a plane crashes into our house, the computer explodes, and so on. The chance of anything going wrong on any *particular* minute is tiny, but the chance that nothing goes wrong for weeks on end is even smaller. For example, if the chance that something goes absurdly wrong is 1/10000 each minute, then the chance that nothing goes wrong for two weeks is $(9999/10{,}000)^{(60\times24\times14)} = 0.13$. In other words, there is an 87% chance that one of those 1-in-10,000 things will happen to you in any two-week period.

**16. A striking deal.** See Mindscape IV.37 for the exact answer.

**17. Drop the needle.** Try Buffon's needle experiment.

**18. IBM again.** After the first week, 300 people see a perfect record; second week, 150; third week, 75. Because 75 is odd, we have to break the symmetry (75 = 38 + 37). If we are lucky, 38 people will still see a perfect track record. If we are always lucky, 2 people will see a perfect track record for 9 consecutive weeks: 300-150-75-38-19-10-5-3-2.

**19. The dart index.** Look at stock histories.

**20. Random walks.** Try a random walk on the grid.

**21. Random guesses.** With a 1/4 chance of correctly answering each question, the chance of a perfect score is $(1/4)^{100}$, or approximately $10^{-60}$. With educated guessing, the chances go up to $(1/2)^{100}$, or roughly $10^{-30}$, still vanishingly small. Note that the chances of getting them *all* wrong are similarly small. The first person will score close to 25%, and the second person will score closer to 50%.

**22. Random dates.** The first person will accept with probability 1/2. You'll be rejected by five in a row, with probability $(1/2)^5 = 1/32$.

**23. Random phones.** Forever is a *long* time. You should expect to eventually run across any sequence of numbers infinitely often. The odds of rolling any seven-digit number right off the bat are $1/10^7$. Therefore, you should expect to see your phone number on average about once every $10^7$ digits.

**24. All sixes.** The table gives the probability $P$ of rolling $N$ 6's in a row. As $N$ tends to infinity, $P$ tends to 0, showing that eventually a 6 will not be rolled.

| $N$ | 1 | 2 | 10 | 100 | . . . | Infinity |
|---|---|---|---|---|---|---|
| $P$ | $1/6 = 0.17$ | $1/6^2 = 0.28$ | $0.16 \times 10^{-7}$ | $0.15 \times 10^{-77}$ | | 0 |

**25. Pick a number, revisited.** The human bias toward the number 3 has been documented in several experiments. One such experiment had the following distribution: $1 - 11\%$; $2 - 29\%$; $3 - 46\%$; $4 - 14\%$.

## III. Creating New Ideas

**26. Good start.** The chance of typing C first is 1/26 *and* the chance of then typing an A is also 1/26 *and* the chance of typing the final T is also 1/26. Because all three independent events must occur, we multiply the probabilities: $(1/26)^3 = 0.00005 \ldots$.

**27. Even moves.** Color the even numbers red and the odd numbers blue. Regardless of where you start, your color alternates with each move. So, an odd number of moves puts you on a different color, and an even number of moves puts you on the same color. To get back to where you started, your color has to change an even number of times; so, you must have made an even number of moves.

**28. Playing the numbers.** The chance of missing the number on any given day is 999/1000. The chance of guessing wrong every day for three years is $(999/1000)^{365 \times 3}$. Thus, the chance of getting at least one correct number is $1 - 0.999^{1095} = 0.6$.

**29. Random results.** It is very strange indeed, because the chance of such a string of 17 consecutive numbers in a list of 10,000 is roughly $10,000/10^{32} = 1/10^{16}$. But keep in mind that we computed the probability *after* looking at the numbers. The chance of seeing any particular sequence of 16-digits (54-63-12-84 . . .) is equally unlikely, despite the fact that we see lots of 16-digit sequences. A reasonable response is the following: Let's compute the expected number of two-digit, three-digit, and four-digit consecutive sequences and compare with another generated list of 10,000 digits. If there are any major discrepancies, we'll look for the bug. (We should see roughly 100 consecutive sequences of length 2, 1 consecutive sequence of length 3, and none of any longer length.)

**30. Monkey names.** Because EDWARDBURGER has 12 letters and MICHAELSTARBIRD has 15 letters, EDWARDBURGER will appear much more frequently (roughly $26^3$ times more frequently). The chance of typing MICHAELSTARBIRD right off the bat is $1/26^{15}$, and because you effectively start over with each key stroke, you will see Michael's full name roughly once every $26^{15}$ digits.

**31. The streak.** Both outcomes are equally unlikely; they both occur with probability $1/2^{10} = 0.001$. The second outcome seems more random because it doesn't have an easily describable pattern like HTHTHTHTHT or HHHHHTTTTT. We might also naturally describe the first outcome as (10 heads, 0 tails) and the second as (6 heads, 4 tails). The chance of rolling (6 heads, 4 tails) is 0.205, while the chance of rolling (10 heads, 0 tails) is still 0.001, which gives us a sense of why we think the second outcome should be more common.

**32. Girl, Girl, . . . .** As in Mindscape III.32, the probability of producing any particular sequence is $1/2^8 = 1/256 = 0.004$. However, the probability of producing 5 boys and 5 girls in no particular order is very different from the probability of producing 10 girls.

**33. One mistake is okay.** There are $2^5 = 32$ different ways to make a series of five predictions. Divide the 128 people into 32 groups of 4, and give each group a different set of predictions. Let's rename the predictions G(ood) and B(ad). Only six groups have seen no more than one mistake, and they correspond to the sequence (GGGGG), (GGGGB), (GGGBG), (GGBGG), (GBGGG), and (BGGGG). Because there are four people in each group, we have 24 in all.

**34. Picking and matching.** With no bias, the chance of picking two 3's is $(1/4)(1/4) = 1/16 = 0.062$, whereas the chance of picking the same number is 0.25. Using the odds given in Mindscape II.25, the chances that you both independently picked 3 is $(0.46)(0.46) = 0.21$. The chance that you both picked the same number is $(0.11)(0.11) + (0.29)(0.29) + (0.46)(0.46) + (0.14)(0.14) = 0.33$.

**35. Picking and matching.** Of course, when you repeat the experiment, the odds remain exactly the same, but the odds that you both picked 3's in one of the two experiments are higher than in either experiment separately. For example, there is a 44% chance that during one of the experiments, you both picked the same number (assuming no bias).

## IV. Further Challenges

**36. Death row.** Of the 49 equally likely possibilities, 7 correspond to the celebrities dying on the same day: (Mon Mon), (Tue Tue), . . . , (Sun Sun), giving a chance of 1/7. But there are 12 outcomes that correspond to adjacent days: (Mon Tue) (Tue Mon) (Tue Wed) (Wed Tue) . . . (Sat Sun) (Sun Sat), for a probability of 12/49.

**37. Striking again.** The chance that you will lose all 52 flips is $(51/52)^{52} = 0.3643\ldots$, so the probability of winning is $0.6357\ldots$. (Note, in this case, the result of each flip is independent of the results of the earlier flips. This was not the case in the original wording of the problem, because your chance of matching the second flip depended on whether you matched your first flip. Nevertheless, the probabilities for both games are very similar. Computer experiments indicate a probability closer to 0.634. With two cards, the chance of winning is 1/2; with 3 cards, 2/3; with 4 cards, 15/24. The chance of winning tends to $(1 - 1/e) = 0.632\ldots$ as the number of cards tends to infinity.)

**38. Random returns.** The hard part is showing that we will eventually hit all points at least one time with probability 1. Suppose we start at 0. We know that we will eventually hit 89 (1 time). Now pretend that we are starting from 89. We know that we will eventually hit 0 again. Now we are back to where we started from, which shows that we will hit 0 and 89 infinitely many times.

Alternatively, the hard part is showing that you return back to your starting point with probability 1. (Turns out the chance of not returning to the origin after $2n$ steps is $(2/4^n)(2n$ choose $n)$, which tends to 0 as $n$ goes to infinity.) Once you know that you will return once, you know that you'll return infinitely many times. Now let's show that you'll hit 134 with probability 1. Each time you hit the origin (your starting point), there is a small but finite chance that you will flip 134 heads in a row. Because you get an infinite number of tries, the probability that you never roll the 134 heads in a row is 0. A similar argument shows that you will eventually hit all numbers once. You can then repeat the argument to show that you'll hit all numbers infinitely many times. (Incidentally, the same is true for a random walk in the plane but *not* for a random walk in 3-dimensional space.)

**39. Random natural.** The probability that there is not a 9 in any particular place is 9/10. The chance that no 9's occur in all 50 places is $(9/10)^{50} = 0.005$, so the chance of at least one 9 occurring is about 0.995. Specifically, we can conclude that most large numbers have at least one 9 in them. More generally, large numbers tend to have lots of 9's, lots of 8's, and so on.

**40. Ace of spades.** The probability that you don't win on the first try is 51/52. The chance that you don't win on any of 36 tries is $(51/52)^{36} = 0.497\ldots$. So, your chance of winning is $0.503\ldots$, or just over 50%.

493

### 7.4 Down for the Count

### I. Developing Ideas

**1. Have a heart.** Probability(red) = 26/52 = 1/2; probability(heart) = 13/52 = 1/4; probability (queen) = 4/52 = 1/13; probability(queen of hearts) = 1/52.

**2. Have a head.** There are $2^{10}$ = 1024 possible outcomes. Probability(at least one head) = 1 − probability(no heads) = 1 − probability(all tails) = 1 − 1/1024 = 0.999023 . . . .

**3. Sunny surprise.** Probability(sunny and doughnuts) = prob(sunny) × prob(doughnuts) = 1/20.

**4. Elephant ears.** The total number of elephants will be four times the number with ear tags. Thus, there are $4 \times 60$ = 240 elephants. If you pick an elephant at random, the probability it has a tag is 1/4.

**5. Little deal.** The number of ways to choose three cards from 52 is $(52 \times 51 \times 50)/(3 \times 2 \times 1)$ = 22,100.

### II. Solidifying Ideas

**6. The gym lock.** There are 36 possibilities for the first position, 36 for the second, and 36 for the third. Therefore, there are $36^3$ = 46,656 possible combinations. The chance of picking one is $1/36^3$ = 0.00002 . . . .

**7. The dorm door.** There are $10^5$ possible codes and only 200 active ones, so the chance of guessing an active code is $200/10^5$ = 1/500 = 0.002. You'll probably have to guess for a while.

**8. 28 cents.** Try to find a systematic way of listing all the possibilities. How many ways can you make 28 cents using just pennies? Pennies and nickels? Pennies, nickels, and dimes? And so on. The chart lists all 13 possibilities.

| 1¢ | 28 | 23 | 18 | 13 | 8 | 3 | 18 | 13 | 8 | 3 | 8 | 3 | 3 |
|---|---|---|---|---|---|---|---|---|---|---|---|---|---|
| 5¢ | | 1 | 2 | 3 | 4 | 5 | 0 | 1 | 2 | 3 | 0 | 1 | 0 |
| 10¢ | | | | | | | 1 | 1 | 1 | 1 | 2 | 2 | 0 |
| 25¢ | | | | | | | | | | | | | 1 |

**9. 82 cents.** Try to find patterns to simplify the enumeration. Notice that in Mindscape II.8, there were 6 + 4 + 2 ways to express 28 with only pennies, nickels, and dimes. 6 with 0 dimes, 4 with 1 dime, 2 with 2 dimes, and so on. Similarly, the number of ways to make 82 cents using pennies, nickels, and dimes is 17 + 15 + 13 + · · · + 1 = 81. 81 + 42 + 16 + 2 + 16 + 2 = 159.

| 50¢ | 25¢ | Amount Needed in Pennies, Nickels, and Dimes | Number of Ways to Do It |
|---|---|---|---|
| 0 | 0 | 82 | 17 + 15 + · · · + 1 = 81 |
| 0 | 1 | 57 | 12 + 10 + · · · + 2 = 42 |
| 0 | 2 | 32 | 7 + 5 + 3 + 1 = 16 |
| 0 | 3 | 7 | 2 |
| 1 | 0 | 32 | 7 + 5 + 3 + 1 = 16 |
| 1 | 1 | 7 | 2 |

494

**10. Number please.** You don't know four of the numbers. Because each number has 10 possibilities, there are $10^4$ numbers to choose from, making the odds of guessing the right one $1/10{,}000 = 0.0001$—very small indeed.

**11. Dealing with jack.** There are two very different approaches. First: How many ways can you pick three cards from 52? Like the lottery number, there are $(52 \times 51 \times 50)/(3 \times 2 \times 1)$ different ways. How many sets contain all jacks? Exactly four of them (JH, JS, JC), (JH, JS, JD), (JH, JC, JD), and (JS, JD, JC). So, the total probability is 4 divided by the fraction above. Inverting and multiplying yields $(4 \times 3 \times 2 \times 1)/(52 \times 51 \times 50)$.

Second: Imagine picking one card at a time. The chance of picking a jack on the first card is 4/52, because there are four jacks in the deck. Assuming that you do pick a jack, the chance of pulling another jack is 3/51, because there is one less card and one less jack. Likewise, the chance of pulling the third jack is 2/50. Multiply these three numbers to get the probability that all three events happen: (4/52)(3/51)(2/50).

**12. MA Lotto.** How many different ways can you choose 6 numbers from 36? There are 36 ways to pick the first number, 35 ways to pick the second number, and so on. Thus, there are $36 \times 35 \times 34 \times 33 \times 32 \times 31$ ways to pick 6 numbers in an *ordered* way. But we have overcounted each combination by exactly the number of orderings of 6 elements. For example, (123456), (263415), (35412), and so on are all the same combination. How many different orderings are there? We can use the same counting argument as above: 6 ways to choose the first number, 5 ways to choose the second, and so on, so that there are $6 \times 5 \times 4 \times 3 \times 2 \times 1$ different orderings of 6 numbers. The total number of combinations is then $(36 \times 35 \times 34 \times 33 \times 32 \times 31)/(6 \times 5 \times 4 \times 3 \times 2 \times 1)$. (See Mindscape II.13 for an alternate way to solve this problem.)

**13. NY Lotto.** This is identical to Mindscape II.12, but we'll answer it using a different method. Imagine that we are watching the lottery balls come out one by one. The chance that the first lottery number matches one of your six numbers is 6/40. The chance that the second lottery number matches one of the remaining five numbers is 5/39 (because there are only 39 balls left in the bin). Likewise, the chances that the third, fourth, fifth, and sixth balls match are 4/38, 3/37, 2/36, and 1/35, respectively. Multiply the six individual probabilities to get the chance that all six events happen. Thus, the chance that we win is (6/40)(5/39)(4/38)(3/37)(2/36)(1/35).

**14. OR Lotto.** The chances of winning are $(42 \times 41 \times 40 \times 39 \times 38 \times 37)/(6 \times 5 \times 4 \times 3 \times 2 \times 1)$. (See Mindscapes II.12 and II.13.)

**15. Burger King.** There are eight questions in all: Do I want cheese?, Do I want lettuce?, and so on. Each answer has two possibilities—yes or no. There are $2 \times 2 \times 2 \times 2 \times 2 \times 2 \times 2 \times 2$ different ways to answer all the questions, corresponding to $2^8 = 256$ different possible hamburgers. (Try simplifying the problem by using a fewer number of toppings. Then look for a pattern as you increase the number of toppings.)

**16. More burgers.** There are eight questions to answer: What goes on first?, What goes on second?, and so on. There are eight ways to answer the first question, seven ways to answer the second question, and so on. There is only one way to answer the eighth question, because by this time, there is only one unused topping. The total number of different answers to the questionnaire is $8 \times 7 \times 6 \times 5 \times 4 \times 3 \times 2 \times 1 = 8! = 40{,}320$.

**17. College town.** Assuming that the 100 people you talked to represented a random cross section of the town, you can assume that 60% of the population are students. If $P$ is the total population, then $0.60P = 2000$, or $P = 2000/0.60$, or about 3333 people.

**18. Car count.** 250/800 = 0.3125, so that 31% of all the cars in the area are Hondas. If $C$ represents the total number of cars, then $0.31C = 10,000$. Dividing by 0.31 to solve for $C$ yields $C = 10,000/0.31$, or roughly 32,300 cars. The small number of Rolls-Royces makes it difficult to get an accurate estimate of the percentage of Rolls-Royces in the area. For example, if you stayed an extra minute, you might have seen one more Honda, so that $251/801 = 0.3134 \ldots$ replaces 0.3125 in your estimate. Not a big deal. However, if there are only five Rolls-Royces in the area, you might be replacing $2/800 = 0.0025$ with $3/801 = 0.0037$ for a relative error of 33%.

**19. One die.** The probability of never seeing a 1 is $(5/6)^4$; so, the probability of seeing at least one 1 is $1 - (5/6)^4 = 0.517$.

**20. Dressing for success.** There are $5 \times 10 \times 3 = 150$ different ways to dress yourself, but you can only go three days without doing laundry.

**21. Band stand.** There are six possible choices for the first city, five for the next, four for the next, and so on. Thus, there are $6 \times 5 \times 4 \times 3 \times 2 \times 1 = 720$ possible choices in all.

**22. Monday's undies.** The important thing here is that the order in which you pick the underwear does not matter. This is the lottery problem in disguise: How many ways can you choose 3 numbers out of 10? There are $(10 \times 9 \times 8)/(3 \times 2 \times 1)$ different possibilities.

**23. Counting classes.** As in Mindscape II.22, the order doesn't matter. So we are looking at a lottery problem. How many ways can we choose 4 numbers from a set of 200? $(200 \times 199 \times 198 \times 197)/(4 \times 3 \times 2 \times 1)$.

**24. Cranking tunes.** This is not the lottery problem because order does matter here. There are nine ways to assign the first button, eight ways to assign the second, seven ways to assign the third, and so on. Thus, there are $9 \times 8 \times 7 \times 6 \times 5 \times 4 = 60,480$ different ways.

**25. The Great Books.** This is the lottery problem. How many ways can you choose 10 numbers from 20? $(20 \times 19 \times 18 \times \cdots \times 11)/(10 \times 9 \times 8 \times \cdots \times 1) = (20!)/(10! \times 10!) = 184,756$.

### III. Creating New Ideas

**26. Morning variety.** The question is really, How many different ways can you choose three things out of five? First, imagine that order counts. There are five ways to pick the first item, four ways to pick the second, and three ways to pick the third. There are $5 \times 4 \times 3$ different ways to pick three items (where [eggs, bagel, coffee] represents a different choice than [bagel, coffee, eggs]). We overcounted each combo by exactly the number of orderings of three things. There are $3 \times 2 \times 1$ orderings of a three-item menu; so, the total number of distinct combinations is $(5 \times 4 \times 3)/(3 \times 2 \times 1) = 10$.

**27. Indian poker.** Any 7, 8, 9, 10, J, Q, K, A (of any suit) will win. Because there are 32 winning cards from a deck of 51, the chances of winning are $32/51 = 0.627 \ldots$. Bet high.

**28. Crime story.** The chance that someone fits all 20 descriptions is $p = (1/2)^{20} = 0.00000095 \ldots$ (less than one in a million). So, if a person is a suspect for other reasons and then fits all these descriptions, the probability that the suspect accidentally had all these characteristics is less than one in a million.

**29. There's a 4.** The number fits the form 4xxx, x4xx, xx4x, or xxx4, where each x represents one of nine possible numbers (not a 4). There are $4 \times 9^3 = 2916$ total possibilities. A random guess will work with probability $1/2916 = 0.0003 \ldots$.

**30. Making up the test.** Since we're not considering the In Your Own Words category, Sections 7.2, 7.3, and 7.4 each have 40 available Mindscapes; Section 7.5 has 22; Section 7.6 has 35; and Section 7.7 has 26. There are $(40 \times 39)/(2 \times 1)$ different ways to choose 2 Mindscapes from Section 7.2. Similarly, there are $(22 \times 21 \times 19)/(3 \times 2 \times 1)$ ways to choose 3 Mindscapes from Section 7.5. Compute similar quotients for all six sections and multiply the results to get $[(40 \times 39)/(2 \times 1)][(40 \times 39)/(2 \times 1)][(40 \times 39 \times 38)/(3 \times 2 \times 1)][(22 \times 21 \times 19)/(3 \times 2 \times 1)][(35 \times 34 \times 33)/(3 \times 2 \times 1)][(26 \times 25)/(2 \times 1)]$. This is approximately $1.87 \times 10^{19}$. The chance that this very Mindscape is on the exam is 3/40.

**31. Moving up.** You have the same chance of getting a raise as any other random employee. Because 6 of 20 are chosen for a raise, your chance of getting a raise is 6/20.

**32. Counterfeit bills.** This is yet another lottery problem in disguise. Number the dollars 1–10 and pick 3 of 10. Pick right, win $700; pick wrong and go to jail. The number of possible triplets of bills is $(10 \times 9 \times 8)/(3 \times 2 \times 1)$; thus, the chance of correctly picking the right triplet is $(3 \times 2 \times 1)/(10 \times 9 \times 8) = 1/120 = 0.0083 \ldots$.

**33. Car care.** There are five possible numbers for each of the three digits, so $5^3 = 125$ in all. There is a 1/125 chance of the first entered guess disarming the system. If he doesn't enter the 20 guesses, he gets no additional information.

**34. Coins count.** Because no combination of coins can add up to the face value of any of the other coins, each different combination of coins will give a different total. How many ways can we pick three out of five? $(5 \times 4 \times 3)/(3 \times 2 \times 1) = 10$. With four coins, we have $(5 \times 4 \times 3 \times 2)/(4 \times 3 \times 2 \times 1) = 5$ different totals. With five coins, there is only one total.

**35. Math mania.** Divide the question into two parts. First, how many different ways can 30 of the students be chosen for Section 1? $(90 \times 89 \times 88 \times \cdots \times 61)/(30 \times 29 \times \cdots \times 1)$ Of the remaining 60 students, how many different ways can we choose 30 of them for Section 2? $(60 \times 59 \times \cdots \times 31)/(30 \times 29 \times \cdots \times 1)$. (This also determines who goes in Section 3.) In total, there are $(90 \times 89 \times \cdots \times 32 \times 31)/(30!^2) = (90!)/(30!^3)$.

## IV. Further Challenges

**36. Party on.** There are $(10 \times 9 \times 8)/(3 \times 2 \times 1)$ ways to pick 3 of 10 soccer players, $(20 \times 19 \times 18 \times 17)/(4 \times 3 \times 2 \times 1)$ ways to pick 4 of 20 orchestral folk, and $(30 \times 29 \times \cdots \times 23)/(8 \times 7 \times \cdots \times 1)$ ways to pick 8 of 30 students. Multiply these three fractions to get the total number of different invitation lists.

**37. No dice.** Either you roll an 11 at least once, or you never roll it. The odds of the second case are easier to compute. Of the 36 equally likely ordered outcomes for a pair of dice, only two total 11 (5-6 and 6-5). Thus, the chance of not rolling an 11 is 34/36. The probability of not rolling an 11 during 24 consecutive rolls is $(34/36)^{24}$, which implies that the chance of rolling an 11 at least once is $1 - (34/36)^{24} = 0.746 \ldots$.

**38. Three angles.** Because any three points not lying on a line form the vertices of a triangle, each triplet of points corresponds to a unique triangle. How many ways can we choose 3 vertices from a group of 10? $(10 \times 9 \times 8)/(3 \times 2 \times 1)$.

**39. Four parties.** Divide the 10 men into two groups of 5, the *short* group and the *tall* group. Similarly, divvy up the women. Party 1: *short men* meet *short women*. Party 2: *short men* meet *tall women*. Party 3: *tall men* meet *short women*. Party 4: *tall men* meet *tall women*. You can also show that you need *at least* four parties. For the 10 men to meet each of the 10 women, 100 matches must take place. If you only had three parties, then some party would have to have more than 33 matches (Pigeonhole principle). With 10 people, the maximum number of matches is 25 and occurs precisely when there are 5 men and 5 women.

**40. Making the cut.** There are $5 \times 5 \times 5 \times 5 \times 5 = 5^5 = 3125$ different possible keys and, therefore, a 1/3125 chance that a key will unlock a random Escort. The probability that a specific key opens a random Escort is the same as the probability that a certain Escort is opened with a random key. In the second case, all we need to know is that there is 1 correct key among 3125 possible ones. The number of cars is extraneous. Take the situation to extremes, what if there were only two keys? Then, any key would open half the Escorts and thus would open any random Escort with probability 0.5.

## 7.5 Collecting Data Rather than Dust

### I. Developing Ideas

**1. Embarrassing data.** Suppose there were $H$ noncheaters in your group. The only No's will come from these $H$ people when they flip tails. Because roughly half of them will flip tails, you expect $H/2$ No's. Because there are 46 No's, solve $H/2 = 46$ to estimate the number of honest students. $H = 92$, leaving eight cheaters in the group.

**2. Hugging both parents.** Only the nonhuggers will contribute to the No's in the survey. If there are $X$ nonhuggers, then we'll see roughly $0.5X$ No's. Because 83 answered Yes, there are 67 No's. Solving $0.5X = 67$ gives $X = 2 \times 67 = 134$. This leaves an estimated 16 people that still hug both parents.

**3. Short test.** The probability that an individual student will get the question right is 0.8 (80%), so the probability that an individual student will get the question wrong is 0.2 (20%). Because there are 100 students in the class, we can expect approximately $0.20 \times 100$, or 20, to get the question wrong and fail the test.

**4. Computer polls.** It is safe to assume that all the people that were sent surveys have access to a computer and the Internet, (they all have e-mail addresses). We are likely to get 100% of the respondents to say that they use the Web. This would be a biased sample of the population to survey considering we are only surveying people with computer and Internet access and not including any noncomputer users in our survey.

**5. Voluntary grade inflation.** Because the students who received A's in the class are most likely happy with their grades, the dean will probably get close to 0% of the A students to respond to her questionnaire, making the sample biased.

### II. Solidifying Ideas

**6. Pornography.** Because 100 students were surveyed using the one-coin method, we would expect half of them, or 50 students, to flip heads and the other half, 50 students, to flip tails. The 50 students who flipped heads would have answered Yes, leaving 48 of the 98 students who answered Yes to have flipped tails and answered honestly. So, we estimate that 48 of the roughly 50 students who flipped tails have pornography. We can expect roughly the same proportion—that is, 48 out of 50 students—who flipped heads to also have pornography in their room. This gives us an estimate of $48 + 48$, or 96, students who have pornography.

**7. Cartoons.** If we use the one-coin method to survey 1000 students, we expect that 500 of these students will flip heads and answer Yes. Of the 380 students who like to watch cartoons, we can expect half, or 190 students, to flip heads and be among the 500 students we are already counting as answering Yes. The other 190 students who like to watch cartoons can be expected to flip tails and also answer Yes. Therefore, we can expect $500 + 190$, or 690, students to answer Yes.

**8. Bias beef.** The sample of students who attended the football game will be a biased sample of the population of all students at the college. Typically, the food served at football games includes items like hot dogs, hamburgers, and barbeque. Tailgating is also a typical activity surrounding football games and usually involves a grill loaded with meat. Therefore, you will most likely end up with a sample that is biased toward meat eaters.

**9. Drug data.** The expected number of Yes's and No's from the two groups is indicated in the table. ($U$ represents the number of recent drug users.)

|  | **Flipped 2 Heads** | **Flipped at Least 1 Tail** |
|---|---|---|
| **Users ($U$)** | No ($0.25U$) | Yes ($0.75U$) |
| **Nonusers ($250 - U$)** | Yes ($0.25 \times (250 - U)$) | No ($0.75 \times (250 - U)$) |

Set the expected number of Yes's, $0.25(250 - U) + 0.75(U)$, equal to 78, and solve for $U$. Multiply by 4 on both sides to get $(250 - U) + 3U = 4 \times 78$, or $2U + 250 = 312$, implying $U = 31$.

**10. Kissing.** Suppose there were $X$ number of kissers. The Yes's come from the $180 - X$ nonkissers who flipped two heads and the $X$ kissers who didn't flip two heads. We expect one-quarter of the nonkissers to flip two heads and three-quarters of the kissers to flip at least one tail. Together, we expect $1/4 \times (180 - X) + (3/4)X = 50$. Solving yields $X = 10$.

**11. Cheating.** If 60 students cheated, we would expect that one-fourth, or 15 students, would flip HH, HT, TH, and TT. That leaves 140 students, of the 200 surveyed, who didn't cheat. We would expect one-fourth, or 35 of them, to flip HH, HT, TH, and TT. We know that the 35 students who didn't cheat and flipped HH will answer Yes. The only other students who will answer Yes are the students who did cheat and flipped HT, TH, and TT. This is $3 \times 15$, or 45, students. Therefore, we would expect $35 + 45$, or 80, students to answer Yes and the other 120 students to answer No ($15 + (35 \times 3)$).

**12. Ask them.** Answers will vary.

**13. Dental hygiene.** The question asked on this survey is one in which many people would lie when they respond. We are told by the dentist that we should brush our teeth in the morning and before we go to bed and sometimes even after every meal. Many people, whether they actually do this, will answer that they brush their teeth as often as the dentist tells them to because this is what they are *supposed* to do, when in fact it is probably more likely that they brush less than that.

**14. More homework.** A professor who wanted to give more homework probably conducted several surveys, each containing two people apiece, and stopped when he finally found a student who wanted more homework. Ignoring all but the last survey, he states his finding. He won't mention that the sample size was 2 because it doesn't help his cause.

**15. Amazing stats.** Answers will vary.

### III. Creating New Ideas

**16. PBS.** There might be different results when people are asked to report their TV-watching habits than when they are monitored, because when people are asked about their TV-watching habits and are not monitored they can adjust the truth. People might say they watch less TV than they do because some people consider it a lazy activity. People might also say that they don't watch "trashy" shows, only shows like the news and *Jeopardy*.

**17. 9:00 a.m. versus 9:00 p.m.** The very cooperative Saturday morning group will, on average, study far more hours than the crowd of students at the party. The mean, median, and mode will be shifted toward 0 in the 9:00 p.m. group.

**18. Four out of five.** Here are some possible questions: Now that you're off TV and you have your money, do you *really* like/use/recommend that product? How difficult was it to find people with a favorable opinion of your product? How did you select these people? How much money did they get? Did the people know they were being filmed?

**19. Sleazy survey.** When the survey results came in, the mayor should have considered that only 8% of the people surveyed responded. These would most likely have been those who felt very strongly about shutting down the pornography vendors. The mayor cannot assume that the feelings of the majority of the 8% that responded are the feelings of the 100,000 households in the city.

**20. Bread winners?** If you take your survey in the middle of the day, you will find only the people who are home during the day and who most likely don't work outside the home. These people would have more time to bake their own bread than those people who work during the day. Therefore, your data set would reflect a higher proportion of bread bakers than actually exist in the whole population. If you survey in the evening, you will find people who don't work and people who do work during the day, so you will get a less biased sample and find that fewer people surveyed during this time bake their own bread than those you surveyed in the afternoon.

## IV. Further Challenges

**21. Lincoln.** The percentage of people who own Lincolns is quite small, so a random sample of 100 may very well find no Lincoln owners at all. Alternatively, it might accidentally find several Lincoln owners when in reality only a tiny fraction of people own Lincolns. When trying to find the percentage of people who own a rather rare item, the number of people in the sample must be considerably larger than 100. There may also be other difficulties with the answers. Although people will most likely not lie when asked if they own a Lincoln, they still may not give the answer that would be expected by the question. The issue of "ownership" is a tricky one for cars. People who are married might feel like they own half of their car and their spouse owns the other half, so will they answer yes or no? Some people have a loan from the bank to pay for their car and might think that they own half of their car and the bank owns the other half, so how will they answer? Also, some people lease their cars, so how will they answer? With so many issues involved in how to answer what seems like a simple question, your results may not reflect reality.

**22. Coffee, tea, or milk?** Even if the students drink coffee, you still do not know enough to determine whether to open a coffee house near campus. Some other information you might want to seek includes, whether the students usually buy their coffee or make it at home, whether they enjoy coffee houses, whether they already have a favorite coffee house near campus, what time of day they tend to drink coffee and if that coincides with when they are on campus.

## 7.6 What the Average American Has

### I. Developing Ideas

**1. Internet costs.**

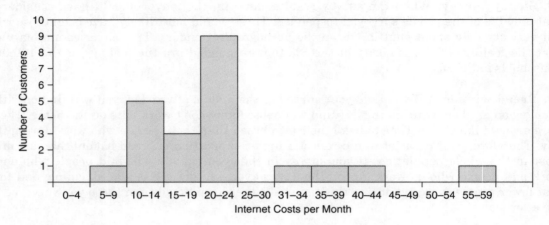

The histogram shows the data skewed toward the left side. Less than one-fourth of the data is in the right side of the histogram. The histogram shows that the interval with the most data points is between $20 and $24, with the second highest being between $10 and $14.

**2. Internet costs—summaries.** The median of the data on the cost of Internet access is $20 and the mean is $20.87, or $21. Both of these numbers are close together; however, more than half of the people, approximately 78%, paid $21 or less, which explains why the mean is higher than the median. The high data points pulled the mean higher. Therefore, the median would give a better answer to the question, In 2001, how much did a typical person with Internet access pay for that access?

**3. Internet costs—more summaries.** The five-number summary includes the minimum value, which is 5; the first quartile, which is 12; the median, which is 20; the third quartile, which is 21; and the maximum value, which is 58. The spread of the data, which is given in the five-number summary, gives an impression similar to that in the histogram. It shows that the data in the lower half is pretty equally spread from 5 to 20, but the data in the third quartile only covers numbers from 20 to 21, and the fourth quartile involves the greatest spread of all, from 21 to 58.

**4. How variable.** The average score is about 87%. Because the highest score is 13 higher and the lowest score is 15 lower, and there are only two scores that are less than 5 away from 87, it seems more likely that the standard deviation is 10.3 rather than 2.7. If the standard deviation were 2.7, we would expect the scores to be closer to 87% than they are.

**5. Gas guzzlers.** It appears as though, for the most part, there is a relationship between the weight of a car and its miles per gallon. As the weight increases, the miles per gallon decrease. The only exception seems to be the weight of 1850, which gets more miles per gallon than the weight of 1700.

## II. Solidifying Ideas

### 6. Who's the best?

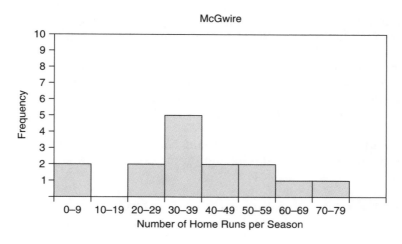

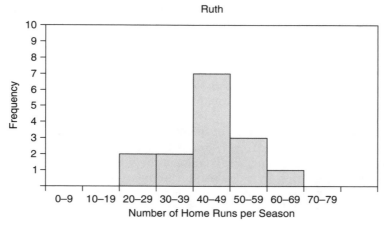

Ruth's histogram shows that most of his data is spread over a smaller interval than McGwire's. McGwire's histogram shows that the interval that had the most seasons represented was 30–39 home runs; however, Ruth's histogram shows that the interval that had the most seasons represented was 40–49. Also, from 40 to 79, Ruth appears to have about twice as much area as McGwire does in their respective histograms. Therefore, you can conclude that overall Ruth is a better home-run hitter.

### 7. Who's the best—short summary. Ruth and McGwire's five-number summaries are below.

|  | Minimum | Q1 | Median | Q3 | Maximum |
|---|---|---|---|---|---|
| **McGwire** | 9 | 29 | 39 | 52 | 70 |
| **Ruth** | 22 | 35 | 46 | 54 | 60 |

You can see that Ruth's data seem to be much closer together than McGwire's because Ruth's minimum is 13 higher than McGwire's, but McGwire's maximum is 10 higher than Ruth's.

**8. Is today different?** In 2001, most home Internet access was through dial-up, which is generally less expensive than other methods. AOL and other dial-up service providers offer many free six-month trials, which would lower the costs of access. Today, many people connect to the Internet at home through DSL or their cable company, which generally costs more than dial-up access does. This would cause the histogram from Mindscape I.1 to be skewed to the right rather than to the left or slightly more normal in shape.

**9. How different are they?** The mean score is approximately 70%. The standard deviation is more likely to be closer to 12, because about half of the scores are within 10 of 70 and the other half are farther than 10 but less than 30 away. Therefore, it seems more likely that the standard deviation would average out to close to 12 rather than 25.

**10. What's normal?** The histogram for Amount has an approximately normal distribution, with a mean between 70 and 80. The histogram for Cost is skewed, and the histogram for Price is bimodal.

**11. Heavy car.** The relationship between weight and miles per gallon seems to be a line with a negative slope. If you continue this line until you reach 6000 on the $x$-axis, it seems like the miles per gallon should be close to 0. Considering 6000 pounds is a much larger value than the values in the data set, these data aren't very useful in predicting the miles per gallon for that great a weight. These data would most likely give a much better prediction for a car that weighs 2400 pounds.

**12. Easy measurement.** Answers will vary. In general, if your three measurements are relatively similar, then the mean is a good estimate. However, if your three measurements vary quite a lot, then the median would probably be a better estimate.

**13. Types of answers.** Some examples of questions that ask you to put a measurement in a category are questions 2, 3, 5, 8, 10, 11, and 12. Some examples of questions that ask for actual measurements are 18, 19, and 20. To determine which set of questions and responses would provide the most meaning by calculating the mean and median, you need to know for what purpose you intend to use the mean and median. Both sets of questions can give you a mean and median; however, with questions that offer categories, your mean may not give you an exact category number to use but rather a decimal between two categories, making the mean of data that are actual numbers rather than numerical categories much easier to work with. The median of the category data will give you an interval median, whereas that exact numerical data will give you an actual number for the median. Either of these could be useful, depending on your intent for analysis of the data.

**14. Is there meaning in summaries?** Question 3 asks about height. If the median response is 3, or between 5' and 5'5", then we can assume that the halfway point of all students' heights falls between 5' and 5'5". However, we don't know how many fall outside or inside this range and in which direction they may fall. Looking at a graph of this data would give you a much better understanding of how the heights of students in your class are distributed. Question 4 asks about favorite body parts, and the median response is 3, or arms. Because this data is qualitative, rather than quantitative, a mean and median doesn't really give you a meaningful summary of the responses. The numbers assigned to the body parts are arbitrary. A bar graph, however, would give a much better picture of how your class's favorite body parts are distributed among the choices.

**15. Your class.** Answers will vary.

**16. Surprising class.** Answers will vary.

**17. Summaries.** Answers will vary.

**18. Relationships.** Answers will vary. Some examples of pairs of possibly correlated variables include weight and height; hours surfing the Internet and number of e-mails per week; number of dates and hours of study per week.

**19. College data.**
Data from a Student Survey

| Student | Hours on Web/Week | Number of E-mails Received/ Week | Number of dates/ Month | Number of Hours Spent Studying/ Week | Number of Hours Participating in Sports/ Week | Number of Parties Attended/ Month |
|---|---|---|---|---|---|---|
| 1 | 16 | 45 | 3 | 25 | 10 | 3 |
| 2 | 5 | 17 | 7 | 50 | 6 | 1 |
| 3 | 10 | 26 | 11 | 21 | 15 | 6 |
| 4 | 13 | 30 | 9 | 25 | 8 | 2 |
| 5 | 3 | 10 | 7 | 30 | 18 | 4 |
| 6 | 7 | 19 | 9 | 14 | 5 | 8 |
| 7 | 9 | 24 | 8 | 35 | 5 | 1 |

The number of hours studied per week is inversely related to the number of parties attended per month (the more you party, the less time you have to study). The weekly hours of sports activity is positively correlated to the partying tendency (outgoing people are likely attracted to both activities). See the following scatterplots. Note that only two graphs seem well represented by a line.

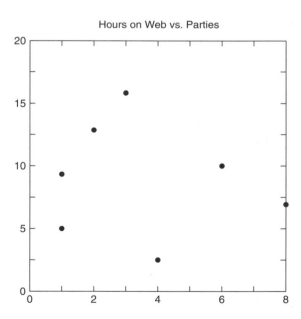

Hours on Web vs. Parties

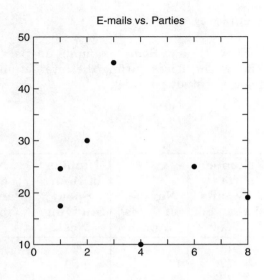

E-mails vs. Parties

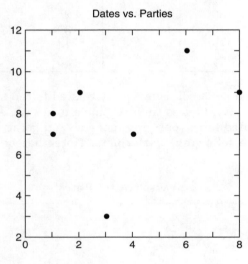

Dates vs. Parties

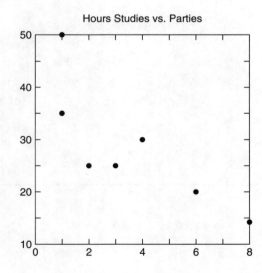

Hours Studies vs. Parties

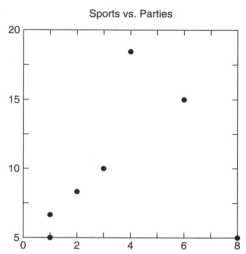

Sports vs. Parties

**20. Web use.** See the following scatterplots. There is a very strong positive correlation between e-mail use and Web surfing time (which makes sense). There's a slight positive correlation between the number of dates and time spent surfing (I can't explain this one). And finally, there may be a slight negative correlation between surfing and studying (again, the more you surf, the less time you have to study).

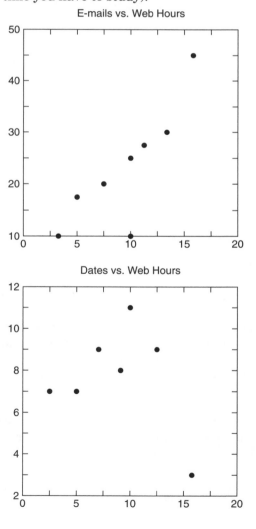

E-mails vs. Web Hours

Dates vs. Web Hours

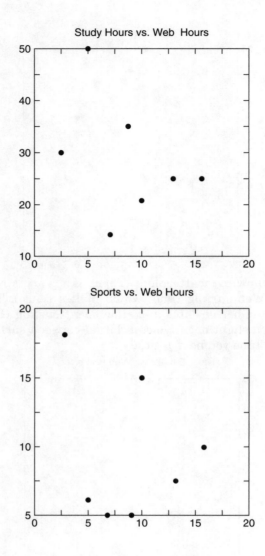

Study Hours vs. Web Hours

Sports vs. Web Hours

## III. Creating New Ideas

**21. News data.** Answers will vary.

**22. Which half?** The spirit of the statement is shockingly true. We can't all be like the children in Lake Woebegone. Implicit in the statement is that the mean and the median statistics are the same. The average refers to the mean, while, by definition, half of the sample size is below the median. Though the mean and median are close in normal-like distributions (height, weight, etc., of a *large* male population), they are far apart for others (salary, number of kids per family, etc.). Because of Microsoft, the average salary is higher than the median salary.

**23. Grades.** The mean is 84.125. The median is halfway between 84 and 87, or 85.5. Both statistics move a little in the direction of the added score, so they will both go down. With the extra score, the mean is 83.66 . . . and the median is 84.

508

**24. Raising scores.** The mean is 79.375. If his next quiz is an 80, the new mean will go up to 79.44 . . . . But if he scores a 79, the mean will go down to 79.33 . . . . With each newly added score, the mean moves a little toward the new score. The amount of movement decreases as the sample size gets large.

**25. Mean and median.** The following is an example of a data set where the mean is greater than the median.

    50,000   75,000   95,000   125,000   135,000   150,000   175,000   200,000   1,500,000

This data set has a mean of 278,333 and a median of 135,000. In this data set, most of the data are somewhat close together and one outlier is much higher than the other values.

**26. Median and mean.** The following is an example of a data set where the median is greater than the mean.

    75,000   265,000   268,000   282,000   350,000   352,000   368,000   375,000   380,000

This data set has a mean of 301,667 and a median of 350,000. In this data set, most of the data are somewhat close together and one outlier is much lower than the other values.

**27. Mean and median.** The mean of the data can be calculated by (25 + 36)/43, which gives 1.42 and would help you determine the percentage of males in the class. The median is 1 and tells you that there are more males in the class than females because at least half are males.

**28. Mode.** Some data sets have no mode, some have one mode, and some are bimodal. The mode is a useful summary statistic for the questions on the survey that ask things like what a person's favorite body type is, gender, and the question that asks about a favorite topic is from a book. These are questions where the choices offered are limited and you are interested in the most popular choice.

**29. Politics as usual.** Nixon wants his listeners to make the following jump, "Humphrey's policies came before a rising crime wave; therefore, Humphrey's policies caused the crime wave," which is an example of *post hoc, ergo propter hoc*. The second statement is weaker, but has the same fallacious flavor, "His policies came before the crime wave; therefore, his policies do nothing to reduce crime." Were it not for his policies, the crime might have increased 20 times as fast. There just isn't enough information here to make *any* conclusions.

**30. Whoops.** Answers will vary.

**IV. Further Challenges**

**31. Further whoops.** The corrected mean is $35,130. Because the mean of 100 salaries is $35,000, the sum of all the salaries originally reported must equal $3,500,000. With the corrected salary, this sum is $3,513,000, giving a mean of $35,130. (Note that this answer is also obtained by adding 1/100 of the salary difference, $13,000, to the orginal mean.)

**32. Mean wins.** Suppose that you and a few friends had bought all the lottery tickets for the past 10 years at a total cost of $100 billion. The state pockets $30 billion and doles out the rest. Because your group wins *all* the cash prizes, you get $70 billion. Thus, the expected return on each dollar was 70 cents. This is meaningless, because the median and mode return is 0 cents. The 70 cents comes from averaging the few cash prizes over all the people who buy tickets, which is similar to the mean salary of graduates from Lakeside School.

**33. Average the grades.** We can compute the mean grade but not the median. Let $M$ represent the sum of the scores of the men's exams and $W$ represent the corresponding sum for the women. We can compute $M$ and $W$ from the separate mean scores. Solving the equations $M/23 = 0.83$ and $W/25 = 0.89$ gives $M = 23 \times 0.83$ and $W = 25 \times 0.89$. So the mean score is $(M + W)/48 = ((23 \times 0.83) + (25 \times 0.89))/48 = 0.86125$, or 86%. The median can't be computed. Two extreme examples: (1) All 23 men make 83. All 25 women score 89. The median score is 89. (2) All 23 men score 83. Two women score 83, two women score 95, and the remaining 23 women score 89. Now the median is 83. There isn't enough information.

**34. Taxes.** A useful summary of the data would be a scatterplot that plots the valuation of the home versus the tax. When each data point is plotted, however, it will be drawn a different color or with a different letter signifying the location. Therefore, you can observe if there is any correlation between tax and valuation, as well as between location and valuation and location and tax.

**35. Doing surveys.** If your only concern is the general amount of time that people spend on the Web, then using categories will give you fewer possible data points and makes it easier to do a basic analysis of the data. If, however, you are looking for a more exact average amount of time spent on the Internet, or maximums and minimums, or even calculating an actual variance for the data, then asking for numerical answers will make this possible. The last three questions on the survey ask for exact numbers. With a class of 120 students, this would be more time consuming to analyze. Simply sorting answers into categories would be much quicker.

## 7.7 Parenting Peas, Twins, and Hypotheses

### I. Developing Ideas

**1. Mendel's snapdragons.** There is only one gene outcome: a red gene from the red parent and a white gene from the white parent. The result is a pink-flowered snapdragon.

**2. Telephone/soda twins.** When we look at the possible combinations of numbers for the last two digits of any phone number, we have 10 choices for the second-to-last digit and 10 choices for the last digit. This means there are 100 possible combinations of numbers for the last two digits of any phone number. Because there are 1000 students, we know that at least 10 of them must have the same last two digits of their phone numbers (there were only 100 possibilities and 10 times as many student). Among the students who have the same last two digits, it is somewhat likely that at least one pair like the same brand of soda, considering that there aren't that many possibilities for soda brands.

**3. Sorority grades.** A sample of only five students is quite small, especially when we don't know how randomly they were selected. A sample of 150 is more likely to be representative of the grades of the sorority, especially if those 150 were chosen randomly from the whole sorority population. Therefore, the evidence from the survey of 150 sorority members is more persuasive of the idea that sorority members have higher-than-average GPAs. Although this is true, it is also important to have a sample that is representative of the population for which you are drawing inferences. If you can say that the 150 sorority members are representative of the population, then the conclusions made are statistically valid; if not, then they are not.

**4. Election up for grabs.** The statement that Jones leads with 51% preference, with margin of error of 3% at the 95% confidence level, means that we can be 95% confident that the actual preference percentage for Jones lies between 48% and 54%. This is a relatively high confidence level. However, one-third of the interval lies below the 50% mark and the other two-thirds lie above 50% mark. Therefore, you would be slightly more confident that Jones will win than lose.

**5. U.S. samples.** The level of confidence we can deduce from a sample is determined by the size of the sample rather than by what proportion of the population the sample represents. So, a sample of 500 randomly chosen people from the United States will give about the same level of confidence as a sample size of 500 from Missouri. A sample size of 10 would give a much less accurate sense of Missouri's population.

### II. Solidifying Ideas

**6. More Mendel.**

|  | **Dad Gives White Gene** | **Dad Gives Red Gene** |
|---|---|---|
| **Mom Gives Red Gene** | (red, white) → Pink | (red, red) → Red |
| **Mom Gives White Gene** | (white, white) → White | (white, red) → Pink |

The third-generation flower is red with probability 1/4, white with probability 1/4, and pink with probability 1/2. You should see twice as many pink flowers as either red or white flowers.

### 7. Oedipus red.

| | Pink Gives White Gene | Pink Gives Red Gene |
|---|---|---|
| **Red Gives Red Gene** | (red, white) → Pink | (red, red) → Red |
| **Red Gives Red Gene** | (red, white) → Pink | (red, red) → Red |

The new flower is equally likely to be red or pink. It won't be white.

### 8. Oedipus white.

| | Pink Gives White Gene | Pink Gives Red Gene |
|---|---|---|
| **White Gives White Gene** | (white, white) → White | (white, red) → Pink |
| **White Gives White Gene** | (white, white) → White | (white, red) → Pink |

The new flower is equally likely to be white or pink. It won't be red.

**9. Pure white.** Yes; second-generation plants are pink and have both a red and a white gene. Third-generation plants can be red, pink, or white. If the second-generation plant mates with either a pink or white flower, there is a chance that the resulting offspring will be white. (The total chance of a white offspring is 1/4).

**10. Random person.** Devise a test.

**11. Another random person.** Survey another random person.

**12. Astrology.** This experiment in no way shows that astrology works. The 100 elderly patients had the experience of working with a "vibrant" experimenter who, no doubt, improved their mood. The other 100 patients had no one to help improve their mood, and therefore it is not surprising that they weren't as happy.

**13. Bigger confidence?** A 99% confidence interval for the same set of data would be wider, because we are now saying that if the actual population value were to fall outside of the range given by the margin of error, then the chance of getting the sample outcome we found is 1% rather than 5%. Therefore, for all else to stay the same, the margin of error would need to increase.

**14. Overlapping confidence.** The percentage of people in the whole population who like green ketchup is more likely to be closer to 41%. Although both confidence intervals are reported at the 95% confidence level, one is done on a much larger sample, which will tend to make the results more accurate. Also, the 45% sample outcome with a margin of error of 12% gives a much bigger range within which the actual population percentage can fall than does the 41% sample outcome with a margin of error of 1%. So, although both say that if the actual population percentage were to fall out of each respective range then only 5% of the time would you find a sample with a different sample outcome, the 41% outcome has a much smaller range for this same confidence statement.

**15. Flipping possibilities.** Because we are flipping a coin 12 times and each flip has two possible outcomes, we come up with $2^{12}$, or 4096 total possible outcomes for the 12 flips. There is only one outcome in which there are no heads—the outcome when all 12 flips come out tails, or TTTTTTTTTTTT. The probability of this outcome is 1/4096. There are 12 possible outcomes in which there is one head: HTTTTTTTTTTT, THTTTTTTTTTT, TTHTTTTTTTTT,

TTTHTTTTTTTT, TTTTHTTTTTTT, TTTTTHTTTTTT, TTTTTTHTTTTT, TTTTTTTHTTTT, TTTTTTTTHTTT, TTTTTTTTTHTT, TTTTTTTTTTHT, and TTTTTTTTTTTH. The probability of any specific flip, regardless of the number of heads or tails involved, is 1/4096 because any one sequence is one of 4096 equally likely outcomes.

**16. Lucky charms.** When the soccer player first realized the necklace was lucky, the reasoning was, "The necklace came before the blocked shots; therefore, the necklace caused the blocked shots." This is a clear case of *post hoc, ergo propter hoc*. Once the player feels lucky, however, the placebo effect may cause him to continue to play well. So, the statement may be true after all.

**17. Read my lips.** The economy is even less predictable than the weather, so the reasoning is likely. "Tax breaks came before the recession; therefore, tax breaks caused the recession." Again, *post hoc, ergo propter hoc*.

**18. Penny luck.** Unlike the situation in Mindscape II.16, the placebo effect is not going to help anyone win the lottery. This is a clear case of *post hoc, ergo propter hoc*.

**19. The lick.** No; although it is possible that Golden gave you a cold, you could have picked it up virtually anywhere. Just because it could be the cause (such as the dog lick occurred before the cold) doesn't mean that it is the cause.

**III. Creating New Ideas**

**20. Mendel genealogy.** One-quarter of the seeds will result from a red-red mating, one-quarter will result from a pink-pink mating, and one-half will result from a pink-red mating. All the red-red seeds will be red flowers. The pink-red flowers will be half red and half pink. And, of the pink-pink seeds, one-quarter will be red, one-quarter white, and one-half pink. This is summarized below: 56.25% Red, 6.25% White, and 37.5% Pink.

| | | |
|---|---|---|
| red-red → red | 0.25 | Red |
| pink-red → red | 0.25 | Red |
| pink-red → pink | 0.25 | Pink |
| pink-pink → red | $0.25 \times 0.25 = 0.0625$ | Red |
| pink-pink → white | $0.25 \times 0.25 = 0.0625$ | White |
| pink-pink → pink | $0.5 \times 0.25 = 0.125$ | Pink |

**21. Martian genetics.** The offspring will have purple eyes (with probability 1). Each parent can only give one kind of gene, so the offspring will have exactly one red, one blue, and one white gene.

**22. More martians.** The two white-eyed parents will donate one white gene each, and the purple-eyed parent will donate either a red, white, or blue gene. The three equally likely outcomes—RWW (pink), BWW (sky blue), and WWW (white)—occur with probability 1/3.

**23. Politics as usual.** Nixon wants his listeners to make the following jump: "Humphrey's policies came before a rising crime wave; therefore, Humphrey's policies caused the crime wave"; this is an example of *post hoc, ergo propter hoc*. The second statement is weaker, but has the same fallacious flavor, "His policies came before the crime wave; therefore, his policies do nothing to reduce crime." Were it not for his policies, the crime might have increased 20 times as fast. There just isn't enough information here to make *any* conclusions.

**24. Up and down.** The example in the chapter explains how this situation could possibly occur. If both the GPAs of the men and women in the college went up, we need to look at the number of men and women who contributed to the original GPAs of the college and the number who contributed to the new GPAs. If, for instance, the women had higher GPAs, on average, than the men and if the number of women in the college decreased as the number of men increased, the amount of effect from the men's lower GPAs on the overall college GPA would be increased as the effect of the women's GPAs was lowered. This could lead to a lower overall college GPA, even as the men and women both show increases in their average GPAs.

## IV. Further Challenges

**25. Modified Mendel.** The table includes an extra row and column of red genes, reflecting the fact that red genes are twice as likely to be passed on as white genes. (P = Parent)

|  | P Gives R Gene | P Gives R Gene | P Gives W Gene |
|---|---|---|---|
| **P Gives R Gene** | (R, R) → Red | (R, R) → Red | (R, W) → Pink |
| **P Gives R Gene** | (R, R) → Red | (R, R) → Red | (R, W) → Pink |
| **P Gives W Gene** | (W, R) → Pink | (W, R) → Pink | (W, W) → White |

4/9 chance of being red, 4/9 chance of being pink, and 1/9 chance of being white.

**26. Sorority grades.** The sample of five is a very small sample and, thus, is not ideal for drawing the broad conclusions that sorority members have higher than average GPAs. The sample of 150 is a much better size sample to use for drawing the same conclusions. However, both samples were taken from only one sorority. This is not necessarily a good representation of all sorority members at P.U. Therefore, neither sample is ideal for supporting the idea that sorority members have higher than average GPAs. If the first sorority has a small number of members, then that data set supports the idea that the members of that sorority have a higher-than-average GPA. Also, the sample of 150 supports the idea that the members of the second sorority have higher-than-average GPAs. However, you must be careful how you define your population and ensure that your sample is representative of the entire population that you defined. Sampling from one sub-group is not a good way to do this.

## 8.1  Great Expectations

### I.  Developing Ideas

**1. What do you expect?** Multiply the value of each outcome with its probability of occurring and then add up all those products.

**2. The average bite.** The probability that your sister receives $1 is 1/2. The probability that she receives $0.50 is 1/2. So the expected value of her Tooth Fairy payoff is ($1)(1/2) + ($0.50)(1/2) = $0.75.

**3. A tooth for a tooth?** The Tooth Fairy payoff has expected value ($1)(1/3) + ($0.50)(1/3) + ($0.80)(1/3) = $0.77 (or $0.76 if the Tooth Fairy rounds down).

**4. Spinning wheel.** The probability of spinning red is 1/2. The probability of spinning blue or green is 3/8.

**5. Fair game.** A game is *fair* if its expected value equals zero.

### II.  Solidifying Ideas

**6. Cross on the green.** The two green slots give the house an edge. The chance of winning on red is 18/38, thus, the expected value is $2000(18/38) = $947.

**7. In the red.** Because the chance of winning on red is 18/38, a fair game would pay back $38/18 = $2.11 \ldots$ for each dollar bet. The expected value of this game is then $38/18 \times 18/38 = $1.

**8. Free Lotto.** Because there are $(36 \times \cdots \times 31)/(6 \times \cdots \times 1)$ possible lottery numbers, the chance of winning is $(6 \times \cdots \times 1)/(36 \times \cdots \times 31) = 0.000000513 \ldots$. The expected value of the ticket is $8,000,000 \times 0.000000513 \ldots = $4.11.

**9. Bank value.** Assuming the bank is a sure thing, then with probability 1, you will have $103 at the end of the year. So, the expected value is $103.

**10. Newcomb your neighbor.** Explain Newcomb's paradox as suggested.

**11. Value of money.** Ordinarily, picking an empty Zero-or-Million-Dollar Box has 0 value, but if you envision yourself starving to death, the perceived value may correspond to a very large negative number—so large that the computed expected value of picking the one box is less than 0. In this situation, you would always pick both boxes, because this clearly has a positive value. On the other hand, if you have $800,000 to your name, you may value the thrill of the unknown more than the measly $1000.

**12. Die roll.** Imagine getting $1 for each dot that appears. The expected number of dots is $1(1/6) + 2(1/6) + 3(1/6) + 4(1/6) + 5(1/6) + 6(1/6) = (1/6)(1 + 2 + 3 + 4 + 5 + 6) = 21/6 = 3.5$. The term "expected value" is a precise definition of the term "average" in the realm of probability.

**13. Dice roll.** It turns out that the expected value of rolling two dice is the sum of the expected values of each die. So from Mindscape II.12, we should anticipate that the answer is $3.5 + 3.5 = 7$. But let's derive this from first principles.

| 2 | 3 | 4 | 5 | 6 | 7 | 8 | 9 | 10 | 11 | 12 |
|---|---|---|---|---|---|---|---|---|---|---|
| 1/36 | 2/36 | 3/36 | 4/36 | 5/36 | 6/36 | 5/36 | 4/36 | 3/36 | 2/36 | 1/36 |

$2(1/36) + 3(2/36) + 4(3/36) + \cdots =$
$(2 + 6 + 12 + 20 + 30 + 42 + 40 + 36 + 30 + 22 + 12)/36 = 252/36 = 7.$

**14. Fair is foul.** The value of this game is $(-\$1)(2/3) + (\$1.50)(1/3) = -2/3 + (3/2)(1/3) = -1/6$. Don't play, because you will lose 17 cents on average when you do.

**15. Foul is fair.** Suppose you get $D$ dollars if you flip tails. The value of the game is then $-5(2/3) + (D - 5)(1/3) = D/3 - 5$. Because the value of a fair game is 0, set $D/3 - 5 = 0$ and solve for $D$. Here $D = \$15$.

**16. Cycle cycle.** The best way to approach insurance problems is from the insurance company's viewpoint. What is the company's expected value from this deal? If the policy costs $P$ dollars, then 80% of the time they will make $P$ dollars and 20% of the time they will make $\$P - \$700$. (Here, making negative money means *losing* money.) The expected value is $(0.2)(P - 700) + (0.8)(P) = P - 140$. The insurance company would break even, on average, if the policy is a fair $140.

**17. What's your pleasure?** Yikes! [TV: 4.0], [Movie: $0.5(11) + 0.5(-2) = 4.5$], [Date: $0.3(21) + 0.7(-2) = 4.9$]. It's difficult to measure some pleasures on the same scale as others. Go on the date.

**18. Roulette expectation.** The chance that the ball lands on 9 is 1/38. The expected value for the game is $(1/38)(\$3600) = \$95$, which is less than the price you pay to play.

**19. Fair wheeling.** If you were paid $3800, the expected return would be $(1/38)(\$3800) = \$100$. This makes it a fair game, because the expected value of the game is equal to the price you paid to play. Alternatively, you could include the price of the game into the expected value computation: expected value = $(1/38)(3700) + (37/38)(-100) = 0$.

**20. High rolling.** The expected value (or average return) is the sum of the individual probabilities of each roll multiplied by the corresponding payback. E.V. = $(1/6)(25) + (1/6)(5) + (1/6)(0) + (1/6)(-10) + (1/6)(-10) + (1/6)(-15) = (1/6)(25 + 5 - 10 - 10 - 15) = -5/6 = -\$0.83$. So, on average, you lose 83 cents per game.

**21. Fair rolling.** You need to be paid more than 83 cents per game to make it worth your while. If you are given $0.83 per game, then don't expect to make or lose money on this game in the long run, which is characteristic of a fair game.

**22. Spinning wheel.** Pick the bottom right spinner, and you'll profit an average of $0.25 per spin.

Top left:     E.V. = $(1/2)(8) + (1/2)(2) = 5.00$
Top right:    E.V. = $(1/2)(5) + (1/4)(0) + (1/4)(10) = 20/4 = 5.00$
Bottom left:  E.V. = $(1/2)(2) + (1/4)(4) + (1/8)(8) + (1/8)(14) = 4.75$
Bottom right: E.V. = $(1/4)(7) + (1/4)(3) + (1/4)(8) + (1/4)(3) = 21/4 = 5.25$

**23. Dice.** Suppose you've already paid $B$ dollars to play. The table represents how much you can get back.

| Roll | 2 | 3 | 4 | 5 | 6 | 7 | 8 | 9 | 10 | 11 | 12 |
|---|---|---|---|---|---|---|---|---|---|---|---|
| Probability | 1/36 | 2/36 | 3/36 | 4/36 | 5/36 | 6/36 | 5/36 | 4/36 | 3/36 | 2/36 | 1/36 |
| Return | 0 | 0 | 2 | 2.5 | 3 | $B$ | 4 | 4.5 | 5 | $B$ | 0 |

The expected return is $(3/36)2 + (4/36)2.5 + (5/36)3 + (6/36)B + (5/36)4 + (4/36)4.5 + (3/36)5 + (2/36)B = (6 + 10 + 15 + 6B + 20 + 18 + 15 + 2B)/36 = (84+8B)/36 = 7/3 + (2/9)B$. It's a fair game if the expected value of the return is equal to the price of the game. Solving $7/3+(2/9)B = B$ gives $B = \$3.00$.

**24. Uncoverable bases.** You could lose big if the jackpot gets split among too many winners. For example, suppose the jackpot for a pick-5-of-40 lottery is \$2,000,000. The total number of different lottery numbers is $(40 \times 39 \times 38 \times 37 \times 36)/(5 \times 4 \times 3 \times 2 \times 1) = 658,008$. If you bought every possible combination and there were no other winners, you'd have a net profit of $2,000,000 - 658,008 = \$1.3$ million! If three other people got lucky that week, then your portion of the jackpot reduces to \$500,000. You would end up losing \$150 thousand (and that's *before* taxes!).

**25. Under the cap.** The expected value of the bottle itself is \$1 because there is a one in a million chance of getting \$1 million. The expected value of *purchasing* the bottle is –\$1, or the expected value of the bottle less what you paid for it.

### III. Creating New Ideas

**26. From a penny to a million.** Suppose spinning yields heads 30% of the time and balancing yields heads 60% of the time. If your friend picks heads, then the expected value of your spinning scam is $0.3(-1) + 0.7(1) = 0.40$. If your friend picks tails, then the expected value of your balancing scam is $0.4(-1) + 0.6(1) = 0.20$. You make either a 20% or 40% return on your investment . . . not bad!

**27. Three coins in a fountain.**

| | 0 Heads | 1 Heads | >1 Heads |
|---|---|---|---|
| Probability | 1/8 | 3/8 | 4/8 |
| Profit | \$15 | \$0 | –\$5 |

Expected value = $(1/8)15 - (1/2)5 = -\$0.625$. In the long run, you should expect to lose an average of 62.5 cents per game.

**28. Insure.** Look at this game through the eyes of the insurance company. The value of the car insurance to the company is $0.02(\$200 - \$9000) + 0.98(\$200) = \$20$. The expected value of the bike insurance is $0.1(\$75 - \$850) + 0.9(\$75) = -\$10$. The company makes enough money on the car insurance to cover its losses on the bike insurance.

**29. Get a job.** An equivalent question: Suppose you knew 100 kids who were in a similar situation. After all the offers were accepted, what would their average salary be? 50 kids would have \$21,000, 30 would have \$32,000, and a lucky 10 would be making \$45,000. The average salary is $((50 \times 21,000 + 30 \times 32,000 + 10 \times 45,000)/100) = \$24,600$.

**30. Take this job and . . . .** Answer: Take the offer . . . always! Suppose you get an offer of \$21,000 from the first company. If you accept the offer, you have a concrete salary of \$21,000.

Before accepting, compute the expected value of your salary if you *reject* the offer. You'll have 0.3 chance of getting \$32,000 and 0.1 chance of getting \$45,000 for an expected return of $0.3(32,000) + 0.1(45,000) = \$14,100$. For this reason, it's better to stick with the bird in hand.

To see this problem in a different light: Suppose you are a cult leader of 200 kids in exactly this situation. No matter what, all the money from all the kids' salaries will go to you. Half the kids (100) get an offer from the first company. How do you advise them? If you told them to all accept the offer, then you'll have \$2,100,000 coming in each year. If you told each of them to reject the offer, then roughly 30 would end up making \$32,000 and 10 would make \$45,000 for a smaller total of \$1,410,000. It is in everyone's interest to take the first job offer they get.

**31. Book value.** $\$70(1.1)^{40} = \$3168$, which is still far less than the estimated \$22,890 that the authors are confident you'll get after reading this book.

**32. In search of . . . .** You will get \$1 million only if these three events happen: They find the shipwreck (0.6), the treasure is worth \$2 million (0.5), and they can retrieve it (0.7). The probability of all three events happening is the product of the individual probabilities $0.6 \times 0.5 \times 0.7 = 0.210$. The expected return of the dive is $0.210(1,000,000) = 210,000$, so that your expected profit is \$10,000.

**33. Solid gold.** After the price of gold changes, the expected profit for each ounce of gold is $0.5(25) - 0.4(40) + 0.2(0) = \$0.50$. If gold costs \$375 per ounce, then your total return on investment is 0.133%. If the price change takes place over a day, then this translates to a yearly return rate of 50%, which is great. But if it takes more than a week, the corresponding yearly return rate is less than 7%, which is not so reasonable.

**34. Four out of five.** The expected value of picking both boxes is $0.8(1000) + 0.2(1,001,000) = 201,000$. The expected value of picking the mystery box is $0.8(1,000,000) + 0.2(0) = 800,000$.

**35. Chevalier de Méré.** The chance of not rolling a 4 on a single throw is 5/6, and the chance of not throwing a 4 on four consecutive throws is $(5/6)^4$. So, the Chevalier will roll at least one 4 with probability $1 - (5/6)^4 = 0.517 \ldots$ . His expected profit is $(0.517 \ldots)(1000) + (0.482 \ldots)(-1000) = 35.5$ euros.

## IV. Further Challenges

### 36. The St. Petersburg paradox.

| Outcome | H | TH | TTH | TTTH | ... |
|---|---|---|---|---|---|
| Probability | $2^{-1}$ | $2^{-2}$ | $2^{-3}$ | $2^{-4}$ | ... |
| Payoff | \$2 | $\$2^2$ | $\$2^3$ | $\$2^4$ | ... |

The expected value of this game is $(2^{-1})(\$2) + (2^{-2})(\$2^2) + (2^{-3})(\$2^3) + (2^{-4})(\$2^4) + \cdots = 1 + 1 + 1 + 1 + \cdots = $ infinite. This means that the longer you play, the larger your averaged winnings become. Your chance of making money on a \$1000 bet is $2^{-10} + 2^{-11} + 2^{-12} + \cdots = 2^{-9} = 1/512 = 0.00195$. Thus, you would expect to win only once for every 512 games, which doesn't seem characteristic of a profitable game. (You win seldom, but when you do, you win big, but you need deep pockets to keep you in the game until the next win.)

**37. Coin or god.** If the odds are 50-50, then the expected value for picking both boxes becomes $0.5(1000) + 0.5(1,001,000) = 501,000$, while the expected value for picking the mystery box is only $0.5(1,000,000) + 0.5(0) = 500,000$. In this case, there is no paradox . . . go for both boxes. If the odds are 100-0, then the expected values become 1.0(1000) = 1000 for picking both boxes and 1.0(1,000,000) = 1,000,000 for picking the mystery box alone.

**38. An investment.** Suppose the riskier venture pays off with probability $p$. The expected profit for the venture is $p(0.25 \times 1000) = 250p$, compared with an assured profit of $(1.0)(0.05 \times 1000) = 50$. The two choices have equal values when $250p = 50$, or when $p = 50/250 = 0.20(20\%)$.

**39. Pap test.** E.V. = $(1/3000)(1,200,000 - 30) + (2999/3000)(-30) = 370$. Again, view these problems from the point of view of an insurance company that has to shell out \$1.2 million for the death of each customer. Is it worth it for the company to pay for all the women to have a pap smear? For every 3000 pap-smeared women, one life is saved. The 3000 pap smears cost \$90,000, which is far less than \$1.2 million. The net savings are \$1,110,000 for an amortized cost of \$370 per pap smear.

**40. Martingales.** Suppose you play until you win once. The table lists all the ways in which you can win. You can win after one flip, two flips, and so on.

| Outcome | H | TH | TTH | TTTH | ... |
|---|---|---|---|---|---|
| Probability | $2^{-1}$ | $2^{-2}$ | $2^{-3}$ | $2^{-4}$ | $2^{-n}$ |
| Profit | 1 | $2 - 1 = 1$ | $4 - 2 - 1 = 1$ | $8 - 4 - 2 - 1 = 1$ | 1 |

Each time, the winnings cancel the prior losses, so that you end up with a profit of \$1. The expected value of the game is $(2^{-1})(1) + (2^{-2})(1) + (2^{-3})(1) + (2^{-4})(1) + \cdots = 2^{-1} + 2^{-2} + 2^{-3} + \cdots = 1$. It's a sure thing, but you may lose arbitrarily large amounts of money before recovering your losses.

## 8.2 Risk

### I. Developing Ideas

**1. Remarkably risky.** Some examples of activities that are more risky than they may first appear include being unmarried or not getting a good education, either of which reduces your life expectancy more than does living near a nuclear plant.

**2. Surprisingly safe.** As discussed in the section, traveling to Israel, rock climbing, and living near a nuclear power plant are all less risky than they may first appear.

**3. Infectious numbers.** To calculate the number of people expected to be infected, we take the U.S. population, 280,000,000, and multiply by 0.02 (for 2%). This gives us 5,600,000 people who we can expect to be infected with this disease.

**4. SARS scars.** To determine the total cost per life saved for the vaccine, you need to multiply the cost per injection by the number of shots you would have to give to save just one life. In the case that the injection would save 1 in 15,000,000 lives, the cost per life would be $150,000,000. In this case, it seems unlikely that people would support this action because of the huge cost per life. However, in the case when the injections would save 1 in 1000 lives and the cost per life is $10,000, it is much more reasonable to support a program to vaccinate each citizen.

**5. A hairy pot.** To determine the loss of life expectancy of taking a swig from the old hairy "potter," you need to first determine the additional years of life this boy is expected to live and multiply this by the chance of death due to drinking from the pot. This gives you $(80 - 12)(0.07)$, which gives 4.76 years of loss of life expectancy.

### II. Solidifying Ideas

**6. Blonde, bleached blonde.** Suppose there are B blondes in the world (fake or real). How many will you suspect being fake? You'll mistake 20% of the 90% that are real blondes $(0.2)(0.9B) = 0.18B$, and you'll correctly identify 80% of the 10% that are fake $(0.8)(0.1B) = 0.08B$. Of the $(0.18B + 0.08B)$ suspects, only 0.08 were really fakes, so the chance that your suspicions are correct is just $0.08B/(0.18B + 0.08B) = 0.08/0.26 = 4/13 = 31\%$. So, even suspecting that Chris's hair color is fake, the probability that Chris is a true blonde is 69%. Might try it.

**7. Blonde again.** Use the reasoning in Mindscape II.6. There are B blondes. You'll correctly identify 85% of the bleached blondes (0.30B) and mistake 15% of the real blondes (0.70B). If you judged every blonde, you would be suspicious of $[(0.85)(0.30B) + (0.15)(0.70B) = 0.36]$ 36% of all the blondes. Since you correctly suspected only 0.255B people, the odds of your suspicions being correct is $0.255B/0.36B = 0.708\ldots$ or 71%. So suspecting that Chris's hair color is fake, the probability that Chris is a true blonde is 29%. Keep waiting.

**8. Bleached again.** If I flipped a coin to make my decision, I'd detect dyed hair 50% of the time. A 50% success rate corresponds to *zero* additional information, so the chance that Chris is blonde is still 80%. Give it a go Mr. Copperfield. (The analogous calculation from the previous Mindscapes gives your chance of being right in suspecting the fake . . . $(0.5 \times 0.20B)/(0.5 \times 0.20B + 0.5 \times 0.80B) = (0.5 \times 0.20B)(0.5 \times B) = 0.20$)

**9. Safety first.** People who drive sports cars tend to be young and tend to drive fast. These traits also make the drivers likely to get into accidents. People who buy the-safest-car-ever-made do so primarily because they are concerned about safety. And such people are likely to be the safest drivers!

**10. Scholarship winner.** Suppose every student calls about the scholarship. Of the 200 winners, the professor will correctly identify 90%, or $0.9 \times 200 = 180$ students. Of the 99,800 *losers*, the professor will incorrectly tell 10% that they won ($0.10 \times 99,800 = 9980$). In summary, the professor tells $9980 + 180 = 10,160$ students that they received the scholarship when really only 180 of those students had won. So your chance of winning the scholarship is $180/10,160 = 0.017$ . . . just under 2%. Sorry! (see also Mindscape II.6.)

**11. Less safe.** Before the safety improvements, there was an average of 183 people killed per year. Afterward, the death rate is 20% of its previous value and 99% of the people still fly. The new annual death toll becomes $(0.99 \times 183)/5 = 36$ airplane deaths per year, for a saving of $183 - 36 = 147$ lives. Of the 1.7 billion miles flown each day, 1% or 17 million miles will be driven instead. With one automobile death per 100 million miles, we can expect ($365 \times 17)/100 = 62$ extra automobile deaths. So the net saving is $147 - 62 = 85$ lives per year.

**12. Aw, nuts!** To determine the loss of life expectancy in this group for eating 1 to 2 tablespoons of peanut butter a day, you need to follow the method outlined in the section. All 100 people eat at least 1 tablespoon of peanut butter a day, and therefore the loss of life expectancy for them is 1.1 days. We let $P$ be the loss of life expectancy for those who eat at least 1 tablespoon of peanut butter a day but less than 2 tablespoons of peanut butter a day, and we let $Q$ be the loss of life expectancy for those who eat at least 2 tablespoons of peanut butter a day. We have the following equations:

$$(75P + 25Q)/100 = 1.1 \quad \text{and} \quad 25Q/25 = 3.$$

By multiplying the equation on the left by 100 and the equation on the right by 25 and subtracting these two new equations, we get the following:

$$75P = 35.$$

Solving for $P$, we find that the loss of life expectancy for those people that eat at least 1 tablespoon a day but less than 2 tablespoons a day is 0.473 years.

**13. Don't cell!** To determine the cost per life saved, you would need to multiply the cost per device by the number of vehicles that would need to have the device installed to save one life. You are given the cost per device, so all you need to know is the number of vehicles that would need to have the device installed to save one life. This number would involve computing the increased risk of an accident for people who use cell phones while driving and how often cell phones are used in cars.

**14. Risk to order.** Answers will vary.

**15. Buy low and cell high.** To determine how much the government is willing to spend on increasing the safety of each cell phone, you can set up the following equation:

$$\$12,500 = C(100,000),$$

where $C$ is the cost the government is willing to spend on increasing the safety of each cell phone. Solving for $C$, you find that the government is willing to spend $0.125, or 12.5 cents, on increasing the safety of each cell phone.

### III. Creating New Ideas

**16. Taxi blues.** If you believe that the witness can be right 100% of the time, then the probability that the car was blue is 1.00 (or 100%). In Mindscape II.8 we saw that 50% accuracy corresponded to zero information and didn't change the original odds. Here, 100% corresponds to perfect information, and the original odds become irrelevant. The analogous calculation becomes: The witness claims that 0% of the green cabs are blue, or $0.00 \times 0.85 \times C = 0$ cabs, and 100% of the blue cabs are blue, or $1.00 \times 0.15 \times C = 0.15C$. Of the 0.15C blue cabs that he *claims* are blue, only $0.15 \times C$ of them actually are. So the chance that the cab is really blue is $0.15C/0.15C = 1.00$ (100%).

**17. More taxi blues.** Let's interpret the statement, "The witness is 80% sure . . ." as "With 80% accuracy, the witness sees a blue cab as blue and a green cab as green." Suppose that the witness were to look at all the cabs in the city, say a total of C cabs. Of the $0.85 \times C$ green cabs, she'll identify 20% of them as blue and of the $0.15 \times C$ blue cabs, she'll identify 80% correctly. She'll claim that $0.20 \times 0.85 \times C + 0.80 \times 0.15 \times C$ cabs are blue when in fact only $0.80 \times 0.15 \times C$ of those cabs were blue. Her chance of correctly identifying a blue cab in *this* city is $0.80 \times 0.15 \times C/(0.20 \times 0.85 \times C + 0.80 \times 0.15 \times C) = 12/29 = 0.41 \ldots = 41\%$.

**18. Few blues.** Use the same reasoning as in all the previous Mindscapes for this section. The probability that the cab was blue is $(0.80 \times 0.05)/(0.80 \times 0.05 + 0.20 \times 0.95) = 0.1739 = 17\%$.

**19. More safety.** Before the safety improvements, there was an average of 183 people killed per year. Afterward, the death rate is 10% of its previous value and 95% of the people still fly. The new annual death toll becomes $0.10 \times 0.95 \times 183 = 17.385$, let's say 17 deaths, for a saving of $183 - 17 = 166$ lives. The 5% fewer flyers now collectively drive the 85 million miles per day that they would have otherwise flown (questionable assumption). Since there's an average of 1 death per 100 million miles $(365 \times 85)/100 = 310$ extra lives would be lost in automobile accidents per year. Net loss of life is $310 - 166 = 144$ lives.

**20. Reduced safety.** Ordinarily 183 people died in plane accidents each year. The 5% increase in death rate and the 15% increase in air travel correspond to $1.15 \times 1.05 \times 183 = 221$ deaths per year, an increase of 38 deaths. The 15% increase in air travel corresponds to an additional $0.15 \times 1.7$ billion = 225 million miles flown that would have ordinarily been driven each day. Since 1 person dies per 100 million automobile miles, there is a saving of 2.25 people per day, or $365 \times 2.25 = 821$ people. There is a net saving of $821 - 38 = 783$ deaths.

### IV. Further Challenges

**21. HIV tests.** Suppose the population of the U.S. is $P$, so that there are roughly 0.001P people with HIV and 0.999P without it. There are three essential questions to ask: How many noninfected people will pass both tests? $0.05 \times 0.01 \times 0.999P = 0.0004995P$. How many infected people will pass both tests? $0.95 \times 0.99 \times 0.001P = 0.0009405P$. And finally, of all the suspected people, what portion are really infected? $0.0009405P/(0.0009405P + 0.0004995P) = 0.6531$. Quite surprising!

**22. More HIV tests.** Of the 0.999P noninfected people, $0.05 \times 0.99 \times 0.999P$ will have a positive Test A and a negative Test B. Of the 0.001P infected, $0.95 \times 0.01 \times 0.001P$ will have the same results. In this group, only $0.95 \times 0.01 \times 0.001P$ are really infected, so the tests indicate a chance of infection at $0.95 \times 0.01 \times 0.001P/(0.95 \times 0.01 \times 0.001P + 0.05 \times 0.99 \times 0.999P) = 0.00019$, a fiftieth of a percent!

## 8.3 Money Matters

### I. Developing Ideas

**1. Simple interest.** To calculate the amount of money that you have at the end of the year, you multiply $P(1 + r/100)$. By substituting \$500 for $P$ and 2 for $r$, you get \$510 after one year. If you are then told that there was an \$11 processing fee, chances are you will be pretty upset, because you will end up with \$1 less than you originally deposited.

**2. Less simple interest.** To calculate the amount of money Sam has after 10 years, you use the following equation: $P(1 + r/100)^y$ and substitute \$500 for $P$, 3 for $r$, and 10 for $y$. Doing this, you find Sam has \$671.96 after 10 years.

**3. The power of powers.** $3^2$ is larger than $2^3$, but $4^5$ is larger than $5^4$. In general, when taking large numbers $n$ and $m$, where $n > m$, $m^n$ will be larger than $n^m$. A large exponent has a much greater effect on the result than a large base. Because of this, you would much rather have \$$2^{100}$, which is more money than there is on the whole Earth, rather than \$$100^2$, which is \$10,000.

**4. Crafty compounding.** To determine how many pieces of gold the minister of finance would receive, you need to compute the sum $2^0 + 2^1 + 2^2 + 2^3 + \cdots + 2^{62} + 2^{63}$, which would give you all the gold pieces on the 64 squares (the exponents 0 through 63 above represent each square), which is equal to $2^{64} - 1$. This gives you approximately 37,000,000,000,000,000,000 pieces of gold. Should you be surprised that the king felt tricked and had the minister executed?

**5. Keg costs.** By taking part in a Thursday night keg party, some potential opportunity costs will include the opportunity to go to a Friday night keg party, if you have spent all your money on the Thursday night keg party, getting to take someone out on a date, buying tickets to a last-minute concert, or anything else exciting that may happen on the upcoming weekend that costs money that you no longer have. The biggest opportunity cost is the cost resulting from doing less well in school by partying on school nights.

### II. Solidifying Ideas

**6. You can bank on us (or them).** If you invest your money at Happy Bank, you will receive $1000(1 + 0.03)$, or \$1030, after a year. If you invest at Glee Bank, you will receive $1000(1 + 0.025/4)^4$, or \$1025.24, after one year. So, although Glee Bank compounds interest quarterly whereas Happy Bank compounds interest only once over the year, you will end up with more money from Happy Bank after one year. Therefore Happy Bank should earn your business.

**7. The Kennedy compound.** With Bobby, you will have $1000(1 + 0.03)^5$, or \$1159.27, after five years. With Jack, you will have $1000(1 + 0.03/365)^{365 \times 5}$, or \$1161.83. If you invest your money with Jack, you will have \$2.56 more than if you invest your money with Bobby.

**8. Three times a lady.** To determine what your Aunt Edna will pay for her three equal payments, you can use the formula developed in this section for the Three-Times-a-Year Savings and Loan Company. When you fill in the numbers for Aunt Edna, you get the following:

$$P = \frac{2000(1+0.07/3)^3}{(1+0.07/3)^2 + (1+0.07/3) + 1}$$

When you calculate this amount, you get $P = \$698.02$.

**9. Baker kneads dough.** Assuming that the Bread Bank compounds interest monthly, you once again can use the formula developed in the section. When you plug in the numbers for Adrian, you get the following:

$$\$10,000((1 + 0.06/12)^{5\times12})(1 + 0.06/12)$$
$$= P((1 + 0.06/12)^{5\times12} - 1),$$

where $P$ is the amount of each monthly payment. When you solve for $P$, you get \$193.33 for each monthly payment.

**10. I want my ATV!** Use the same formula as Mindscape II.9 and assume that the bank compounds interest monthly. But use \$29,000 as the amount borrowed, 4% as the interest rate, and 120 as the number of times the interest is compounded to get the following equation:

$$\$29,000((1 + 0.04/12)^{120})(1 + 0.04/12) = P((1 + 0.04/12)^{120} - 1).$$

Solving for $P$ gives each equal monthly payment is \$288.13.

**11. Lottery loot later?** If you choose the five-payment option, remember that only four years will elapse between the date of your first payment and the date of your fifth. For example, if you get your first payment on January 1, 2005, then your fifth payment will arrive four years later on January 1, 2009. Therefore, if you calculate the amount that you would have after four years if you took the \$45,000 from Jimmy, you find you have $45,000(1 + 0.03)^4$, or \$50,647.90. If you take the five payments of \$10,000 and invest each one in the same savings account, you will have \$10,300 after the first year. At the beginning of the second year, you invest this \$10,300 plus the \$10,000 payment you receive, so at the end of year 2, you have \$20,909. At the beginning of the third year, you invest the \$20,909 plus the third annual payment of \$10,000, so at the end of the year you have \$31,836.27. At the beginning of the last year, you invest the \$31,836.27 plus the fourth annual payment of \$10,000, so at the end of the year you have \$43,091.36 plus the final \$10,000 payment. This gives you a total of \$53,091.36, which exceeds the amount you would have if you took Jimmy's offer by approximately \$2500. Jimmy made you an offer you can definitely afford to refuse.

**12. Open sesame.** After 50 years, Bert will have $1.00(1 + 0.05/365)^{365\times50}$, or \$12.18, whereas Ernie will have $1.00(1 + 0.06/365)^{365\times50}$, or \$20.08. If their initial deposits were \$10,000, the only thing that would change in each formula is the 1.00 to 10,000. All this does is multiply the final amount you found earlier by 10,000, leaving Bert with \$12,180 and Ernie with \$20,080.

**13. Jelly-filled investments.** To determine how much money you would have after three years if you had invested your money in a three-year CD, you need to calculate $2000(1 + 0.05/12)^{12\times3}$, which gives you \$2,322,94. Therefore, had you invested the \$2000 in a CD rather than in the stock of Krispy Kreme Donuts, you would have made \$22.94 more.

**14. Taking stock.** To determine the annual rate of a savings account, you can set up the following equation and solve for $r$:

$$\$2000 + \$155 = \$2000(1 + r/100)^3.$$

Solving for $r$ gives you 2.52%.

**15. Making your pocketbook stocky.** As in Mindscape II.14, you can set up the following equation and solve for $r$ to determine the annual rate of return of the investment:

$$\$3500 + \$248 = \$3500(1 + r/100)^2.$$

Solving for $r$ gives you 3.48%.

### III. Creating New Ideas

**16. Money-tree house.** Assuming that the loans are compounded monthly, you can determine the cost of a monthly payment for the 30-year loan at a 7.5% interest rate using the following equation:

$$\$250{,}000((1 + 0.075/12)^{30\times12})(1 + 0.075/12 - 1) = P((1 + 0.075/12)^{30\times12} - 1)$$

By solving for $P$, each monthly payment for the 30-year loan is \$1748.04, which makes a total of \$629,293.06 paid for the loan of \$250,000. To find the cost of a monthly payment for the 15-year loan at a 7% interest rate, you can use the following equation:

$$\$250{,}000((1 + 0.07/12)^{15\times12})(1 + 0.07/12 - 1) = P((1 + 0.07/12)^{15\times12} - 1)$$

By solving for $P$, each monthly payment for the 15-year loan is \$1458.33, which makes a total of \$262,500 paid for the loan of \$250,000.

**17. Future value.** To determine the future value of the \$3000, you need to calculate $3000(1 + 0.045/4)^{6\times4}$, which will give you \$3923.97.

**18. Present value.** The amount that you are trying to find is your original investment, or $P$. By using the following equation

$$\$20{,}000 = P(1 + 0.057/12)^{4\times12},$$

and solving for $P$, $P$ = \$15,931.082. Usually you would round this amount to \$15,931.08; however, when you plug that number back in for $P$, you will find that you are just short of \$20,000. Therefore, your initial investment needs to be \$15,931.09.

**19. Double or nothing.** Use the following equation and solve for $t$ to find the amount of time it will take to double your \$1500 deposit.

$$\$3000 = \$1500(1 + 0.03/365)^{365t}$$
$$2 = (1.000082192)^{365t}$$

By using logs, you can bring the $t$ down from the exponent to find that $t = 23.1$ years. We also notice that in the second line, the actual amount you invested and the amount you are trying to get are gone and replaced simply with 2, which represents doubling your money. Therefore, it would take the same amount of time to double your money if you started with \$5000 because the same thing would happen with the \$5000 and \$10,000.

**20. Triple or nothing.** Set up the following equation and solve for $t$ to determine how long it will take to triple your initial $3500.

$$\$10{,}500 = \$3500(1 + 0.025/52)^{52t}$$

$$3 = (1.00480769)^{52t}$$

By using logs, you can bring the $t$ down from the exponent to find that $t = 43.96$ years. Once again, you should notice that on the second line, the actual amount you invested and the amount you are trying to get are gone and replaced simply by 3, which represents tripling your money. Therefore, it would take the same amount of time to triple your money if you started with $500 rather than $3500.

## IV. Further Challenges

**21. Power versus product.** When you determine the amount of money that is made when interest is compounded, you notice that the first time you compound interest, the initial investment is multiplied by 1 + the interest per interval. When this amount is used and the interest is compounded again, you multiply by 1 + the interest per interval. This process continues the same number of times as the number of intervals over which you are compounding. Observe the pattern in the following formula:

$P(1 + \text{the interest per interval})(1 + \text{the interest per interval}) \cdots (1 + \text{the interest per interval})$

Same number of this term as the number of intervals over which you are compounding.

Notice that you end up with the amount (1 + the interest per interval) multiplied by itself the same number of times as intervals you have. This leads to using an exponent, which simplifies multiple multiplications. By multiplying the amount (1 + the interest per interval) by the number of intervals, it is as if we added the amount to itself the same number of times as the number of intervals. This doesn't make sense. Therefore, you use the number of intervals as an exponent when dealing with compounding interest.

**22. Double vision.** You can start this process by setting up the following equation:

$$2P = P(1 + (r/100)/t)^{ty}.$$

If you divide both sides by $P$, you will end up with the following:

$$2 = (1 + (r/100)/t)^{ty},$$

which no longer depends on $P$. To most, this seems quite surprising—to double $500, you have to wait just as long as if you wanted to double $500,000.

## 8.4 Peril at the Polls

### I. Developing Ideas

**1. Landslide Lyndon.** LBJ won the statewide election, thus beginning a 12-year career in the U.S. Senate. This victory was by a mere 87 votes, which earned LBJ the nickname "Landslide Lyndon." The vote totals for Precinct 13 were originally counted as 765-60. The 203 last-minute voters changed the vote count for Precinct 13 to 967-61 in favor of Johnson.

**2. Electoral college.** The Electoral College is the voting method employed by the United States to elect the president and vice president. The Electoral College is made up of electors selected by each state. Each state has one elector for each of their two senators and one for each representative. When the population votes, whichever candidate has the most votes in that state wins all of the electors. The only exceptions to this are Maine and Nebraska, where the statewide election results choose the two electors based on the two senators and the other electors are chosen based on congressional district voting. The electors place their votes for president and vice president in December, and the candidates with an absolute majority (more than half of the electoral votes) are declared president and vice president. Several times in American history candidates who did not win the popular vote won the electoral vote. This occurred in 1876 between Samuel J. Tilden, who won 51% of the popular vote, and Rutherford B. Hayes, who only won 48%. However, Hayes had 185 electoral votes and Tilden had 184. In 1888, Benjamin Harrison won only 47.8% of the popular vote and received 233 electoral votes, whereas Grover Cleveland had 48.6% of the popular vote but only received 168 electoral votes, making Harrison the victor. Most recently was the result of the 2000 election, when Al Gore won 48.38% of the popular vote and George W. Bush won only 47.87% of the popular vote. However, Gore received 266 electoral votes and George W. Bush received 271 electoral votes, making the latter our president.

**3. Voting for voting.** Plurality voting allows each voter to vote only for his or her top choice. This can result in a three-candidate situation where no one candidate would win in a head-to-head vote against any of the other candidates. The Borda count method allows voters to voice an opinion on all candidates. However, in this case, even someone who is ranked last by one group of voters can win. The approval voting method allows each voter to determine how many candidates he or she wants to vote for. Because there is no regulation as to how many candidates each voter is to vote for, once again a candidate who may lose in a head-to-head competition can be victorious if some voters vote for just one candidate while others vote for more than one.

**4. Voting for sport.** There is a long-standing competition between the football teams of Texas A&M University (Texas A&M), The University of Texas at Austin (UT at Austin), and Oklahoma University (OU). In general, OU is ranked higher than both UT at Austin and Texas A&M. These teams might be ranked against each other, Oklahoma University #1, UT at Austin #2, and Texas A&M #3. However, in the 2002–2003 season, OU defeated UT at Austin, UT at Austin defeated Texas A&M, and Texas A&M defeated OU. This led to a ranking in which #1 is better than #2, #2 is better than #3, and #3 is better than #1—an example of the Condorcet Paradox!!

**5. The point of the arrow.** Every nondictatorial voting scheme must fail to satisfy one of the following reasonable principles—go along with consensus, ignore the irrelevant, or better is better.

527

## II. Solidifying Ideas

**6. Dictating an election through a dictator.** One way to determine a true winner would be to have a run-off among the four tied candidates and then have that winner go against Ellen in a run-off. A possible problem with this suggestion is that the people who voted for Ellen, possibly all the girls, might vote for Mary in the run-off, causing her to beat the three boys. Then, in the run-off between Mary and Ellen, all the boys who are bitter that Mary beat them in the first run-off will vote for Ellen. Ellen will win even though there are probably more boys than girls, because there were three boys to the two girls originally running who all had pretty much equal vote totals.

**7. Pro- or Con-dorcet?** This situation is not an example of the Condorcet Paradox because six of the nine prefer A over B, six out of nine prefer B over C, and six out of nine prefer A over C. Therefore, this is an example of transitive ranking.

**8. Where is Dr. Pepper?** Given the voting data from Mindscape II.7, using the Borda count method, A would receive 15 votes (3 from the first three voters, 6 from the second voters, and 6 from the third group of voters), B would receive 18 votes (6 from the first three voters, 9 from the second voters, and 3 from the last group of voters), and C would receive 27 votes (9 from the first three voters, 3 from the second voters, and 9 from the third group of voters). This would leave Austin Arctic as the winner because A had the smallest total number of votes.

**9. Approval drinking.** Given that the voters in the first column approve of A and B, they are likely to cast their votes for both if you employ approval voting. However, the voters in the second column will only vote for C and those in the third column will only vote for B. This leaves A with 3 votes, C with 3 votes, and B with 6 votes. Therefore, Boston Breeze will win if approval voting is the method of voting used.

**10. Long live Acid Burn!** If you use the plurality method of voting, Acid Burn Baby Burn will receive 2 votes and Billie Hooker, Cool KK, and Doctor DoDo will receive only 1 vote apiece, leaving Acid Burn Baby Burn as the winner.

**11. The Hooker rules!** If you use the Borda method, then Acid Burn Baby Burn will receive 14 votes, Billie Hooker will receive 11 votes, Cool KK will receive 12 votes, and Doctor DoDo will receive 13 votes, which makes Billie Hooker the winner.

**12. A Cool win, oh-KK?** If you use the voting method where each voter can vote for three candidates, Acid Burn Baby Burn will receive 2 votes, Billie Hooker will receive 4 votes, Cool KK will receive 5 votes, and Doctor DoDo will receive 4 votes, making Cool KK the winner.

**13. Gotta love DoDo!** A method that will result in Doctor DoDo being victorious is the bullying method. If Voter 4 beats up Voters 1, 2, 3, and 5 so that they are no longer able to vote, Voter 4's vote for Doctor DoDo will be the only vote cast and will win him the Wrapper.

**14. The Hooker scandal.** Voters 1, 3, 4, and 5 will not change their vote using the plurality method if Billie Hooker drops out. However, Voter 2 will change his vote from B to C. This leaves you with a tie between Acid Burn Baby Burn and Cool KK.

**15. Borda without Billie.** Once again, if Billie Hooker were out of the running and you used the Borda count method, Acid Burn Baby Burn would receive 11, Cool KK would receive 9, and Doctor DoDo would receive 10. This gives you Cool KK as the winner.

## III. Creating New Ideas

**16. What's it all about, Ralphie?** The situation described is an example of the principle "ignore the irrelevant," as discussed in this section.

**17. Two, too.** It is not possible for there to be two different results using the Borda count method and the plurality method in an election between just two candidates. If you let $P_A$ be the number of plurality votes that candidate A receives and $P_B$ be the number of plurality votes that candidate B receives, you can determine the number of Borda count votes that each candidate would receive in terms of one of these variables. The number of Borda count votes that candidate A and candidate B receive can be represented by $B_A$ and $B_B$, respectively. Therefore, the following table and equations show how the Borda count votes are determined based on the plurality votes.

|  | $P_A$ | $P_B$ |
|---|---|---|
| **First Choice** | A | B |
| **Second Choice** | B | A |

$$B_A = P_A + 2P_B \quad \text{and} \quad B_B = P_B + 2P_A.$$

Consider the case of candidate A winning the plurality vote by the smallest margin possible, 1. You can write $P_B$ as $P_A - 1$. If you substitute this into both equations, you get:

$$B_A = 3P_A - 2 \quad \text{and} \quad B_B = 3P_A - 1.$$

This also gives A the win for the Borda count method. Therefore, even if A's plurality victory were bigger, A would still win using the Borda count method. Thus, there is no way for an election between just two candidates to produce two different victors using the Borda count method and plurality voting.

**18. Two, too II.** Yes, it is possible to get two different outcomes when using the Borda count method and the approval voting method in an election with just two candidates. For example, let A and B be the two candidates running. Assume that there are 5 people who approve of A over B and 10 people who approve of B over A (see table). This gives A 25 votes and B 20 votes, using the Borda count method, giving the victory to B. However, in approval voting method, suppose that the 5 people who approve of A only vote for A and the 10 people who approve of B also think A is okay and vote for both A and B. This would give A 15 votes and B only 10 votes; thus, A would be victorious in contradiction to the result from the Borda count method.

|  | 5 | 10 |
|---|---|---|
| **First Choice** | A | B |
| **Second Choice** | B | A |

**19. Instant runoffs.** If you number the candidates 1 through 4, then a voter would be asked to cast a vote for the following choices—an overall choice, 1 or 2, 1 or 3, 1 or 4, 2 or 3, 2 or 4, and 3 or 4. This would be seven total votes that a voter would need to cast.

**20. Run runoff.** Yes, it is possible for there to be a runoff because of the runoff results. Let's say that candidate 1 gets 20 votes, candidate 2 gets 20 votes, candidate 3 gets 11 votes, and candidate 4 gets 5 votes. Then assume that the candidate 1 voters vote 1 over 2 in the runoff votes and candidate 2 voters vote 2 over 1 in the runoff votes, giving each of the two candidates 20 votes. Then assume that 8 of the 11 voters who voted for candidate 3 prefer 1 over 2 and the other 3 prefer 2 over 1. Candidate 1 gets 28 votes and candidate 2 gets 23 votes. Then further assume that all of the candidate 4 voters prefer candidate 2 over 1 in the runoff votes. The results from the runoff leave us with a tie between candidate 1 and 2, with 28 votes apiece.

## IV. Further Challenges

**21. Coin coupling.** Yes, it is possible to get all six pairs in your bank. The following is one possible solution for your moveable coins columns and your bank additions.

| D | D | N | N | P |
|---|---|---|---|---|
| P | N | D | P | N |
| N | P | P | D | D |

| P | D | D | N | N | P |
|---|---|---|---|---|---|
| N | N | P | P | D | D |

This is true because you have taken three choices of coins and created the six possible orderings of all three, which results in the six possibilities for ranking just two of the coins.

**22. From money-mating to cupid's arrow.** Each addition to your bank is based on the ranking from your moveable coins column. By accepting that the top coin is better than the bottom coin and then maintaining that ranking when you move from your bank back to your moveable coins column, you end up with a dilemma in which all possible rankings end up like the argument behind Arrow's theorem. By maintaining a rule in which the ordering in your bank is based on the ranking in the coin column, which is based on what happens in your bank, you end up with all possible rankings that all followed the established rules.

530

## 8.5 Cutting Cake for Greedy People

### I. Developing Ideas

**1. You-cut you-lose.** The cutter cuts the cake so that the two pieces are equally valuable in the cutter's mind. As the chooser, you get to choose the piece that appears more valuable to you. In the case that your preferences are *exactly* that of the cutter, then you both believe that the two pieces are equal in value. However, chances are your preferences are different. Therefore, as the chooser, you can choose a piece that you deem more valuable, whereas the cutter only gets a piece that he or she views as having the same value as yours.

**2. Understanding icing.** This person is missing the idea that some people prefer different parts of the cake based on features rather than size. Some are icing fanatics and are willing to take a smaller piece to maximize the amount of icing they get. Some prefer the writing; some prefer the candy rose. Therefore, dividing the cake based on size alone would not necessarily leave everyone satisfied.

**3. Liquid gold.** This method works for gold because every part of gold is equally valuable. There are no parts with more icing or a candy rose. Therefore, if the weights are equal, then the division is fair. However, it is quite difficult to melt down a cake. As in Mindscape I.2, it is important to realize that different people view different parts of a cake with different levels of value. Therefore, this method of using equal weights will not necessarily be consistent with the preferences of the people involved, resulting in an unfair division.

**4. East means West.** As you move toward the east vertex, the east piece and the north piece are getting smaller, because you are moving in a northeasterly direction. Therefore, your preference would be for the west piece, because that is the only piece that increased in size, which only increases its value. The north and east pieces lost size and value.

**5. Two-bedroom bliss.** Consider the following:

At the far left of the line, your (and your roommate's) preference is for the big bedroom, because it is free. At the far right of the line, your (and your roommate's) preference is for the small bedroom. There exists a point on the line for each of you, as you move from left to right, at which your preference changes from the big bedroom to the small bedroom. At this point, you equally prefer either the big bedroom for the given price or the small bedroom for $1000 minus that price. The same is true for your roommate. For instance, say you and your roommate's preference points are as shown here:

In this case, you can choose your preference point where you are as happy to pay $600 for the big bedroom as you are to pay $400 for the small bedroom. At these prices, your roommate's preference is for the big bedroom, so he gets that room for $600. Because you were equally happy with either, you get the small bedroom for $400. You are always able to do this because either your preference points are the same or one is to the left of the other. In both situations, you can find a solution that makes both you and your roommate happy to pay the price for the room you get.

## II. Solidifying Ideas

**6. Your preference.** If the cake were cut into three pieces from point $A$ by straight lines to the vertices, I would prefer the east piece. If the cake were cut into three pieces from point $B$ by straight lines to the vertices, I would prefer the west piece. If the cake were cut into three pieces from point $C$ by straight lines to the vertices, I would prefer the west piece. If the cake were cut into three pieces from point $D$ by straight lines to the vertices, I would equally prefer the north or east piece.

**7. Bulk.** Your preference diagram will choose the piece with the largest area, because that will lead to the biggest volume from any point on the cake. Your preferences will fall along the imaginary lines that extend at a right angle from the halfway point of each side, because each point along these lines is equidistant from two vertices, making them equal in volume and equally valuable in your eyes. Therefore, your preference diagram should look like this:

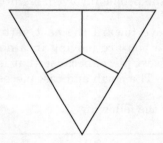

**8. Don't move that knife.** Consider the situation where Julie, Bob, and Stacy are sharing Julie's birthday cake. All Julie wants is her name, whereas Bob and Stacy want as much cake and icing as they can get. Let's say the cake looks something like this:

As the knife moves from the left side of the cake to the right side of the cake, Bob yells stop right after the rose. Julie then yells stop right after her name is passed. Stacy is happy because she has quite a lot of cake and icing. Julie is happy with her piece; however, Bob feels that his piece, although equal to one-third of the value of the cake in his eyes, is not as good a piece as Stacy's piece. Bob envies Stacy's piece of cake.

**9. On the edge.** To cut the cake, cut from the chosen point to each vertex of the triangle. When you choose a point on an edge, two of the three vertices lie on that edge. Therefore, when you cut to those the three vertices, you produce only two pieces of cake. Here are some examples:

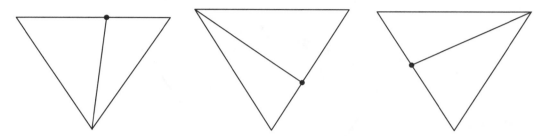

**10. Just do it.** This method works because you are guaranteed that the three preference diagrams, when superimposed onto each other, will contain a point where the three people all have three different preferences. You are guaranteed to have two people's preference borders cross each other, which gives a point at which two people have two equally valuable preferences. This allows the third person to choose his or her preferred piece and leaves the other two with either one or two choices. For example, if the cut point is located on person 1's northeast border and person 2's northwest border, then person 3 can pick his or her favorite piece. If that piece is north, then person 1 gets east and person 2 gets west. If that piece is east, then person 1 gets north and person 2 gets west. If that piece is west, then person, 1 gets east and person 2 gets north. You are guaranteed, by the Greedy Division Theorem, that two borders from two different preference diagrams will intersect, therefore, you can always find this point.

**11. The real world.** Some other examples, which are also given in the section, include dividing rooms and rent in an apartment or dividing up land.

**12. Same tastes.** If all three people have the exact same preference diagrams, you would cut at the point where all three preference diagrams branch out. At this point all three people will feel that all three pieces are equally valuable and therefore will be equally happy with any of the three pieces.

**13. Crossing the line.**

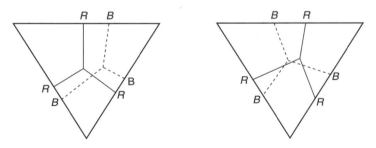

The fewest number of times that the lines must intersect in the first diagram is once; in the second diagram, twice.

**14. Cutting up Mass.**

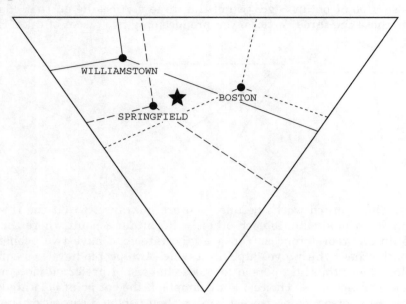

The preference diagrams for you, Joan, and John are drawn. Your preference diagram is the solid line, Joan's is the dashed line, and John's is the dotted line. All three preference diagrams have their branch points on the preferred city for each person. If you cut at the star, John will pick the east region (because the star lies in his east preference region), Joan will pick the west region, and you will pick north. Therefore, all three of you will have a region that contains your city of preference.

**15. Where to cut.** If you cut the cake at point $x$, then the dashed line will choose east, the solid line will choose west, and the dotted line will choose north, because in the region containing $x$, the preferences of dashed, solid, and dotted are all different. Thus, each takes the piece each prefers.

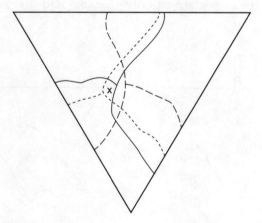

**III. Creating New Ideas**

**16. Land preference.** Your preference diagram would roughly look like the one below. Your western preference region is as shown, because anywhere in that region the west portion can contain at most 23. As you move toward the branch point, the 6 becomes part of the north; the 5, part of the east; and the 8 stays in the west. The big change happens when

534

you reach the 4; the western portion changes from a worth of 12 to a worth of 8 as the north or east become greater, depending on where in the triangle you are. The other preferences are drawn in a similar manner, with the branch point contained in the triangle with a value of 4.

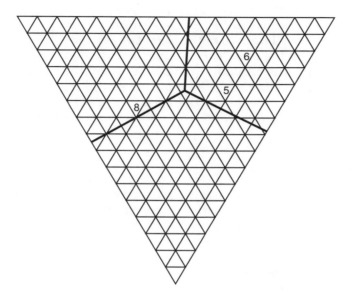

**17. Uneven pair.** To divide the cake two-thirds/one-third, have each person draw a preference diagram as though the cake were being divided among three people. You are guaranteed to get one intersection point that lies on a preference border for each person. At this point, the first person will prefer two pieces equally and the second will also prefer two pieces equally. However, one of these preferred pieces must be the same for both people and one will be different. Therefore, give the person who gets one-third the piece he prefers that is not one of the two preferred by the second person. Then give the second person the two pieces that she prefers. Therefore, the first person feels that his piece is equally valuable to one of the second person's pieces and more valuable than her other piece, leaving him feeling as if he got more than one-third the worth of the cake. The second person gets the two pieces that she found more valuable than the third piece, leaving her feeling like she got more than two-thirds the worth of the cake. Everybody wins.

**18. Diversity pays.** If everyone has the same preferences, then the cake is cut from everyone's branch point. This means that when each person gets a piece they each believe that the piece is as equally valuable as the other pieces. No one feels that his or her piece is better. If someone did feel this way, then it wouldn't have been the branch point of his or her preference diagram. By having different preferences, you get a situation in which each person believes, his or her piece is more valuable than the other pieces (or at least as valuable) because each preference diagram chose that piece over the other pieces.

**19. Be fair.** For four people, you can employ the same technique. Move the knife until someone yells stop. At this point, that person feels that the piece to the left of the knife is worth one-fourth of the whole cake, and the other three people feel that what is left is worth at least three-fourths of the whole cake. Then, use what is left and the knife-moving method for three people, as described in the section. For five people, do the same thing and continue with the four person method. This method works in most situations; however, as in the case of Mindscape II.8, everyone might not always get his or her favorite piece.

**20. Nuclear dump.** The preference diagram given would be considered inconsistent if not for the negative value of the nuclear dump. The only preference regions that are potentially inconsistent are the four in the northern portion of the triangle. Moving from the western vertex toward the eastern vertex, the preferences change from east to west, back to east, and then back to west. However, although these seem inconsistent, they all make sense. The first preference region is east, because although the eastern portion when cut from this region contains the dump, it is much larger than the other two portions. Thus, the negative value of the dump is outweighed by the size. This is also the case in the preference region by the eastern vertex, except in this case the preference is for the western portion. The two preference regions in the middle change because the size of the portions created no longer outweigh-the negative value of the dump. Therefore, the preference is for the piece that does not contain the dump. This preference diagram is self-consistent due to the location of the nuclear dump.

### IV. Further Challenges

**21. Disarming.** When the nuclear superpowers make their own assessments of their arsenals, they give each piece the valuation equal to their true feeling of its worth. Each country determines its own feeling of the value of the items in the other country's arsenal. So, if Country A feels that Country B has undervalued an item, then Country A can include that item in those to be destroyed. Each country will feel that it loses 3% of its arsenal (because it is losing 3% by its own valuation scheme), whereas it can choose items that, in its eyes, are worth more than 3% of the other country's arsenal. So, for example, Country A may feel that it is giving up 3% of its arsenal while making Country B give up 5% of its arsenal, whereas Country B feels that it is giving up 3% of its arsenal while making Country A give up 6% of its arsenal. Both countries feel as though they have made the better deal.

**22. Cupcakes.** It is possible to divide the 100 cupcakes and cut only three because you are able to use any size triangle for your preference diagram. If you choose a large enough triangle, you can arrange your cupcakes so that for every possible line from each vertex you will only go through one cupcake. This means if you choose a vertex and draw every possible cut line from it to the opposite side, you can arrange the cupcakes such that once you hit one, you will not hit another. Because you can do this with all three vertices, you can divide the cupcakes and only cut three. In fact, if you take the east and west vertices and do this and then use the intersection point to arrange the cupcakes so that a cut line from this point to the south vertex will not cut any cupcakes, you can reduce the number of cupcakes cut from three to two.

**23. Barely consistent.** If you let the value of the shaded region be $S$, then at point $W$ the western portion is equal to 9, the north is equal to $3 + S$, and the east is equal to 8. Because you chose the western portion at $W$, then $3 + S$ has to be less than 9. At point $E$, your preference changes to east, which now has a value of $7 + S$. The north is now equal to 2 and the west is equal to 11. Therefore, $7 + S$ has to be greater than 11. You have the following two inequalities:

$$3 + S < 9 \quad 7 + S > 11$$
$$S < 6 \quad S > 4$$

By simplifying both inequalities, you find that $S$ is bounded by 4 and 6; therefore, $S$ could equal 5. So, if the value of the shaded region is 5, the preferences would be consistent.